Elements of Scientific Inquiry

Elements of Scientific Inquiry

Eric Martin and Daniel Osherson

A Bradford Book
The MIT Press
Cambridge, Massachusetts
London, England

This book was set in Times Roman by Windfall Software using ZzTEX and was printed and bound in the United States of America.

Library of Congress Cataloging-in-Publication Data

Martin, Eric.
  Elements of scientific inquiry / Eric Martin and Daniel N.
  Osherson.
    p.   cm.
  "A Bradford book."
  Includes bibliographical references and index.
  ISBN 0-262-13342-3 (hardcover : alk. paper)
  1. Research—Methodology. 2. Logic, Symbolic and mathematical.
I. Osherson, Daniel N. II. Title.
Q180.55.M4M36   1998
001.42'01'5118—dc21   97-41019
                                                              CIP

Contents

Preface

The aim of this book is to articulate a model of scientific inquiry based on the idea of theory acceptance. For technical development we use the tools of mathematical logic. We hope in this way to exhibit inductive logic as part of the intellectual tradition identified with Frege, Tarski, Carnap, and others.

We are not, of course, the first to attempt to forge induction from logic, and can only hope that the present initiative offers a few insights not already available from the earlier literature. In particular, the philosophy of inquiry has been so skillfully analyzed by others (notably [Kelly, 1996]) that we shall enjoy the luxury of remaining at the more detailed level of model development.

The first chapter of the book indicates the character of the theory we intend to develop. Chapter 2 prepares the way for our logical approach by reviewing parts of "Formal Learning Theory," which investigates induction within a purely numerical framework. The latter theory has developed prodigiously in recent years, and Chapter 2 is not intended as a survey.[1] Rather, we investigate just those aspects of the numerical framework that facilitate understanding of what follows. Chapter 3 offers a model of inquiry built from the concepts of first-order logic. The last chapter specializes the model in view of exhibiting successful hypothesis selection as issuing from the rational revision of belief.

It is a pleasure to acknowledge Noam Chomsky, Clark Glymour, Kevin Kelly, Michael Stob, and Scott Weinstein for shaping our views of induction (although we do not expect all of them to approve all of what follows). We are also indebted to Franco Montagna and Piergiorgio Odifreddi for reading early versions of the manuscript. We bear responsibility, of course, for the blunders that remain, and look forward to hearing all about them from our readership (write to emartin@cse.unsw.edu.au or to osherson@rice.edu).

The following institutions have helped provide the material setting necessary for the realization of our project.

• The Office of Naval Research (contract #N00014-87-K-0401 to D. Osherson and S. Weinstein).

• The Swiss National Science Foundation (contract #21-32399.91 to D. Osherson).

• CogniScience, France.

• Istituto San Raffaele, Milan.

---

1. One such survey will soon be available in [Jain *et al.*, forthcoming].

• Laboratoire d'Automatique et de MicroInformatique Industrielle, Université de Savoie.

We thank them.

How to Read This Book

After a remark about exercises and another about cross references, we propose some itineraries through the book.

Many lemmas and nonessential topics have been relegated to the exercises. Solutions to all of them appear at the end of the book. Exercises headed by card suits are more challenging than others in the same section. We've used the symbols ♣, ♦, ♥, ♠ to rate the degree of difficulty; a ♠ signifies the greatest difficulty, at least in our opinion.

To refer, for example, to Proposition (4) of Chapter 2, we write "Proposition (4)" in Chapter 2 and "Proposition 2.(4)" in other chapters.

A minimal voyage across these pages would be as follows.

• Section 1.1: Modeling inquiry
• Section 3.1: Fundamentals
• Section 3.2: Solvability characterized
• Section 3.3: Four applications
• Section 3.4: Efficient discovery—however, without the optional subsection 3.4.3
• Section 4.1: Belief revision
• Section 4.3: Augmenting the background theory
• Section 4.4: Efficient inquiry via belief revision
• Section 4.5: Closure

There are several ways to supplement the bare minimum. In particular, for the newcomer to acceptance-based theories of inductive logic, the balance of Chapter 1, and all of Chapter 2 might be helpful preparation for the paradigm discussed in Chapters 3 and 4. We note as well that probability enters our discussion only in Sections 2.7 and 3.6; either or both can be omitted.

But you needn't make up your mind right now. Along the way we indicate which sections of the book are essential to the sequel and which can be skipped on a first reading.

Elements of Scientific Inquiry

1 Introduction

1.1 Modeling Inquiry

Scientific inquiry has become so essential to modern life that scholars have been led to analyze it, study its history, and even reassemble it in the form of computer programs. Inquiry itself has become the object of inquiry.[1] To justify such self-reflection it may be hoped that better understanding of science will help us do it more accurately, or else faster, or maybe automatically. Alternatively, we can recognize in scientific inquiry one of the purest expressions of human intelligence, and seek to perceive more clearly the latter through analysis of the former. Or again, we may desire insight into the scope and limits of rational inquiry as a way of evaluating our epistemological position within an immense universe.

Whatever the motives for studying inquiry, we must begin by appreciating the complexity of the subject matter. As a form of human behavior, science involves a wide range of activities, both in and out of laboratories. Surely such a phenomenon cannot be understood without substantial idealization. By limiting attention to just a few, salient aspects of science we may hope to understand their interaction within the larger scheme, and eventually illuminate further variables that can be added to our model at a later stage. The insight to be gained from such an approach depends, of course, upon which aspects of science are idealized away, and which are kept for study. A poor choice at the beginning dooms the enterprise to sterility.

For those who dislike concentrating risk, it is fortunate that different students of inquiry have different opinions about fruitful idealization. One influential view focusses on the credibility that scientists attach to alternative theories, and on the evolution of these credibilities under the impact of data. Interpreting credibility as probability leads to the Bayesian analysis of inquiry, which has greatly illuminated diverse aspects of scientific practice.[2] It is not the Bayesian program that we will pursue in this book, however, since our starting point is a different set of intuitions about the variables to be retained in a model of inquiry. Disagreement with the Bayesian picture concerns hypothesis "acceptance." Bayesians typically find this concept to be epistemologically inert.

---

1. For an introduction to the analysis of scientific reasoning, see [Gustason, 1994]. For a case study in the history of science, see [Earman & Glymour, 1980]. Computer programs that carry out scientific investigation are discussed in [Langley *et al.*, 1987].

2. For illustrations, see [Earman, 1992, Horwich, 1982, Howson & Urbach, 1993, Myrvold, 1996, Rosenkrantz, 1977]. A classic example of Bayesian reconstruction of scientific history is provided in [Dorling, 1979].

Others (including the present authors) are struck by the fact that inquiry often makes headway even when the only credible claim is that all our theories are wrong.[3] In this kind of circumstance propounding a theory seems not to reflect an underlying sensation of credibility but rather a voluntary postulation for the sake of further exploration. Such is the hallmark of acceptance, insofar as the latter is distinguished from belief.[4]

Within the acceptance perspective, scientific success consists not in gradually increasing one's confidence in the true theory, but rather in ultimately accepting it, and then holding on to it in the face of new data. This way of viewing inquiry is consonant with the philosophy of Karl Popper [Popper, 1959]. It is also consistent with the use of the probability idiom by scientists evaluating competing theories. For, high credibility is no doubt an important factor in theory acceptance. Other factors include the scientist's impression that one theory is more interesting than another, as well as her suspicion that postulating a certain theory will promote inquiry leading to a better one.

The preceding remarks are intended neither as a decisive objection to the Bayesian model, nor as definitive motivation for the acceptance perspective. Their purpose is merely to justify exploring the mathematics of acceptance-based discovery. If the resulting theory is elegant and surprising, the background, epistemological issues may come to be seen in new light.

In any event, all the models of inquiry to be examined in this book are based on the idea of acceptance. Other details vary, so it will be useful to describe the structure common to each paradigm. Throughout our discussion, inquiry is analyzed into the following components:

(1) **Components of our models of inquiry:**

(a) potential realities, or "worlds," along with a question that can be asked about any particular world;

(b) a scientific problem;

(c) for each world, a set of potential data-streams, or "environments," which provide information about the world;

(d) (formal) scientists;

---

3. Thus, the astronomer Simon Lilly says of current cosmological theory: "We have these simple-minded ideas and won't be surprised if it's all different than we think" (reported in *Science*, 7 June 1996, vol. 272, p. 1436).

4. Detailed discussion of the distinction is provided in [Cohen, 1992].

(e) a criterion of success that stipulates the conditions under which a scientist is credited with solving a given problem.

Each way of specifying the foregoing concepts may be conceived as a game played between Nature and a scientist. The scientist is of the kind envisioned in (1)(d). To start the game, a class **P** of potential realities is announced to both players. The members of **P** are drawn from (1)(a), and **P** itself constitutes the problem evoked in (1)(b). It will typically happen that **P** does not include every possible reality. Intuitively, this is the case if the possibilities are constrained from the outset in a way already known to the scientist (such constraints constitute the scientist's "background knowledge"). To continue the game, Nature chooses one member w of **P** to be the actual world. Her choice is not revealed to the scientist. Nature also chooses an environment for w, in the sense of (1)(c). The scientist is then progressively shown the data in the environment. She reacts to each datum in turn, and wins the game just in case her evolving conjectures stabilize to the correct answer to the question about w formulated in (1)(a). Of course, it is possible that the scientist would win the game if Nature makes one choice of world and environment, but lose the game for other choices. If the scientist would win for "many" of the choices that Nature might make, then she is credited with solving **P**. The requirements for winning any particular game, and the precise sense given to "many," are spelled out in the success criterion (1)(e).

Let us note that the locution "scientist Ψ solves the problem **P**" should be understood in the sense of Ψ *being a solution* to the problem posed by **P**. The latter problem is to exhibit a strategy for choosing hypotheses that is capable of answering the question envisioned in (1)(a) for a wide class of possible choices of actual world.

This should all become clearer with an illustration, so a simple paradigm is discussed in the next section. Then we return to general remarks. Presentation of our illustrative paradigm relies on the following notation, which will be ubiquitous in the sequel. The set $\{0, 1, 2 \ldots\}$ of natural numbers is denoted N. The symbol ω is used to denote N under its standard order, whereas ω^* denotes N ordered backwards $(\ldots 3, 2, 1, 0)$.

1.2   The Order Paradigm

The model of inquiry now to be presented will be called the *order paradigm*. It illustrates features of the more complex paradigms discussed later. Since the order paradigm is only for illustrative purposes, it will not arise in later chapters (except in exercises). We present the paradigm by stepping through the components listed in (1). This is the purpose of the next subsection. Then we prove some simple facts about solvable and unsolvable problems within the paradigm. In Section 1.3 the paradigm is exploited to discuss some general issues about inquiry.

1.2.1 Components

Worlds. The potential realities, or "worlds," are all the strict total orders $\prec$ over N.[5] For brevity, they are referred to in this chapter simply as "orders." Here are some of the possibilities, where left-to-right position is used to represent $\prec$.

(2) *Example:*   Some ways of ordering N:

(a) 0 1 2 3 4 5 6 7 8 ...

(b) 2 1 0 5 4 3 8 7 6 ...

(c) ... 8 7 6 5 4 3 2 1 0

(d) 0 2 4 6 8 ... 1 3 5 7 9 ...

(e) ... 11 9 7 5 3 1 0 2 4 6 8 10 ...

(f) 0 ... 11 9 7 5 3 2 4 6 8 10 ... 1

Up to isomorphism, both (a) and (b) look like ω, (c) looks like ω^*, (d) looks like $\omega \, \omega$ (two copies of N, one after the other), (e) looks like $\omega^* \, \omega$, and (f) looks like $0 \, \omega^* \, \omega \, 1$.[6] In case you are not familiar with these kinds of orders, consider $\omega \, \omega$. It has a least element, namely, the number 0 in the bottom copy of ω. This least element is followed by 1, then by 2, etc. Only after traversing all the natural numbers in this way do we begin again with the top ω. It starts off with its own copy of 0, and proceeds again through all remaining members of N. The point about the upper 0, is that it is preceded by infinitely many other numbers. This is not true of the bottom 0, since it is first in the entire order. The isomorphism of (2)d to $\omega \, \omega$ works this way: the even members of N are used in place of the bottom ω, the odd members of N in place of the top ω.

---

5. A binary relation is a strict total order iff it is irreflexive, connected, transitive, and antisymmetric. For a review of relations, see [Partee *et al.*, 1990].

6. $\prec_0$ is isomorphic to $\prec_1$ just in case there is a one-one map h from N onto N such that $i \prec_0 j$ iff $h(i) \prec_1 h(j)$ for all $i, j \in N$. For example, (2)b is isomorphic to (2)a.

There is just one question that we will ask about an order, namely, does it have a least element? The answer is Yes for orders (2)a,b,d,f, and No for the others.

Problems.   A *problem* in our paradigm is a collection of orders. For example, the set of all six orders shown in (2) is a problem.

Environments.   By an *environment* for a given order $\prec$ is meant any enumeration of all facts of the form $i \prec j$, with $i, j \in N$. To express such facts in a neutral vocabulary, the expression $R(i, j)$ is used in place of $i \prec j$. (Writing "$i \prec j$" in the environment would give the game away since "$\prec$" names the order chosen secretly by Nature.) No restriction is placed on how the facts are enumerated, and repetitions are allowed. In sum, enumeration e is an environment for order $\prec$ just in case the elements of e form the set: $\{R(i, j) \mid i, j \in N$ and $i \prec j\}$. Here are sample environments.

(3)   *Example:*   Two environments:

(a) $R(2, 3), R(1, 2), R(0, 2), R(1, 4), R(0, 3), R(0, 4), R(0, 5), R(1, 3) \ldots$

(e) $R(5, 10), R(0, 4), R(11, 1), R(0, 2), R(11, 9), R(0, 6), R(1, 0), R(11, 7) \ldots$

The first is for (2)a, the second is for (2)e.

By "environment" (without qualification) is meant an environment for some particular order. Observe that there are uncountably many environments for any given order. Also, notice the following.

(4)   LEMMA: Suppose that an environment is for two orders $\prec_0$, $\prec_1$. Then $\prec_0$ is identical to $\prec_1$.

Thus, environments provide complete information about the orders they are for.

Scientists.   Scientists in our paradigm examine an environment in piecemeal fashion, seeing only some finite initial segment of it at any one time. In response to each segment, they say either "yes" or "no" (meaning "yes it has a least element" or "no it doesn't"). Officially, let *SEQ* be the collection of all finite initial sequences of any environment. Then a *scientist* is any mapping from *SEQ* to $\{\mathbf{y}, \mathbf{n}\}$. To illustrate, *SEQ* includes all the finite initial sequences arising from the environment (3)a. Some of these sequences are shown here:

(5) *Example:*   Sample members of *SEQ*:

∅ (namely, the finite sequence of length 0)

$\langle\, R(2,3)\, \rangle$

$\langle\, R(2,3), R(1,2)\, \rangle$

$\langle\, R(2,3), R(1,2), R(0,2)\, \rangle$

$\langle\, R(2,3), R(1,2), R(0,2), R(1,4)\, \rangle$

$\vdots$

$\langle\, R(2,3), R(1,2), R(0,2), R(1,4), R(0,3), R(0,4), R(0,5), R(1,3)\, \rangle$

A scientist might map the first two of these sequences into **y**, and every other member of *SEQ* into **n**. Let us note that *SEQ* is a countable set.

Success.   Suppose we are given a scientist Ψ and an environment e for order $\prec$. For $k \in N$, the finite initial sequence of length k in e is denoted $e[k]$. For example, if e is the environment (3)a, then (5) shows $e[0], e[1], e[2], e[3], e[4] \dots$ $e[8]$. Providing Ψ with longer and longer initial segments of e produces the sequence of hypotheses $\Psi(e[0]), \Psi(e[1]), \Psi(e[2]), \dots$. [The first conjecture $\Psi(e[0])$ can be thought of as a "wild guess" issued prior to examining any data.] For example, the sequence might begin: **y, n, n,** . . . .. Now we can define success within our paradigm.

(6)   DEFINITION: We say that scientist Ψ *solves* problem **P** just in case for every $\prec\, \in \mathbf{P}$ and every environment e for $\prec$, the following conditions hold:

If $\prec$ has a least element then for cofinitely many k, $\Psi(e[k]) = \mathbf{y}$.

If $\prec$ has no least element then for cofinitely many k, $\Psi(e[k]) = \mathbf{n}$.[7]

If some scientist solves **P**, then it is called *solvable*, otherwise *unsolvable*.

Suppose for example that **P** consists of the orders (2)a,e. Then for **P** to be solvable there must be a scientist Ψ that issues **y** on cofinitely many initial segments of any environment for (2)a [such as (3)a] and issues **n** on cofinitely many initial segments of any environment for (2)e [such as (3)e].

---

7. By "cofinitely many k" is meant: "for all $k \in N$ except for a finite set (possibly empty)."

Lemma (4) allows us to make sense of the idea that a scientist solves a given environment. Since every environment is for just one order, a successful scientist must end up with an unbroken string of **y**'s in case the underlying order has a least element, and with an unbroken string of **n**'s otherwise. So, scientist Ψ solves problem **P** just in case Ψ solves every environment for every order in **P**. Let us stress that solvability requires a *single* scientist to behave appropriately on all the environments for all the orders in **P**. If **P** includes both kinds of orders (those with a least element and those without), then a scientist Ψ must exercise some ingenuity to succeed. This is because no initial segment of an environment for an order in **P** directly informs Ψ of the right answer, yet Ψ's local responses to the incoming data must end up globally appropriate in the sense of converging to the correct choice between **y** and **n**. In contrast, the solvability of **P** would be a trivial affair if it allowed different scientists to succeed on different environments. For, every environment is solved in a brain-dead sort of way by one of the two constant scientists.[8]

By requiring a single scientist to solve all the environments for all the orders in **P**, we leave the door open to unsolvable problems. Such problems tax the ingenuity required of scientists beyond what is theoretically possible, and thus shed light on the limits of inquiry—or at least, of inquiry as modeled by the order paradigm. In fact, the order paradigm engenders a rich set of unsolvable problems, along with a rich set of solvable ones. The next subsection provides an example of each.

1.2.2   Problems, Solvable and Unsolvable

The following proposition exhibits a solvable problem.

(7)   PROPOSITION: Let **P** consist of every order that is isomorphic either to ω or to ω^*. Then **P** is solvable.

Proof: Let $\sigma \in SEQ$ be given. By the "L-score" of σ is meant the smallest natural number that does not appear as the second argument of R in σ (this number is 0 if $\sigma = \emptyset$). Intuitively, the L-score of σ is the best candidate for the least element of $\prec$, if it has one. Similarly, the "G-score" of σ is the smallest natural number that does not appear as the first argument of R in σ. So, it is the best candidate for the greatest element of $\prec$, if there is one. It is easy to see that every member of *SEQ* has a well defined L-score and G-score.

---

8. A "constant scientist" is either the mapping of *SEQ* uniformly to **y**, or the mapping of *SEQ* uniformly to **n**.

Suppose that $\prec$ is isomorphic to ω, and let e be an environment for $\prec$. Then, the least element of $\prec$ appears somewhere in e, and never appears as the second argument of R. It follows that the L-scores of the $e[k]$, $k \in N$, are bounded. In contrast, since $\prec$ has no greatest element, every natural number appears as the first argument of R somewhere in e. It follows that the G-scores of the $e[k]$, $k \in N$, are unbounded. The situation is reversed for $\prec$ isomorphic to ω^*. In this case, the G-scores are bounded in e whereas the L-scores are unbounded. Observe also that for $k' < k$, the L-score of $e[k]$ cannot be smaller than that for $e[k']$, and likewise the G-score of $e[k]$ cannot be smaller than that for $e[k']$. As a consequence of these facts, we have:

(8)   Let environment e be for order $\prec$.

(a) If $\prec$ is isomorphic to ω then for cofinitely many k, the L-score of $e[k]$ is lower than the G-score of $e[k]$.

(b) If $\prec$ is isomorphic to ω^* then for cofinitely many k, the G-score of $e[k]$ is lower than the L-score of $e[k]$.

Define scientist Ψ as follows. For all $\sigma \in SEQ$,

$$\Psi(\sigma) = \begin{cases} \mathbf{y} & \text{if the L-score of } \sigma \text{ is lower than its G-score;} \\ \mathbf{n} & \text{otherwise.} \end{cases}$$

We claim that Ψ solves $\mathbf{P}$. Indeed, let environment e be for $\prec \in \mathbf{P}$. If $\prec$ is isomorphic to ω then (8)a implies that $\Psi(e[k]) = \mathbf{y}$ for cofinitely many k. If $\prec$ is isomorphic to ω^* then (8)b implies that $\Psi(e[k]) = \mathbf{n}$ for cofinitely many k. So, in either case, Ψ solves e.   ∎

The next proposition shows that there are also unsolvable problems.

(9)   PROPOSITION: Let $\mathbf{P}$ consist of every order that is isomorphic either to ω or to $\omega^* \omega$. Then $\mathbf{P}$ is unsolvable.

Proof: Choose a scientist Ψ such that:

(10) Ψ solves every order isomorphic to ω.

It suffices to exhibit an environment e_0 for some order isomorphic to $\omega^* \omega$ such that $\Psi(e_0[k]) = \mathbf{y}$ for infinitely many k. To define e_0 we rely on the following notation. Given $\sigma \in SEQ$, we denote by $D(\sigma)$ the set of numbers appearing in σ. Let $\sigma, \tau \in SEQ$ and $n \in N$ be given. We say that σ is "complete through n" just in case:

(a) $D(\sigma) = \{0 \ldots n\}$, and

(b) for all $i < j \le n$, either $R(i, j)$ or $R(j, i)$ appears in σ.

We say that τ is a "completion of σ through n" just in case:

(a) $\sigma \subseteq \tau$, and

(b) τ is complete through n.

Intuitively, a completion of σ through n is consistent with σ, contains all information about the ordering of $\{0 \ldots n\}$, and contains nothing more. We say that τ is an "upward completion of σ through n" just in case τ is a completion of σ through n and:

for all $i \in D(\sigma)$ and $j \in D(\tau) - D(\sigma)$, $R(i, j)$ appears in τ.

Similarly, we say that τ is a "downward completion of σ through n" just in case τ is a completion of σ through n and:

for all $i \in D(\sigma)$ and $j \in D(\tau) - D(\sigma)$, $R(j, i)$ appears in τ.

It is evident that for every $\sigma \in SEQ$ and $n > \max(D(\sigma))$, there is an upward completion of σ through n, and there is a downward completion of σ through n. We now demonstrate:

(11)   For every $\sigma \in SEQ$, there is $\tau \in SEQ$ and $n > \max(D(\sigma))$ such that:

(a) τ is an upward completion of σ through n, and

(b) $\Psi(\tau) = \mathbf{y}$.

Proof of (11):   Suppose for a contradiction that $\sigma \in SEQ$ does not conform to (11). We define by induction a sequence $\{\gamma^i \mid i \in N\} \subseteq SEQ$, as follows. $\gamma^0 = \sigma$. Suppose that γ^i has been defined. Then γ^{i+1} is an upward completion of γ^i through $1 + \max(D(\gamma^i))$. It is clear that $\gamma^i \subseteq \gamma^{i+1}$ for all $i \in N$, and that $\bigcup_{i \in N} \gamma^i$ is an environment e for an order isomorphic to ω. By our reductio hypothesis, $\Psi(\gamma^i) = \mathbf{n}$ for all $i > 0$. Hence, for infinitely many k, $\Psi(e[k]) = \mathbf{n}$, contradicting (10).   ∎

We use (11) to construct a sequence $\{\sigma^i \mid i \in N\} \subseteq SEQ$. It will be the case that:

(12)   (a) for all $i \in N$, $\sigma^i \subseteq \sigma^{i+1}$;

(b) for all $i \in N$, σ^i is complete through some $n \ge i$;

(c) for all $i \in N$, $\Psi(\sigma^{2i+1}) = \mathbf{y}$;

(d) $\bigcup_{i \in N} \sigma^i$ is an environment for some order isomorphic to $\omega^* \, \omega$.

Conditions (12)a,b imply that $e_0 = \bigcup_{i \in N} \sigma^i$ is an environment, and Conditions (12)c,d imply that it has the promised properties.

Construction of the σ^i

Stage 0: Set $\sigma^0 = \emptyset$. So σ^0 is complete through 0.

Stage $2i + 1$: Suppose that σ^{2i} is defined and complete through $n_0 \geq 2i$. Then $n_0 = \max(D(\sigma^{2i}))$. By (11) let $\tau \in SEQ$ and $n > n_0$ be such that:

(a) τ is an upward completion of σ^{2i} through n, and

(b) $\Psi(\tau) = \mathbf{y}$.

We take $\sigma^{2i+1} = \tau$. So σ^{2i+1} is complete through some $n \geq 2i + 1$.

Stage $2i + 2$: Suppose that σ^{2i+1} is defined and complete through $n_0 \geq 2i + 1$. Let $\tau \in SEQ$ and $n > n_0$ be such that τ is a downward completion of σ^{2i+1} through n. We take $\sigma^{2i+2} = \tau$. So σ^{2i+2} is complete through some $n \geq 2i + 2$.

End construction

It is immediate that Conditions (12)a–c are satisfied. Condition (12)d follows easily from the definitions of upward and downward completions.   ■

   With examples of solvable and unsolvable problems in hand, we could go on to seek an illuminating characterization of solvability within the order paradigm. Let us reserve this enterprise, however, for the more substantial paradigms to be introduced later. In fact, it will be seen in Exercise 3.(22) that the order paradigm is a special case of the one advanced in Chapter 3. So the characterization of solvability to be offered for the latter paradigm will apply as well to the present one. A more pressing claim on our attention is to identify aspects of the order paradigm that will be conserved in later work. This is the topic of the next section.

1.2.3   Exercises

(13)   EXERCISE: Show that every countable problem is solvable.

(14)   EXERCISE: Scientist Ψ *half solves* problem **P** just in case for all environments e for $\prec \in$ **P**, $\prec$ has a least element iff $\Psi(e[k]) = \mathbf{y}$ for cofinitely many k. In this case, **P** is called *half solvable*. Show that the entire collection of orders is half solvable. In view of Proposition (9), conclude that half solvability is strictly weaker than solvability as a criterion of success.

(15)   EXERCISE: Let scientist Ψ, environment e, and $X \in \{\mathbf{y}, \mathbf{n}\}$ be given. We say that Ψ *gradually converges* to X on e just in case:

$$\liminf_{k \to \infty} \frac{\mathrm{card}(\{k' \leq k \mid \Psi(e[k']) = X\})}{k} > 0.5.$$

Scientist Ψ *gradually solves* environment e for order $\prec$ just in case the following is true:

if $\prec$ has a least element then Ψ gradually converges to $\mathbf{y}$ on e, and

if $\prec$ has no least element then Ψ gradually converges to $\mathbf{n}$ on e.

Problem $\mathbf{P}$ is *gradually solvable* if there is a single scientist that gradually solves every environment for every order in $\mathbf{P}$. Prove that a problem is solvable iff it is gradually solvable.

(16)   EXERCISE: ♣   Fix an enumeration $\{(x_i, y_i) \mid i \in N\}$ of $N \times N$ (all pairs of numbers). Given order $\prec$, the *canonical environment* for $\prec$ is the list generated by the following loop:

Set $i = 0$.

LOOP: if $x_i \prec y_i$ then put out $R(x_i, y_i)$. Increment i by 1. Goto *LOOP*.

We say that a problem $\mathbf{P}$ is solvable *on canonical environments* just in case there is a single scientist that solves the canonical environment for each order in $\mathbf{P}$.

   Show that a problem is solvable iff it is solvable on canonical environments.

(17)   EXERCISE: ◆   By an *imperfect environment* for an order $\prec$ is meant the result of removing from any environment for $\prec$ a finite number of entries [of the form $R(i, j)$]. So, every environment for $\prec$ is also an imperfect environment for $\prec$. Scientist Ψ solves problem $\mathbf{P}$ *on imperfect environments* just in case for every $\prec \in \mathbf{P}$ and every imperfect environment e for $\prec$, Ψ issues $\mathbf{y}$ cofinitely often on e if $\prec$ has a least element, and issues $\mathbf{n}$ cofinitely often on e otherwise. In this case $\mathbf{P}$ is said to be solvable *on imperfect environments*.

(a) Show that the problem $\mathbf{P}$ of Proposition (7) is solvable on imperfect environments.

(b) Exhibit a solvable problem that is not solvable on imperfect environments.

1.3   Discussion of the Paradigm

Several features of the order paradigm will recur throughout the book. They
are highlighted in the present section.

1.3.1   Convergence versus Knowledge Thereof

Suppose that scientist Ψ solves environment e for an order with least element.
Then there is a least $k_0 \in N$ such that $\Psi(e[k]) = \mathbf{y}$ for all $k \geq k_0$. In other words,
Ψ begins to *converge* on e starting at $e[k_0]$. Imagine that k_0 is substantially
greater than 0, and consider Ψ's behavior prior to $e[k_0]$. There is no guarantee
of the tidy situation in which $\Psi(e[k]) = \mathbf{n}$ for all $k < k_0$. Rather, Ψ will likely
meander within $\{\mathbf{y}, \mathbf{n}\}$ prior to reaching $e[k_0]$, occasionally issuing $\mathbf{y}$, only
to withdraw it later. It might even happen that Ψ announces a long string
of $\mathbf{y}$'s over some stretch of e prior to $e[k_0]$. Examining Ψ's behavior on this
stretch might fool us into thinking that Ψ has converged prior to $e[k_0]$, with
disillusionment following Ψ's subsequent change of heart at, say, $e[k_0 - 1]$.
This kind of case reveals that nothing in the behavior of Ψ distinguishes the
stable $\mathbf{y}$ issued at $e[k_0]$ from the vacillatory $\mathbf{y}$ issued earlier. The general point
is that Ψ is not required to recognize or signal in any way that its conjectures
have begun to converge. In this respect our paradigm is faithful to the situation
of real scientists, whose theories remain open to revision by new, unexpected
data. It is, of course, possible to define paradigms that require scientists to
signal convergence. The prospects for success, however, are then diminished
[see Exercise (19), below].

1.3.2   Reliability

Consider scientist Ψ defined as follows. For all $\sigma \in SEQ$, $\Psi(\sigma) = \mathbf{y}$ if σ begins
with either $R(0, 1)$ or $R(1, 0)$; otherwise, $\Psi(\sigma) = \mathbf{n}$. It is easy to verify that
for every order $\prec$—whether or not $\prec$ has a least point—Ψ solves uncountably
many environments for $\prec$. Yet Ψ reveals not the slightest intelligence thereby,
and there is no temptation to credit it with discovery. The virtue that Ψ lacks
is reliability, since its success is a hit-or-miss affair, depending on accidental
features of the environment. Such is the motivation for requiring scientists to
succeed on all the environments for $\prec \in \mathbf{P}$ in order to be credited with solving
$\mathbf{P}$. Later paradigms preserve the idea that success on a problem requires success
on "many" environments. Often, as in the present paradigm, we take "many"
to be *all*. Perhaps this is excessive, since probable success on a "random

environment" might be enough to distinguish genuine from spurious discovery. This idea is explored in Sections 2.7 and 3.6.

Similar considerations apply to the constant scientist who issues **y** in response to every $\sigma \in SEQ$. This scientist succeeds on a wide class of orders, namely, all those with a least point! Yet, once again, no scientific competence is revealed by such an accomplishment since it does not rest upon an inductive strategy that is reliable in the right sense.

Now suppose, by way of contrast, that scientist Ψ solves every environment for a rich and varied collection **P** of orders. Let e be an environment for a member of **P**, and let $k_0 \in N$ be such that $\Psi(e[k]) = \Psi(e[k_0])$ for all $k \geq k_0$. Then Ψ's conjecture at $e[k_0]$ is true, stable, and the result of a reliable process. These conditions are often considered central to the analysis of knowledge, in particular, to its separation from mere, true belief. Accordingly, Ψ might be considered at $e[k_0]$ to know the answer to the question "Does the underlying order have a least point?". This does not entail that Ψ knows he knows the answer, since (as observed above) Ψ may lack any reason to believe that his hypotheses have begun to converge. Nonetheless, to the extent that the reliability perspective on knowledge can be sustained, our paradigms concern scientific discovery in the sense of *acquiring knowledge*.

1.3.3 Idealization

Our scientists are idealized agents in many respects. For one thing, they are able to examine every initial segment of an infinite environment whereas this is not possible for actual scientists. Such unbounded inquiry is justified by the desire to study scientific strategy apart from its material realization. The strategy so isolated is a formal object, hence not subject to the practical limitations that afflict its embodied counterpart. Less justifiable is the liberality of the success criterion, which allows arbitrary behavior on all $\sigma \in SEQ$ of length less than, say, a million. The liberality in question is expressed in the following lemma, which is straightforward to verify.

(18) LEMMA: Suppose that scientist Ψ solves problem **P**. Let Ψ' be any scientist such that for all $\sigma \in SEQ$ with length greater than a million, $\Psi'(\sigma) = \Psi(\sigma)$. Then Ψ' also solves **P**.

So, a scientist may relax for the first million data without compromising its prospects for ultimate success, according to the criterion introduced in Section 1.2.1. To distinguish intelligent use of data from dawdling, the solvability criterion must therefore be reinforced to require rapid convergence. Later paradigms incorporate a condition of this kind.

This is a good place to distinguish our enterprise from an alternative tradition that is equally devoted to modeling inquiry, namely, the "PAC" paradigm. Within the latter it is enough for the scientist to reach, with high probability, an hypothesis that is approximately predictively accurate using a randomly composed data set whose size is bounded by a slow-growing function of the allowed error and the chance of failure.[9] The PAC tradition has offered profound results about the feasibility of learning, but its focus seems distinct from the kind of inquiry common to science. Under mild idealization, scientists are not limited to a finite number of data (fixed in advance), and their goal is to state the underlying truth, not merely to predict correctly with high probability. Even the PAC idea of randomly drawing independent, identically distributed samples of data is discrepant with the bulk of scientific practice—even though we ourselves shall later exploit a similar idea to measure "how many" environments lead a given scientist to success (see Section 3.6, below). For these reasons, the pages that follow do not discuss the many important findings about PAC learning; our idealizations are fundamentally different.[10]

In addition to the idealization of longevity, scientists in our framework benefit from unlimited choice among the maps from SEQ to $\{y, n\}$; any of them can be used to convert data into conjectures about the presence of a least point. Greater realism consists in limiting attention to scientists that implement computable functions. Observe in this connection that both the domain and range of scientists—namely, SEQ and $\{y, n\}$, respectively—are countable sets of finite objects, easily conceived as the inputs and outputs of Turing Machines.

Later paradigms will address computability issues. Uncomputable scientists nonetheless remain central to the discussion in view of our goal of understanding the logic of inquiry independently of its implementation. Similarly, Bayesian analyses generally ignore the problem of actually performing the recommended calculations. It is well known that even elementary problems involving probability cannot be computably decided in the context of an expressive language (see [Gaifman & Snir, 1982]).

---

9. See [Kearns & Vazirani, 1994] for an introduction to the PAC framework. The word "PAC" means "probably approximately correct."

10. For preliminary attempts to unite PAC with the kinds of paradigms studied here, see [Osherson *et al.*, 1991a, Maas & Turán, 1996].

1.3.4 Rationality

The class of all maps from *SEQ* to {**y**, **n**} is broad enough to include pathology of various kinds. For example, call scientist Ψ *dotty* if $\Psi(\sigma) = \mathbf{n}$ for every $\sigma \in SEQ$ of length less than one million. As part of understanding rational inquiry, we would like to find a principled basis for separating scientists of the dotty kind from the more sober variety we prefer. This cannot be achieved by invoking the prospects for success since Lemma (18) shows every solvable problem to be solved by a dotty scientist. Indeed, even considerations of rapid convergence cannot be straightforwardly applied to this case, as discussed in Exercise (21).

Considerable attention will be devoted in what follows to the question of rational hypothesis choice. To help formulate and evaluate proposals, we conserve a liberal conception of "scientist," namely, as any map of appropriate domain and range. Special subsets thereof will then be isolated for further analysis.

1.3.5 Exercises

(19) EXERCISE: A scientist that signals its convergence makes but a single, genuine conjecture per environment; for, the only conjecture that really counts is the one given at the signal. To show how limited is the competence of signaling scientists, it is convenient to offer them somewhat more flexibility. Fix $m \in N$. Scientist Ψ is said to "m-solve" problem **P** if Ψ solves **P**, and for every environment e for an order in **P**, card($\{k \mid \Psi(e[k]) \neq \Psi(e[k+1])\}$) $\leq$ m.[11] Intuitively, to m-solve **P**, Ψ must "change its mind" no more than m times on any environment for an order in **P**. Show that for all m, the problem described in Proposition (7) is not m-solvable.

(20) EXERCISE: ◆ Give an example of a solvable problem that is solved by no computable scientist.

(21) EXERCISE: ♣ If scientist Ψ solves environment e, then we let CP(Ψ, e) denote the least $k_0 \in N$ such that $\Psi(e[k]) = \Psi(e[k_0])$ for all $k \geq k_0$ (the "convergence point" of Ψ on e"). Let **P** be the problem described in Proposition (7). Show that there is a dotty scientist Ψ with the following properties.

(a) Ψ solves **P**.

---

11. We use card(D) to denote the cardinality of the set D.

(b) Let scientist Ψ' also solve **P**, and let environment e for an order in **P** be such that $CP(\Psi', e) < CP(\Psi, e)$. Then there is $\prec \in$ **P** and $k \in N$ such that:

- some environment for $\prec$ begins with $e[k]$, and
- $CP(\Psi, e') \leq k < CP(\Psi', e')$ for every environment e' for $\prec$ that extends $e[k]$.

The second property can be read this way: If Ψ' is faster than Ψ on some environment for an order in **P**, then Ψ is faster than Ψ' on many environments for a different order in **P**.

1.4 Notes

The Bayesian analysis of science is not without its critics, notably, [Glymour, 1980]. A review of the issues is provided by [Earman, 1992]. The contrast between Bayesianism and the present model emerges more clearly in Sections 3.1.4 and 3.6.9 below.

The decomposition of inquiry into the components listed in (1) is based on a discussion found in [Wexler & Culicover, 1980].

The role of reliability in the definition of knowledge is discussed in [Goldman, 1986, Kornblith, 1985, Pappas, 1979]. The more specific relation between reliable inquiry and scientific knowledge is elucidated in [Kelly, 1996].

The order paradigm is a species of "learning-in-the-limit," whose first modern appearance seems to be [Solomonoff, 1964, Putnam, 1975, Gold, 1967]. The idea of "convergence point" in Exercise (21) comes from [Gold, 1967]. We shall meet it again in later chapters.

2 A Numerical Paradigm

2.1 Fundamentals

2.1.1 The Numerical Perspective on Inquiry

Scientists employ rich languages, both for stating theories and for describing data. As a small step in the same direction, the principal paradigm examined in this book is embedded in the framework of first-order logic (see Chapters 3 and 4). Many of the issues to be confronted, however, already arise in the simpler context of a purely numerical paradigm, which is the subject of the present chapter. The paradigm is a variant of an influential model of language acquisition due to E. M. Gold [Gold, 1967]. A considerable literature has been devoted to analyzing and elaborating Gold's model, and it is not our aim to provide a survey of results.[1] Instead, we focus on just those aspects that presage later chapters. In order to maximize overlap with the first-order framework, our treatment of the numerical paradigm will not be the standard one found in most studies. For example, we highlight the learning of languages rather than functions, and introduce computability towards the end rather than the beginning of the discussion.

Some readers will have already encountered the numerical paradigm in other contexts, and not be eager to inch their way through it once again. To them we say: Fine, jump right to Chapter 3; the material in the present chapter is meant only to level the road for newcomers.

Gold's paradigm was motivated by the problem that every human infant faces, namely, to discover an accurate theory of its linguistic environment. By conceiving natural numbers as codes for a fixed, denumerable stock of potential sentences, subsets of N can be understood as potential human languages. The paradigm then allows mathematical formulation of the question: Which collections of languages are learnable by agents with child-like inductive powers? The utility of this perspective for linguistic theory has been much discussed.[2] Following suit, nonempty subsets of N will be called *languages* throughout the chapter. We emphasize, however, that our concern is with inquiry in the general sense, not with the child's specific discovery problem.

---

1. For surveys, see [Angluin & Smith, 1983, Osherson *et al.*, 1986c] and [Jain *et al.*, forthcoming].

2. For example, in [Kanazawa, 1994, Matthews & Demopoulos, 1989, Matthews, 1990], [Gibson & Wexler, 1994, Osherson *et al.*, 1984, Osherson & Weinstein, 1995]. An introduction to the empirical issues surrounding first-language acquisition, is available in [Gleitman & Liberman, 1995].

This is the place for some remarks about notation and terminology. By a "sequence over a set S" is meant a function with domain N and range equal to a nonempty subset of S. Intuitively it helps to think of a sequence e as an infinite list of elements drawn from S. Then, range(e) is the set of elements appearing in the list. Suppose that e begins this way: $5, 6, 9, 5, 0, 25 \ldots$. Then $e(0) = 5$, $e(1) = 6$, and $e(5) = 25$. To denote the restriction of a sequence e to $\{0, 1, \ldots, k - 1\}$, we use: $e[k]$. In terms of lists, $e[k]$ is the finite initial segment of e with length k. Thus, our example yields: $e[0] = \emptyset$, $e[1] = \langle 5 \rangle$, $e[2] = \langle 5, 6 \rangle$, and $e[6] = \langle 5, 6, 9, 5, 0, 25 \rangle$. Note that $e(k)$ shows up in $e[k + 1]$ but not before (except in the case of fortuitous repetition). Given a finite sequence σ of objects, length(σ) denotes the length of σ. Thus, length$(e[k]) = k$ for all $k \in N$. Finally, we use $*$ to concatenate sequences in the obvious way. Thus, $\langle 5, 6, 1 \rangle * e$ is the sequence d with $d[3] = \langle 5, 6, 1 \rangle$ and $d(i + 3) = e(i)$ for all $i \in N$.

2.1.2   Components of the Paradigm

To define the numerical paradigm, we consider its components, in the sense of 1.(1).

Worlds.   Any language is a potential world. Recall that languages are nonempty subsets of N, not necessarily recursively enumerable. The question to be answered about a given language is this: which set of numbers is it?

Problems.   We construe a *problem* **P** to be a collection of languages.

Environments.

(1)   DEFINITION: An *environment* is any sequence over N. Let environment e, language L and problem **P** be given. We say that e is *for* L just in case range$(e) = L$. We say that e is *for* **P** just in case e is for some language in **P**.

Thus, an environment for L enumerates L in arbitrary order, and repetitions of L's elements are allowed. If L is finite, repetition is needed in order to fill out the environment to infinite length (and since $L \neq \emptyset$, there are always elements to repeat). There are uncountably many environments for a language with more than one element.

We let *SEQ* denote the collection of proper initial segments of any environment. Thus, *SEQ* consists of all finite sequences over N, and is a countable set. It is helpful to conceive of $\sigma \in SEQ$ as a data-record or "evidential position." Specifically, for any environment e with $\sigma \subseteq e$, σ records the evidence about range(e) available after reading length(σ) elements of e. The set of elements occurring in σ is denoted range(σ). To illustrate, if $\sigma = \langle 5, 9, 1, 9 \rangle$ then range(σ) = $\{1, 5, 9\}$.

Scientists.   Within the numerical paradigm, a *scientist* is an arbitrary function (partial or total) from *SEQ* to pow(N), the power set of N. (It is convenient to allow $\emptyset$ into the range of a scientist, even though it corresponds to no potential language.) Thus, scientists are systems for converting finite data sets into conjectures about the potentially infinite language inscribed in the environment. If for scientist Ψ and $\sigma \in SEQ$, $\sigma \notin$ domain(Ψ), then Ψ offers no conjecture in response to the information available in σ. Given environment e, the sequence $\{\Psi(e[k]) \mid k \in N\}$ denotes the succession of hypotheses that Ψ issues while examining e (where Ψ is defined). Observe that conjectures are typically infinite objects (e.g., the set of prime numbers). Finitization of the scientist's output is considered in Section 2.5, when considerations of computability arise.

Success.   Success is defined at three levels: first, success for an environment; then, success for a language; finally, success for a problem.

(2)   DEFINITION: Let scientist Ψ, environment e, language L, and problem **P** be given.

(a) We say that Ψ *converges* on e to L just in case $\Psi(e[k]) = L$ for cofinitely many k. In this case, we also say that Ψ *converges* on e (without mention of the language to which Ψ converges). We say that Ψ *solves* e just in case Ψ converges on e to range(e).

(b) We say that Ψ *solves* L just in case Ψ solves every environment for L.

(c) We say that Ψ *solves* **P** just in case Ψ solves every member of **P**. In this case, **P** is said to be *solvable*, otherwise *unsolvable*.

Thus, Ψ solves **P** if and only if Ψ solves every environment for **P**. Since *SEQ* is countable, so is range(Ψ). Since Ψ cannot solve what it cannot say, it follows that:

(3)   LEMMA: Every solvable collection of languages is countable.

2.1.3   An Example of Solvability

To solve problem **P**, Ψ must converge to range(e) on any environment e for **P**. This might seem easy inasmuch as Ψ is shown every part of e. However, Ψ

never sees e in its entirety, and this feature of the paradigm renders discovery nontrivial; that is, not every countable problem is solvable. Before discussing unsolvability, however, we give an example in which all goes well.

(4)   PROPOSITION: Let **P** consist of every set of the form $N - \{n\}$, for $n \in N$. Then **P** is solvable.

Proof: Let $f : SEQ \to N$ map $\sigma \in SEQ$ into the least $n \in N$ with $n \notin \text{range}(\sigma)$. For all $\sigma \in SEQ$, define $\Psi(\sigma) = N - \{f(\sigma)\}$. Then it is easy to see that Ψ solves **P**.   ∎

2.1.4   Exercises

(5)   EXERCISE: Show that the following are solvable:

(a) the collection of all nonempty, finite subsets of N;

(b) for fixed $m \in N$, the collection of all sets of the form $N - D$, where $\text{card}(D) = m$.[3]

(6)   EXERCISE: A scientist is called *confident* if it converges on every environment. Show that no confident scientist solves any of the problems mentioned in Exercise (5).

2.2   Solvability Characterized

We now proceed to develop the theory of the numerical paradigm. Our first order of business is to provide a necessary and sufficient condition for the solvability of a problem. Among other things, the condition will allow us to demonstrate interesting cases of unsolvability. In addition, it provides a useful warm-up for a similar condition that governs solvability in the paradigm of Chapters 3 and 4.

As a preliminary, we prove a lemma of independent interest.

2.2.1   Locking Sequences

Some data seem to lock a scientist onto a particular conjecture in the sense that further data consistent with the conjecture do not prompt its revision. This idea is captured in the following definition.

---

3. We use $\text{card}(D)$ to denote the cardinality of the set D.

(7)   DEFINITION: [Blum & Blum, 1975] Let language L, $\sigma \in SEQ$, and scientist Ψ be given. Then σ is a *locking sequence for Ψ and L* just in case range(σ) $\subseteq L$ and for all $\tau \in SEQ$ with range(τ) $\subseteq L$, $\Psi(\sigma * \tau) = \Psi(\sigma)$.

It follows that if Ψ solves L, and σ is a locking sequence for Ψ and L, then $\Psi(\sigma) = L$; for otherwise, any environment for L that extends σ would cause Ψ to converge to the wrong conjecture $\Psi(\sigma)$, contradicting the hypothesis that Ψ solves L. A deeper fact about locking sequences is this:

(8)   LEMMA: [Blum & Blum, 1975] Suppose that scientist Ψ solves language L. Then there is a locking sequence for Ψ and L.

Thus, the existence of a locking sequence for Ψ and L is a necessary condition for Ψ to solve L. The condition is not sufficient, however, as shown by the following example. Let $\Psi(\sigma) = \{0, 1\}$ if σ begins with 0; otherwise, $\Psi(\sigma)$ is undefined. Then $\sigma = \langle 0 \rangle$ is a locking sequence for Ψ and $\{0, 1\}$, but Ψ does not solve $\{0, 1\}$ since Ψ is undefined on every initial segment of an environment for $\{0, 1\}$ that begins with 1.
   Now let us prove the lemma.

Proof of Lemma (8):   We follow [Blum & Blum, 1975]. Let Ψ solve L, and suppose for a *reductio* that there is no locking sequence for Ψ and L. Then Definition (7) implies:

(9)   For every $\sigma \in SEQ$ with range(σ) $\subseteq L$, there is $\tau \in SEQ$ with range(τ) $\subseteq$ L such that $\Psi(\sigma * \tau) \neq \Psi(\sigma)$.[4]

Let e_1 be an arbitrary environment for L. We construct a sequence $\{\gamma_i \mid i \in N\} \subseteq SEQ$ such that for every $i \in N$:

(10)   (a) $\gamma_j \subseteq \gamma_i$ for all $j \leq i$;

      (b) $e_1(i) \in$ range(γ_i);

      (c) range(γ_i) $\subseteq L$;

      (d) $\Psi(\gamma_i) \neq \Psi(\gamma_{i-1})$ if $i > 0$.

Construction of the γ_i

Stage 0: $\gamma_0 = \langle e_1(0) \rangle$. Trivially, γ_0 satisfies the conditions of (10).

---

4. When we write an inequality like $\Psi(\sigma * \tau) \neq \Psi(\sigma)$, we mean that it is not the case that both terms are defined and equal. So, in particular, the inequality is true if either or both of $\Psi(\sigma * \tau)$, $\Psi(\sigma)$ are undefined.

Stage $i + 1$: Suppose that γ_i satisfies (10). Then range(γ_i) $\subseteq L$, so range($\gamma_i *$ $e_1(i + 1)$) $\subseteq L$. If $\Psi(\gamma_i) \neq \Psi(\gamma_i * e_1(i + 1))$, then we define γ_{i+1} to be $\gamma_i *$ $e_1(i + 1)$. Otherwise, by (9) there is $\tau \in SEQ$ with range(τ) $\subseteq L$ and $\Psi(\gamma_i *$ $e_1(i + 1) * \tau) \neq \Psi(\gamma_i * e_1(i + 1)) = \Psi(\gamma_i)$, and we define γ_{i+1} to be $\gamma_i *$ $e_1(i + 1) * \tau$. It is easy to see that γ_{i+1} also satisfies (10).

End Construction

By (10)a, let $e_0 = \bigcup_{i \in N} \gamma_i$.[5] Then by (10)b,c, e_0 is an environment for L, and by (10)d, $\Psi(e_0[k]) \neq L$ for infinitely many k. Hence, Ψ does not solve e_0, so does not solve L, contradicting our hypothesis. ∎

2.2.2   Characterization by Tip-offs

In the present subsection we introduce a necessary and sufficient condition for solvability, relying on Lemma (8). To explain the idea intuitively, let D be a finite subset of language L in collection **P**. The presence of D in the scientist's data does not justify the conjecture of L if there is $L' \in \mathbf{P}$ with $D \subseteq L' \subset L$. For, in this case systematic conjecture of L on the basis of D will lead the scientist to converge to the wrong answer on any environment e for L'. This is because D will eventually show up in e [since $D \subseteq L' = $ range(e)], and the scientist will respond with the superset L. Finite subsets of L without this drawback are called "tip-offs," and defined as follows.

(11)   DEFINITION: [Angluin, 1980] Let problem **P**, $L \in \mathbf{P}$, and finite $D \subseteq L$ be given. Then D is a *tip-off for L in* **P** just in case there is no $L' \in \mathbf{P}$ with $D \subseteq L' \subset L$. If for all $L \in \mathbf{P}$ there is a tip-off for L in **P**, then we say that **P** *has tip-offs.*

To illustrate, let **P** be the collection of all finite languages. For all $n \in N$, $\emptyset$ and $\{n\}$ are tip-offs for $\{n\}$ in **P**. If $X \in \mathbf{P}$ is not a singleton then X is the only tip-off for X in **P**. We now show that tip-offs are sufficient for solvability.

(12)   PROPOSITION: If problem **P** is countable and has tip-offs then **P** is solvable.

Proof: Suppose that nonempty, countable **P** has tip-offs, with D_L being a tip-off for L in **P**. Fix an enumeration $\{L_i \mid i \in N\}$ of **P**. Define scientist Ψ such that for all $\sigma \in SEQ$, $\Psi(\sigma)$ is the $L \in \mathbf{P}$ with least index such that $D_L \subseteq$ range(σ) $\subseteq$

---

5. Thus, e_0 is the unique environment e with $e[\text{length}(\gamma_i)] = \gamma_i$.

L, if such exists; $\Psi(\sigma)$ is undefined otherwise. Let environment e be for $L \in \mathbf{P}$, and let i be the least index for L. We must show that Ψ converges on e to L. For this purpose, let $j < i$ be given. Then by the choice of i:

(13) $L_j \neq L$.

To conclude the proof it suffices to show that for cofinitely many k, either $D_{L_j} \not\subseteq \text{range}(e[k])$ or $\text{range}(e[k]) \not\subseteq L_j$. For a contradiction suppose the contrary, namely, that for infinitely many k, $D_{L_j} \subseteq \text{range}(e[k])$ and $\text{range}(e[k]) \subseteq L_j$. It follows that:

(14) $D_{L_j} \subseteq L$ and $L \subseteq L_j$.

From (13) and (14) we obtain that $D_{L_j} \subseteq L$ and $L \subset L_j$, which by Definition (11) contradicts the choice of D_{L_j} as a tip-off for L_j in $\mathbf{P}$.   ∎

Next, let us demonstrate the converse to Proposition (12), namely, the necessity of tip-offs.

(15)   PROPOSITION: A solvable problem has tip-offs.

Proof: Suppose that scientist Ψ solves problem $\mathbf{P}$. By Lemma (8), for each $L \in \mathbf{P}$, choose locking sequence σ_L for Ψ and L. We claim that $\text{range}(\sigma_L)$ is a tip-off for L in $\mathbf{P}$. To show this, suppose for a *reductio* that there is $L' \in \mathbf{P}$ with $\text{range}(\sigma_L) \subseteq L' \subset L$, and let e be an environment for L' that begins with σ_L. Then by Definition (7), Ψ converges on e to $L \neq L'$, contradicting the assumption that Ψ solves $\mathbf{P}$.   ∎

From Propositions (12) and (15) and from Lemma (3), we deduce our promised condition for solvability.

(16)   THEOREM: [Angluin, 1980] A problem is solvable if and only if it is countable and has tip-offs.

As an application, suppose that $L \subseteq N$ is infinite. Then it is easy to verify that L has no tip-off in either of the collections: $\{L\} \cup \{D \subseteq L \mid D \text{ finite and non-empty }\}$, and $\{L\} \cup \{L - \{n\} \mid n \in L\}$. We thus have the following corollary to Theorem (16).

(17)   COROLLARY: Let $L \subseteq N$ be infinite. Then

(a) [Gold, 1967] $\{L\} \cup \{D \subseteq L \mid D \text{ finite and nonempty }\}$ is not solvable;

(b) $\{L\} \cup \{L - \{n\} \mid n \in L\}$ is not solvable.

2.2.3 Exercises

(18) EXERCISE: Suppose that scientist Ψ solves language L. Show the following.

(a) For every $\sigma \in SEQ$ with range$(\sigma) \subseteq L$, some extension of σ is a locking sequence for Ψ and L.

(b) Environment e is called a *locking environment* for Ψ and L if it begins with a locking sequence for Ψ and L. Does it follow from the assumption that Ψ solves L that every environment for L is a locking environment for Ψ and L?

(19) EXERCISE: There are two ways that a scientist Ψ can perform poorly in an environment e. Ψ can either converge on e to an incorrect conjecture or else fail to converge at all. The former behavior might be considered worse than the latter since in the case of nonconvergence Ψ signals the incorrectness of each conjecture by ultimately abandoning it. With this in mind, call Ψ *reliable* just in case for all environments e, if Ψ converges on e then Ψ solves e. Prove the following, depressing fact about reliability: If Ψ is reliable then Ψ solves no infinite language.

(20) EXERCISE: ♣ Environment e is *ascending* if for all $i \in N$, $e(i) \leq e(i + 1)$. Problem **P** is solvable *on ascending environments* if there is a scientist that solves every ascending environment for **P**. Prove that some unsolvable problem is solvable on ascending environments. Also, give an example of a problem that is not solvable on ascending environments.

(21) EXERCISE: Use Theorem (16) to show that every finite problem is solvable.

(22) EXERCISE: ♦ Show that the qualifier "countable" cannot be omitted from the statement of Theorem (16).

(23) EXERCISE: ♣ A problem is *saturated* if it is solvable but none of its proper supersets are solvable. Show that there is exactly one saturated problem.

2.3 Efficiency and Rationality

As discussed in Section 1.3, our policy is to define scientists in liberal fashion, with no initial attempt to weed out the inefficient and irrational. In a second step, however, it becomes very much our concern to characterize subclasses of scientists who are rational or efficient in various senses. Indeed, rationality and efficient use of data are pivotal issues for the theory developed in this book. These concepts are now introduced for the numerical paradigm.

2.3.1 Efficiency

We consider a scientist to be efficient if it stabilizes as quickly as possible to the correct hypothesis. One way to make this idea precise is as follows.

(24) DEFINITION: Suppose that scientist Ψ solves problem **P**.

(a) Given environment e for **P**, we let $CP(\Psi, e)$ denote the least $k_0 \in N$ such that $\Psi(e[k]) = \text{range}(e)$ for all $k \geq k_0$.

(b) We say that Ψ solves **P** *efficiently* just in case for every scientist Ψ' that solves **P** and for every environment e_0 for **P**, if $CP(\Psi', e_0) < CP(\Psi, e_0)$ then the following holds: there is $L \in \mathbf{P}$ and $\sigma \in SEQ$ with $\text{range}(\sigma) \subseteq L$ such that $CP(\Psi, e_1) < CP(\Psi', e_1)$ for every environment e_1 for L that extends σ. In this case, **P** is solvable *efficiently*.

Intuitively, suppose that scientists Ψ and Ψ' both solve problem **P**, and that Ψ solves it efficiently. Then if Ψ' is faster than Ψ on some environment for **P**, there are many other environments for **P** on which Ψ is faster than Ψ'. The notation $CP(\Psi, e)$ may be read as "the convergence point of Ψ on e." It is a striking fact that reinforcing our success criterion by requiring efficiency does not alter the class of solvable problems. The matter can be expressed as follows.

(25) THEOREM: Every solvable problem is efficiently solvable.

Proof: Let nonempty, solvable problem **P** be given. By Theorem (16), **P** is countable, and for each $L \in \mathbf{P}$ we may choose a tip-off D_L for L in **P**. Fix an enumeration $\{L_i \mid i \in N\}$ of **P** (possibly with repetitions). Let $\sigma \in SEQ$ and $i \in N$ be given. The "σ-score of i" is defined to be the smallest $n \in N$ such that for some extension $\tau \in SEQ$ of σ with $\text{length}(\tau) - \text{length}(\sigma) = n$ the following conditions hold.

(26) (a) $D_{L_i} \subseteq \text{range}(\tau) \subseteq L_i$, and

(b) for every $j < i$, either $\text{range}(\tau) - L_j \neq \emptyset$ or $D_{L_j} - L_i \neq \emptyset$.

If there is no such n, then the σ-score of i is ∞. It is easy to verify:

(27) Let $L \in \mathbf{P}$ be given, and let i be the least index for L. Let environment e for L also be given. Then for cofinitely many k:

(a) the $e[k]$-score of i is 0;

(b) if $j < i$, the $e[k]$-score of j is ∞.

Define scientist Ψ as follows. For all $\sigma \in SEQ$, $\Psi(\sigma) = L_i$, where i is least with minimal σ-score. In view of (27), it is clear that Ψ solves **P**.

For efficiency suppose that scientist Ψ' solves **P**, and that for some environment e_0 for **P**, $CP(\Psi, e_0) > CP(\Psi', e_0)$. Then there is $k_0 \in N$ with $\Psi(e_0[k_0]) \neq \Psi'(e_0[k_0])$. Let $\sigma = e_0[k_0]$ and $L_i = \Psi(\sigma)$. Because e_0 is for **P**, let natural number n be the σ-score of i. Let extension $\tau \in SEQ$ of σ be such that $\text{length}(\tau) - \text{length}(\sigma) = n$ and (26) holds. Because i is least with minimal σ-score, it is clear that $CP(\Psi, e) = k_0$ for every environment e for L_i that extends τ. In contrast, since $\Psi'(\sigma) \neq L_i$, $CP(\Psi', e) > k_0$ for any such environment e. ∎

A strong form of *inefficiency* can be defined as follows. It will be exploited later.

(28) DEFINITION: Suppose that scientist Ψ solves problem **P**. We say that Ψ is *dominated* on **P** just in case there is a scientist Ψ' that solves **P** such that:

(a) for every environment e for **P**, $CP(\Psi', e) \leq CP(\Psi, e)$, and

(b) for some environment e for **P**, $CP(\Psi', e) < CP(\Psi, e)$.

Informally, Ψ' dominates Ψ on **P** if Ψ' is never slower than Ψ on environments for **P**, and is sometimes faster. Unravelling Definitions (28) and (24) yields the following.

(29) LEMMA: Suppose that scientist Ψ solves problem **P**. If Ψ is dominated on **P** then Ψ does not solve **P** efficiently.

2.3.2 Rationality

What is the rational design for a scientist charged with solving a given problem? The next definition introduces properties of scientists that seem essential to their rationality.

(30) DEFINITION: Let scientist Ψ and problem **P** be given. We say that $\sigma \in SEQ$ is *for* **P** just in case some environment for **P** extends σ.

(a) Ψ is **P**-*total* just in case for every σ for **P**, $\Psi(\sigma)$ is defined.

(b) Ψ is **P**-*bound* just in case for every σ for **P**, if $\Psi(\sigma)$ is defined then $\Psi(\sigma) \in$ **P**.

(c) [Angluin, 1980] Ψ is *P-consistent* just in case for every σ for **P**, if $\Psi(\sigma)$ is defined then range$(\sigma) \subseteq \Psi(\sigma)$.

(d) [Angluin, 1980] Ψ is *P-conservative* just in case for all σ for **P** and $n \in N$, if $\Psi(\sigma)$ is defined and range$(\sigma) \cup \{n\} \subseteq \Psi(\sigma)$, then $\Psi(\sigma * n) = \Psi(\sigma)$.

(e) Ψ is *P-rational* if it is **P**-total, **P**-bound, **P**-consistent and **P**-conservative.

Intuitively, we can paraphrase Definition (30) as follows.

- **P**-total scientists never lack an hypothesis in response to data from **P**.

- **P**-bound scientists only announce conjectures relevant to **P** in response to data from **P**.

- **P**-consistent scientists do not announce hypotheses that are contradicted by their current data from **P**.

- **P**-conservative scientists stick with conjectures that are not falsified by data from **P**.

- **P**-rational scientists possess all of the foregoing virtues.

It is easy enough, of course, to award the term "rational" to a set of scientists. Issues of substance arise only when we try to justify the prize. As a start, let us offer the following proposition, which reveals an advantage of rationality as we have defined it.

(31) PROPOSITION: Let problem **P** and **P**-rational scientist Ψ be given. If Ψ solves **P** then Ψ solves **P** efficiently.

Proof: Let problem **P** be given. Suppose that scientist Ψ is **P**-rational and solves **P**. Let scientist Ψ' also solve **P**. Let environment e_0 for **P** be such that $\mathrm{CP}(\Psi', e_0) < \mathrm{CP}(\Psi, e_0)$. Let $k_0 = \mathrm{CP}(\Psi', e_0)$. We show:

(32) $\Psi(e_0[k_0]) \neq \Psi'(e_0[k_0])$.

Proof of (32): By Definition (30)a, $\Psi(e_0[k_0])$ is defined. Suppose for a contradiction that $\Psi(e_0[k_0]) = \Psi'(e_0[k_0])$. Since Ψ' solves **P** and $\mathrm{CP}(\Psi', e_0) = k_0$, it follows that $\Psi'(e_0[k_0]) = \mathrm{range}(e_0) = \Psi(e_0[k_0])$. So by Definition (30)d, $\Psi(e_0[k]) = \mathrm{range}(e_0)$ for all $k \geq k_0$. Hence, $\mathrm{CP}(\Psi, e_0) \leq k_0 = \mathrm{CP}(\Psi', e_0)$, contradicting the assumption that $\mathrm{CP}(\Psi', e_0) < \mathrm{CP}(\Psi, e_0)$. ∎

By Definition (24), to complete the proof we exhibit $L \in \mathbf{P}$ and $\sigma \in SEQ$ with range$(\sigma) \subseteq L$ such that $\mathrm{CP}(\Psi, e_1) < \mathrm{CP}(\Psi', e_1)$ for every environment e_1 for L that extends σ. Let $\sigma = e_0[k_0]$, and let $L = \Psi(\sigma)$. By Definition (30)b, $L \in \mathbf{P}$. By Definition (30)c, range$(\sigma) \subseteq L$. Let environment e_1 extend σ and be for L. By Definition (30)d, $CP(\Psi, e_1) \leq k_0$. On the other hand, $CP(\Psi', e_1) > k_0$ since otherwise, $\Psi'(e_1[k_0]) = \Psi'(e_0[k_0]) = L = \Psi(\sigma) = \Psi(e_0[k_0])$, contradicting (32).   ∎

The results recorded in Theorem (25) and Proposition (31) raise an obvious question. Does rationality offer a *canonical form* for inquiry? In other words, is every solvable problem **P** solved efficiently by some **P**-rational scientist? Canonical forms for scientific inquiry will take center stage in Chapter 4, where many will be exhibited. In the present context our results are negative: rationality within the numerical paradigm is far from canonical. Indeed, the next proposition shows that some solvable problems **P** cannot even be solved in the **P**-rational way, never mind efficiently.

(33) PROPOSITION: There exists a solvable problem **P** that is solved by no **P**-total, **P**-bound, **P**-conservative scientist (hence it is solved by no **P**-rational scientist).

Proof: For $n \in N$, denote $\{n, n + 1, \ldots\}$ by L_n, and set $\mathbf{P} = \{L_n \mid n \in N\}$. Let scientist Ψ be such that for every non-empty $\sigma \in SEQ$, $\Psi(\sigma) = L_n$, where $n \in N$ is smallest with $n \in \text{range}(\sigma)$. It is easy to verify that Ψ solves **P**.

Let **P**-total, **P**-bound, **P**-conservative scientist Ψ' be given. To finish the proof we show that Ψ' does not solve **P**. Since Ψ' is **P**-total and **P**-bound, $\Psi'(\emptyset) = L_n$ for some $n \in N$. Choose environment e for L_{n+1}. Since $L_{n+1} \subseteq L_n$ and Ψ' is **P**-conservative, $\Psi'(e[k]) = L_n$ for every $k \in N$. This proves that Ψ fails to solve e, hence fails to solve L_{n+1}, hence fails to solve **P**.   ∎

By Exercise (40), any scientist that solves a problem **P** and is not dominated on **P** is **P**-total, **P**-bound, and **P**-consistent. Hence Proposition (33) yields:

(34) COROLLARY: There is a solvable problem **P** such that for all scientists Ψ, if Ψ solves **P** and Ψ is **P**-conservative, then Ψ is dominated on **P**.

So the questionable component in the concept of rationality appears to be conservatism. In its defense we may evoke the apparent capriciousness of announcing an hypotheses that will later be withdrawn in the face of confirmatory data. Moreover, conservatism by itself is compatible with solvability. Indeed, the next proposition shows this to be so even in the presence of boundedness and consistency.

(35) PROPOSITION: Every solvable problem **P** is solved by a **P**-bound, **P**-consistent, and **P**-conservative scientist.

Proof: Let nonempty, solvable problem **P** be given. By Theorem (16), **P** is countable, so we may fix a repetition-free enumeration $\{L_i \mid i < \kappa\}$ of **P**, where $\kappa = \text{card}(\mathbf{P})$. For each $i < \kappa$, choose a tip-off D_{L_i} for L_i in **P**. Given $\sigma \in SEQ$ and $i < \kappa$, say that σ is "*i*-complete" just in case the following holds:

(36)   (a) $D_{L_i} \subseteq \text{range}(\sigma) \subseteq L_i$, and

     (b)   for every $j < i$, either $\text{range}(\sigma) - L_j \neq \emptyset$ or $D_{L_j} - L_i \neq \emptyset$.

We now show:

(37)   For all $\sigma \in SEQ$, there is at most one i such that σ is i-complete.

Proof of (37):   Suppose for a contradiction that $\sigma \in SEQ$ is both i-complete and j-complete for $j < i < \kappa$. It follows from (36)a that:

(38)   (a) $\text{range}(\sigma) \subseteq L_j$;

     (b) $D_{L_i} \subseteq \text{range}(\sigma)$;

     (c) $\text{range}(\sigma) \subseteq L_i$.

From (36)b and (38)a we infer that $D_{L_j} - L_i \neq \emptyset$, which together with (38)b implies that $\text{range}(\sigma) - L_i \neq \emptyset$. But this contradicts (38)c.   ∎

    Now define scientist Ψ as follows. For every $\sigma \in SEQ$, $\Psi(\sigma) = L_i$ if σ is i-complete; if σ is i-complete for no $i < \kappa$, then $\Psi(\sigma)$ is undefined. It is easy to see that Ψ solves **P** and is both **P**-bound and **P**-consistent. Let $\sigma \in SEQ$ be such that $\Psi(\sigma) = L_i$ for some $i < \kappa$. Then σ is i-complete, and it is immediately verified that for all $n \in L_i$, $\sigma * n$ is i-complete, hence $\Psi(\sigma * n) = L_i = \Psi(\sigma)$. So Ψ is **P**-conservative.   ∎

    Taking stock, Theorem (25), Corollary (34), and Proposition (35) suggest tension between conservatism and the desire for efficient use of data. This is because every solvable problem is solvable efficiently and also conservatively, but the two qualifiers are sometimes incompatible. We are thus led to the following conclusion, with respect to the numerical paradigm. Although it may seem capricious to announce an hypothesis that will later be withdrawn in the face of confirmatory data, such behavior is sometimes necessary to arrive at the truth as fast as possible.[6]

---

6. An historical example of successful theory change that appears not to be conservative is discussed in [Kuhn, 1957].

2.3.3   Exercises

(39)   EXERCISE: Suppose that scientist Ψ solves problem **P**. Show that Ψ is not dominated on **P** iff the following holds. For every $\sigma \in SEQ$, if σ is for **P** then there exists environment e for **P** such that:

(a) e extends σ, and

(b) for every $k \geq \text{length}(\sigma)$, $\Psi(e[k]) = \text{range}(e)$.

(40)   EXERCISE: Suppose that scientist Ψ solves problem **P**. Show that if Ψ is not dominated on **P**, then Ψ is **P**-total, **P**-bound, and **P**-consistent. On the other hand, show that Ψ can be **P**-total, **P**-bound, **P**-consistent, and dominated on **P**.

(41)   EXERCISE: ♦   Suppose that scientist Ψ solves problem **P**. Show that Ψ solves **P** efficiently iff the following holds. If $\sigma \in SEQ$ is for **P** then there exists $L \in \mathbf{P}$ and $\tau \in SEQ$ such that:

(a) τ extends σ,

(b) $\text{range}(\tau) \subseteq L$, and

(c) for every environment e for L that extends τ, for every $k \geq \text{length}(\sigma)$, $\Psi(e[k]) = L$.

(42)   EXERCISE: Suppose that scientist Ψ solves problem **P**. We say that Ψ solves **P** *strongly efficiently* just in case for every scientist Ψ' that solves **P** and for every environment e for **P**, if $k_0 = \text{CP}(\Psi', e) < \text{CP}(\Psi', e)$ then the following holds: for some $L \in \mathbf{P}$, $\text{range}(e[k_0]) \subseteq L$ and $\text{CP}(\Psi, e') \leq k_0 < \text{CP}(\Psi', e')$ for every environment e' for L that extends $e[k_0]$. In this case, **P** is solvable *strongly efficiently*. Now let problem **P** and scientist Ψ that solves **P** be given. Show that Ψ is **P**-rational if and only if Ψ solves **P** strongly efficiently.

(43)   EXERCISE: ♣   Given a well-ordering $\prec$ over a problem **P**, we define an associated scientist $\Psi_\prec$ as follows. For all $\sigma \in SEQ$, $\Psi_\prec(\sigma)$ is the $\prec$-least member of **P** that includes $\text{range}(\sigma)$; $\Psi_\prec(\sigma)$ is undefined if there is no such. Note that $\Psi_\prec$ is **P**-rational. Problem **P** is *solvable by enumeration* just in case there is a well-ordering $\prec$ of **P** such that $\Psi_\prec$ solves **P**. Show that there exists a problem **P** such that some **P**-rational scientist solves **P**, but **P** is not solvable by enumeration.

2.4 Memory Limitation

2.4.1 Strategies and Canonical Forms

Definition (30) embodies one attempt to define a canonical form for inquiry within the numerical paradigm. The attempt founders on Proposition (33), which shows that scientists of the "rational" kind (as specified by the definition) are unable to solve certain, solvable problems.

A different way to approach the same issue is to single out a subset S of the vast class of scientists, and to determine whether every solvable problem falls into the competence of some scientist in S. [This is not quite the approach of Definition (30) since the concept of rationality was there linked to specific problems **P**.] Subsets of scientists are often called *strategies*, and we say that a strategy is "canonical" if every solvable problem is solved by at least one of its members.[7] As an illustration we consider a strategy that provides an alternative perspective on efficient data use. One formalization of this concept was presented in Definition (24), using time-to-convergence CP(Ψ, e). A different idea is to bound the number of data employed in choosing hypotheses at each stage of inquiry. Let us explain.

Working in environment e, scientist Ψ modifies its hypothesis $\Psi(e[k])$ on the basis of $e[k + 1]$, which includes all data $e(0) \ldots e(k)$ observed since the start of inquiry. In contrast, real scientists seldom have access to the entire record of previous observations, and could not long keep them in mind even if they did. Rather, inquiry often proceeds by modifying current theory in light of recent data only. In the present section we consider the simplest model of this kind; others are developed in the exercises, and an extended study is available in [Kinber & Stephan, 1995]. In a manner reminiscent of the "total evidence requirement" in probabilistic contexts [Carnap, 1950, Sec. 45] it will be seen that the competence of scientists is impaired by limiting their memory. In other words, memory-limited scientists conserve resources in one clear sense (since they do not require data to be archived), but for this same reason they do not yield a canonical form of inquiry.

2.4.2 Definition and Example

Now for the details. The following definition isolates scientists whose current datum and current conjecture determine the next conjecture.

---

7. The "strategy" terminology is from [Osherson *et al.*, 1982]. In place of canonicity it is often said that a strategy is "restrictive" in case some solvable problem is solved by no scientist within it.

(44) DEFINITION: [Wexler & Culicover, 1980] Scientist Ψ is *memory-limited* just in case Ψ is total, and there is a total function $f : \text{pow}(N) \times N \to \text{pow}(N)$ such that for all $\sigma \in SEQ$ and $n \in N$, $\Psi(\sigma * n) = f(\Psi(\sigma), n)$.

On first impression memory limitation might appear to be debilitating. Such is not the case, however, since it is possible temporarily to store past data in the current hypothesis. This tactic is illustrated in the following proposition.

(45) PROPOSITION: Some memory-limited scientist solves the collection of all finite languages.

Proof: Define $f : \text{pow}(N) \times N \to \text{pow}(N)$ as follows. For all $S \subseteq N$ and $n \in N$, $f(S, n) = S \cup \{n\}$. Let scientist Ψ be such that $\Psi(\emptyset) = \emptyset$ and for all $\sigma \in SEQ, n \in N$, $\Psi(\sigma * n) = f(\Psi(\sigma), n)$. It is clear that Ψ is memory-limited. Moreover, given environment e and $k \in N$ it is easy to see that $\Psi(e[k]) = \text{range}(e[k])$. Hence, Ψ solves e if $\text{range}(e)$ is finite. ∎

2.4.3 Competence of Memory-Limited Scientists

Although not debilitating, memory-limitation is nonetheless restrictive. In particular, comparison of Proposition (4) to the following fact reveals that some solvable problems cannot be solved by memory-limited scientists.

(46) PROPOSITION: No memory-limited scientist solves the collection of all languages of the form $N - \{n\}$, for some $n \in N$.

Proof: For a *reductio*, suppose that memory-limited Ψ solves the collection **P** of all sets of the form $N - \{n\}$, for some $n \in N$. By Lemma (8), let $\sigma \in SEQ$ be a locking sequence for Ψ and $N - \{0\}$. Choose $x, y, z \in N$ with $z > y > x > 1$ and $x, y \notin \text{range}(\sigma)$. Let $\tau_x = \langle 1 \ldots x - 1, x + 1 \ldots z \rangle$ and $\tau_y = \langle 1 \ldots y - 1, y + 1 \ldots z \rangle$. By Definition (7):

(47) $\Psi(\sigma * \tau_x) = \Psi(\sigma * \tau_y) = \Psi(\sigma) = N - \{0\}$.

Let $e = \langle 0, z + 1, z + 2, \ldots \rangle$, $e_x = \sigma * \tau_x * e$, and $e_y = \sigma * \tau_y * e$. Then by (47) and Definition (44), $\Psi(e_x[k]) = \Psi(e_y[k])$ for all $k \geq \text{length}(\sigma) + z$. Hence, Ψ fails to solve at least one of e_x and e_y. But both environments are for **P**. ∎

So, the memory-limited strategy is not canonical. A second drawback to memory-limitation concerns efficient discovery. Whereas Theorem (25) shows that every solvable problem is solvable efficiently, the next proposition shows that this is no longer true for the class of memory-limited scientists.

(48) PROPOSITION: There is a problem **P** such that some memory-limited scientist solves **P**, but every memory-limited scientist that solves **P** is dominated on **P**.

Proof: Let $\{E_i \mid i \in N\}$ partition $N - \{0\}$, and be such that E_i is infinite for all i. Let **P** consist of $N - \{0\}$ along with every set of the form $N - E_i$. We will specify a memory-limited scientist that solves **P**, and then show that no **P**-consistent, memory-limited scientist solves **P**. It will then follow from Exercise (40) that every memory-limited scientist that solves **P** is dominated on **P**.

Let function $f : \text{pow}(N) \times N \to \text{pow}(N)$ be defined as follows. For every $S \subseteq N$ and $n \in N$:

(a) if $n = 0$ and $S = N - \{0\}$, then $f(S, n) = N - E_0$;

(b) if $n \in E_i$ and $S \cap E_i = \emptyset$, then $f(S, n) = (S \cup E_i) - E_{i-1}$;

(c) otherwise, $f(S, n) = S$.

Let scientist Ψ be such that $\Psi(\emptyset) = N - \{0\}$ and for all $\sigma \in SEQ$ and $n \in N$, $\Psi(\sigma * n) = f(\Psi(\sigma), n)$. It is clear that Ψ is memory-limited. We show that Ψ solves **P**. Let environment e be for $N - \{0\}$. Then it is easy to verify that $\Psi(e[k]) = N - \{0\}$ for all $k \in N$, and that Ψ solves e. Let $i \in N$ and environment e be for $N - E_i$. Let $k_0 \in N$ be least such that $e(k_0) = 0$. There exists a unique sequence $k_0 < k_1 < \ldots < k_{i+1}$ of integers such that for all $1 \leq j \leq i + 1$, k_j is the least $k > k_{j-1}$ such that $e(k) \in E_{j-1}$. (This sequence exists since every E_j, $j \in N$, is infinite.) It follows that:

(a) for all $k \leq k_0$, $\Psi(e[k]) = N - \{0\}$;

(b) for all $0 \leq j \leq i$, for all $k_j < k \leq k_{j+1}$, $\Psi(e[k]) = N - E_j$;

(c) for all $k > k_{i+1}$, $\Psi(e[k]) = N - E_i$.

Hence Ψ converges on e to range(e), so we have proved that Ψ solves **P**.

Now suppose that memory-limited scientist Ψ solves **P**. We have to show that Ψ is not **P**-consistent. Since Ψ solves $N - \{0\}$, choose $\sigma \in SEQ$ that is a locking-sequence for Ψ and $N - \{0\}$. Choose $i \in N$ with $E_i \cap \text{range}(\sigma) = \emptyset$. Let environment e be such that $\sigma * e$ is an environment for $N - E_i$. Since Ψ solves $N - E_i$, there exists $k > 0$ with $\Psi(\sigma * e[k]) = N - E_i$. Fix $n_i \in E_i$. By the choice of σ, $\Psi(\sigma * n_i) = N - \{0\} = \Psi(\sigma)$. Since Ψ is memory-limited, $\Psi(\sigma * n_i) = \Psi(\sigma)$ implies by induction on $k' \leq k$ that for all $k' \leq k$, $\Psi(\sigma * n_i * e[k']) = \Psi(\sigma * e[k']) = N - E_i$. Hence $\Psi(\sigma * n_i * e[k]) = N - E_i$. Trivially, $\sigma * n_i * e[k]$ is for **P** since it is an initial segment of $N - E_j$ for any j such that $E_j \cap (\text{range}(\sigma * n_i * e[k]) - \{0\}) = \emptyset$. But $\text{range}(\sigma * n_i * e[k]) \not\subseteq N - E_i$ (since $n_i \in E_i$), so Ψ is not **P**-consistent. ∎

2.4.4 Exercises

(49) EXERCISE: ♣ Show that for every problem **P**, if some memory-limited scientist solves **P** then some **P**-conservative, memory-limited scientist solves **P**.

(50) EXERCISE: Scientist Ψ is *gradualist* if for all $\sigma \in SEQ$,

$$\{\Psi(\sigma * n) \mid n \in N \text{ and } \Psi(\sigma * n) \text{ is defined}\}$$

is finite. Thus, gradualist scientists modify their conjectures slowly.

(a) Show that every solvable problem is solved by a gradualist scientist.

(b) Exhibit a problem **P** such that some memory-limited scientist solves **P** but no gradualist, memory-limited scientist solves **P**.

Thus, gradualism is not a restrictive strategy for selecting hypotheses. However, it amplifies the difficulties associated with memory-limitation.

(51) EXERCISE: ♠ Environment e is *fat* if for all $n \in \text{range}(e)$, $\{j \mid e(j) = n\}$ is infinite. Show that a problem **P** is solvable iff some memory-limited scientist solves every fat environment for **P**.

(52) EXERCISE: ♥ Let $m > 0$ be given. Scientist Ψ is *m-memory-limited* just in case Ψ is total and there is a total function $f : \text{pow}(N) \times N^m \to \text{pow}(N)$ such that for all $\sigma \in SEQ$ and $(x_1 \ldots x_m) \in N^m$, $\Psi(\sigma * \langle x_1 \ldots x_m \rangle) = f(\Psi(\sigma * \langle x_1 \ldots x_{m-1} \rangle), x_1 \ldots x_m)$. Thus, memory-limitation in the sense of Definition (44) is 1-memory-limitation. Show that for every m-memory-limited scientist Ψ there is a 1-memory-limited scientist that solves every language that Ψ solves.

(53) EXERCISE: ♣ Given $m \in N$, let D_m be the set of all subsets of N of cardinality at most m. By an "m-data selector" is meant any function g from SEQ to D_m such that $g(\emptyset) = \emptyset$ and $g(\sigma * n) \subseteq g(\sigma) \cup \{n\}$ for all $\sigma \in SEQ$ and $n \in N$. Scientist Ψ is *m-buffer-limited* just in case Ψ is total and there is a total function $f : \text{pow}(N) \times D_m \times N \to \text{pow}(N)$ and an m-data selector g such that for all $\sigma \in SEQ$ and $n \in N$, $\Psi(\sigma * n) = f(\Psi(\sigma), g(\sigma * n), n)$. Definition (44)

is 0-buffer-limitation. Thus, an m-buffer-limited scientist can hold in mind an updated register of m data.

(a) Given $m \in N - \{0\}$, fix a coding of $\{1 \ldots m\} \times N$ into $N - \{0\}$, and denote by $\langle x, y \rangle$ the image of (x, y) under this coding. Let problem $\mathbf{P}_m$ consist of $N - \{0\}$ together with every set of form $\{0\} \cup \bigcup_{1 \le i \le m} \{\langle i, 0 \rangle \ldots \langle i, p_i \rangle\}$, for $p_1 \ldots p_m \in N$. Show that for all $m \in N$, $\mathbf{P}_m$ is solved by some m-buffer-limited scientist, but $\mathbf{P}_m$ is solved by no $(m - 1)$-buffer-limited scientist (in particular, $\mathbf{P}_1$ is solved by no memory-limited scientist).

(b) Exhibit a solvable problem that is solved by no m-buffer-limited scientist, for any $m \in N$.

(c) Let $\mathbf{P}$ be the problem specified in the proof of Proposition (48). Show that for all $m \in N$, there is no m-buffer-limited scientist that solves $\mathbf{P}$ efficiently.

2.5   Computable Scientists

2.5.1   Motives and Goals

Computability is perhaps the most interesting scientific strategy. For one thing, human scientists may be computing engines in a relevant sense, and in this case we would like to know what such devices are good for.[8] Moreover, even if real scientists turn out to escape the confines of the computable, it is important to understand how much of empirical inquiry can in principle be shifted to electronic computers. So, the present section focusses on scientists whose behavior is algorithmic in a sense to be made precise.

When we isolate the computable scientists for separate study the mathematics of inquiry takes on considerable richness and subtlety. For this reason it is worth stressing that our aim is only to introduce the topic, not to explore it in depth. More thorough treatments are available elsewhere, notably in [Odifreddi, 1997, Jain et al., forthcoming]. In what follows we presuppose elementary concepts of recursion theory, as developed in [Rogers, 1987, Davis & Weyuker, 1983, Machtey & Young, 1978, Odifreddi, 1989]. In particular, let us now fix an acceptable indexing $\{W_i \mid i \in N\}$ of the recursively enumerable ($r.e.$) subsets of N.

---

8. How to determine whether a physical device—like a human scientist—instantiates an abstract scheme for computation is a delicate issue that will not be pursued here. For discussion, see [Osherson, 1985].

2.5.2 Definitions and Background Concepts

Computability enters the numerical paradigm in two ways. First, we use computer programs as finite names for subsets of N. Second, to define the computable scientists, we consider the computable functions from SEQ to the set of r.e. indexes. Putting together the two uses of computability yields the following definition, central to our discussion.

(54) DEFINITION: Scientist Ψ is *computable* just in case there is computable $\psi : SEQ \to N$ such that for all $\sigma \in SEQ$, $\Psi(\sigma)$ is defined iff $\psi(\sigma)$ is defined, and when both are defined $\Psi(\sigma) = W_{\psi(\sigma)}$. In this case, we say that ψ *underlies* Ψ. A problem that is solved by a computable scientist is *computably solvable*.

Note that the conjectures of a computable scientist may stabilize without the same being true of the function underlying it. To appreciate the point, let $\{x_i \mid i \in N\}$ be a recursive enumeration without repetition of (some of the) indexes of N. Suppose that computable $\psi : SEQ \to N$ underlies scientist Ψ and is such that for all $\sigma \in SEQ$, $\psi(\sigma) = x_{\text{length}(\sigma)}$. Then for all $k \in N$, $\Psi(\langle 0, 1 \ldots k \rangle) = N$, but $\psi(\langle 0, 1, \ldots, k \rangle) \neq \psi(\langle 0, 1, \ldots, k, k+1 \rangle)$. So Ψ stabilizes to N on any environment for N whereas the output of ψ keeps changing. The volatility of ψ does not prevent Ψ from solving N, in the sense of Definition (2). Requiring both Ψ and ψ to stabilize in the environments Ψ solves yields a stricter criterion of success, considered in Section 2.5.5, below.[9]

Since the languages issued as hypotheses by computable scientists are r.e., it is evident that a problem can be computably solved only if it consists of such languages. [Generalization to a richer class of names for sets is considered in Exercise (77).] Observe also that there are only countably many computable scientists, since there are only countably many computable functions to underlie them.

We emphasize that computable scientists are scientists in the original sense given in Section 2.1.2, namely, functions (total or partial) from SEQ to pow(N). They are distinguished from other scientists only in their simulability via an underlying computable function ψ from SEQ to N. The predicates "**P**-consistent," "efficiently solvable," etc. defined in Section 2.3 thus apply

---

9. The criterion of computable solvability defined above is thus closest to "behaviorally correct (BC) identification" defined in [Case & Smith, 1983, Jain *et al.*, forthcoming], and to "extensional identification" defined in [Osherson *et al.*, 1986c].

straightforwardly to computable scientists. Their interaction with computability, however, can have complex consequences, as will be seen. First, here is a simple example of computable solvability.

(55)   PROPOSITION: Let A be nonempty and $r.e.$ Then $\{A \cup D \mid D \subseteq N$ and D finite$\}$ is computably solvable.

Proof: Since A is $r.e.$, there is a computable function $f : SEQ \rightarrow N$ such that for all $\sigma \in SEQ$, $f(\sigma)$ is an index for $A \cup \text{range}(\sigma)$. It is easy to see that f underlies a scientist that solves every environment for a finite extension of A. ∎

It follows immediately that the collection of finite languages is computably solvable. It can also be easily shown that $\{N - \{n\} \mid n \in N\}$ is computably solvable.

2.5.3   Competence of Computable Scientists

Even when attention is limited to problems consisting of $r.e.$ sets, scientific competence is reduced by the constraint of computability. The following theorem shows that part of the reduction is due merely to the cardinality of the class of computable scientists.

(56)   THEOREM: Let Σ be a countable collection of scientists. Then there is a problem $\mathbf{P}$ such that $\mathbf{P}$ is solvable, every member of $\mathbf{P}$ is recursive, but no scientist in Σ solves $\mathbf{P}$.

Proof: Given $X \subseteq N$, let $\mathbf{P}_X$ denote the collection of all nonempty $L \subseteq N$ such that if n is the smallest member of L, then either $n \in X$ and L is finite, or $n \notin X$ and $L = \{n, n + 1 \ldots\}$. It is easy to verify that for every $X \subseteq N$, $\mathbf{P}_X$ is a solvable problem consisting of recursive sets. Since there are uncountably many $X \subseteq N$ and Σ is countable, the theorem is proved if we show the following fact.

(57)   If X and Y are distinct subsets of N then $\mathbf{P}_X \cup \mathbf{P}_Y$ is not solvable.

To demonstrate (57), let $X, Y \subseteq N$ and $n \in N$ be such that $n \in X$ iff $n \notin Y$. Denote by $\mathbf{R}$ the collection consisting of $\{n, n + 1 \ldots\}$ together with every finite subset of N whose least member is n. It is immediate that $\{n, n + 1 \ldots\}$ has no tip-offs in $\mathbf{R}$, so by Proposition (15) $\mathbf{R}$ is not solvable. Since $\mathbf{R} \subseteq \mathbf{P}_X \cup \mathbf{P}_Y$, $\mathbf{P}_X \cup \mathbf{P}_Y$ is not solvable. ∎

Since no superset of an unsolvable problem is solvable, Corollary (17) implies that the collection of all *r.e.* languages is not solvable, computably or otherwise. Theorem (56) darkens the picture by showing that even among the solvable collections of recursive languages, some are not solvable computably. It is thus surprising to learn that there is a computably solvable problem that contains "almost every" *r.e.* language. To formulate the fact precisely, let us say that set X is a *finite variant* of set Y just in case $(X - Y) \cup (Y - X)$ is finite. Then we have:

(58) PROPOSITION: [Wiehagen, 1977] There is a computably solvable problem **P** such that for every *r.e.* $L \subseteq N$, some member of **P** is a finite variant of L.

Proof: Let $L \subseteq N$ be *r.e.*. Then it is easy to define a total recursive function $f : N \to N$ such that for all $n \in N$, $W_{f(n)} = (L \cup \{n\}) \cap \{m \mid m \geq n\}$. By an application of the Recursion Theorem there is $n \in N$ with $W_{f(n)} = W_n$.[10] It is clear that W_n is a finite variant of L, and that the least member of W_n is n. This proves:

(59)   For every *r.e.* $L \subseteq N$ there is an *r.e.* $L' \subseteq N$ such that (a) L' is a finite variant of L, and (b) the least element of L' is an index for L'.

Now let **P** be the collection of all *r.e.* $L' \subseteq N$ such that the least element of L' is an index for L'. By (59), for every *r.e.* $L \subseteq N$, some member of **P** is a finite variant of L. So it suffices to show that **P** is computably solvable. However, this is trivial since it is accomplished by the scientist whose underlying function maps each $\sigma \in SEQ$ into the least member of range(σ).   ∎

Juxtaposition of Corollary (17) and Proposition (58) reveals that the usual criteria of set complexity are not directly related to solvability. To see this, note that the problem $\{N\} \cup \{D \subseteq N \mid D$ finite and nonempty$\}$ consists of languages whose computational complexity is trivial; yet it is unsolvable. In contrast, the solvable problem **P** of Proposition (58) contains languages with arbitrary complexity among the *r.e.* subsets of N (since finite variation of a language does not affects its computational complexity according to most classifications [Machtey & Young, 1978, Papadimitriou, 1994]). Yet it is trivially solvable.

---

10. The version of the Recursion Theorem used here is as follows. Let f be any recursive function from N to N. Then there exists an n such that $W_n = W_{f(n)}$. See [Rogers, 1987, Cor. I, p. 181]. The surprising character of Theorem (58) derives from the foregoing fact about recursive functions.

Finally, we demonstrate a contrast to Theorem (25). The latter shows solvability and efficient solvability to be coextensive properties of problems. It will now be seen that the theorem breaks down when attention is limited to computable scientists.

(60) PROPOSITION: There is a computably solvable problem **P** such that every computable scientist that solves **P** is dominated on **P**.

Proof: Let $K \subseteq N$ be *r.e.* and nonrecursive. Let **P** consist of every singleton language of the form $\{x + 1\}$ with $x \in K$, along with every doubleton language of the form $\{x + 1, 0\}$ with $x \in \overline{K}$. It is evident that **P** is computably solvable. For every $x \in N$, let e^x be the environment for $\{x + 1\}$. Let computable $\psi : SEQ \to N$ be given. We show:

(61) Either there is $x \in K$ such that for some $k \in N$, $0 \in W_{\psi(e^x[k])}$, or there is $x \in \overline{K}$ such that for every $k \in N$, $0 \notin W_{\psi(e^x[k])}$.

For if (61) were false, then for every $x \in N$, $x \in \overline{K}$ iff there is $k \in N$ with $0 \in W_{\psi(e^x[k])}$; and this would exhibit $\overline{K}$ as *r.e.*, which contradicts the choice of K as nonrecursive.

Now suppose that computable $\psi : SEQ \to N$ underlies scientist Ψ that solves **P**. Choose $x \in N$ to satisfy (61). Then for some $k \in N$, $\psi(e^x[k])$ is not an index for a member of **P**. Hence, Ψ is not **P**-bound, hence by Exercise (40), Ψ is dominated on **P**. ∎

2.5.4 Incomplete Environments

Real scientists almost always face the problem of inaccessible data, so we are led to consider the impact on inquiry of leaving out of environments some of the numbers comprising a language.[11] The issue has particular interest when attention is limited to the computable scientists, as will be the case here. To keep matters simple, we consider the loss of just one point. The relevant definition is as follows.

(62) DEFINITION: Let language L and environment e be given. We call e an *incomplete environment* for L just in case range$(e) \subseteq L$ and card$(L -$ range$(e)) \leq 1$. (Thus, environments for L count as incomplete environments for L.) Scientist Ψ solves problem **P** *on incomplete environments* just in case

---

11. The present subsection may be omitted without loss of continuity.

for every $L \in \mathbf{P}$ and incomplete environment e for L, Ψ converges on e to L. In this case, $\mathbf{P}$ is *solvable on incomplete environments*.

Naturally, solvability of $\mathbf{P}$ in this stricter sense is impossible if any pair of distinct languages in $\mathbf{P}$ share an incomplete environment since it is impossible to converge to different languages on the same environment. However, incompletion can also disrupt inquiry by computable scientists even when all the languages in play are radically distinct. Indeed, we have:

(63) PROPOSITION: There is a problem $\mathbf{P}$ with the following properties.

(a) Each member of $\mathbf{P}$ is infinite, and disjoint from every other member of $\mathbf{P}$ (hence, $\mathbf{P}$ is solvable on incomplete environments).

(b) $\mathbf{P}$ is computably solvable.

(c) No computable scientist solves $\mathbf{P}$ on incomplete environments.

Proof: Let $\langle \cdot, \cdot \rangle$ be a recursive isomorphism between N^2 and N. Given sets X, Y, let $X \times Y$ be $\{\langle x, y \rangle \mid x \in X, y \in Y\}$. Given permutation h of N, let problem $\mathbf{P}_h$ consist of all sets of the form $(N - \{h(m)\}) \times \{m\}$ for $m \in N$. It is easy to verify that:

(64)   For all permutations h of N, $\mathbf{P} = \mathbf{P}_h$ satisfies (63)a,b.

For (63)c, we rely on the following fact.

(65)   Let h and h' be distinct permutations of N. Then no scientist solves both $\mathbf{P}_h$ and $\mathbf{P}_{h'}$ on incomplete environments.

Proof of (65):   Since h and h' are distinct, let $m \in N$ be such that $h(m) \neq h'(m)$. Let $L_h = (N - \{h(m)\}) \times \{m\}$, and let $L_{h'} = (N - \{h'(m)\}) \times \{m\}$. Then L_h and $L_{h'}$ differ only on the set $\{\langle h'(m), m \rangle, \langle h(m), m \rangle\}$. Let environment e be such that range$(e) = L_h - \{\langle h'(m), m \rangle\}$. Then e is an incomplete environment for both L_h and $L_{h'}$. Since $L_h \neq L_{h'}$, no scientist can converge on e to both languages.   ∎

Since there are uncountably many permutations of h and only countably many computable scientists, (65) implies the existence of h such that no computable scientist solves $\mathbf{P}_h$. Together with (64), this concludes the proof.   ∎

The dual deformation of "noise" in environments is taken up in Exercise (82).

2.5.5   Stable Solvability

We now return to a point made just after Definition (54), concerning the stability of the conjectures issued by an underlying scientist. To remind you about the issue, suppose that computable $\psi : SEQ \to N$ underlies scientist Ψ, and that Ψ solves environment e. Then Ψ's conjectures on e ultimately converge to range(e). In contrast (as noted earlier), ψ's outputs need not stabilize to any particular index. It suffices that $W_{\psi(e[k])} = \text{range}(e)$ for cofinitely many k, and this may be achieved by alternating among the infinitely many $r.e.$ indexes for range(e). The desire for more stability in the computer simulation of scientists leads to the following definition.

(66)   DEFINITION: Suppose that computable $\psi : SEQ \to N$ underlies scientist Ψ, and let environment e, language L, and problem **P** be given.

(a)   We say that ψ *converges* to n_0 on e just in case $\psi(e[k]) = n_0$ for cofinitely many k. In this case, we also say that ψ *converges* on e. We say that Ψ solves e *stably* just in case Ψ solves e, and ψ converges on e.

(b) Ψ solves L *stably* just in case Ψ stably solves every environment for L.

(c) Ψ solves **P** *stably* just in case Ψ stably solves every member of **P**. In this case, **P** is said to be *stably solvable*.

Thus, Ψ solves e stably if and only if ψ converges on e to an index for range(e). And Ψ solves **P** stably if and only if Ψ stably solves every environment for **P**.[12]

We shall now show that stability is purchased with a reduction in scientific competence. For the demonstration we rely on the following fact, whose proof is an easy variant of that for Lemma (8).

(67)   LEMMA: Suppose that computable scientist Ψ solves language L stably, and let ψ underlie Ψ. Then there is $\sigma \in SEQ$ with the following properties.

(a)   range(σ) $\subseteq L$,

(b) $W_{\psi(\sigma)} = L$, and

(c)   for all $\tau \in SEQ$ with range(τ) $\subseteq L$, $\psi(\sigma * \tau) = \psi(\sigma)$.

(68)   PROPOSITION: [Osherson & Weinstein, 1982b] There is a computably solvable problem that no computable scientist solves stably.

---

12. The criterion of stable solvability corresponds to "explanatory (EX) identification" in [Case & Smith, 1983, Jain *et al.*, forthcoming], and to "intensional identification" in [Osherson *et al.*, 1986c].

Proof: Let $A \subseteq N$ be *r.e.* and nonrecursive. Let $\mathbf{P} = \{A \cup \{x\} \mid x \in N\}$. Then by Proposition (55), $\mathbf{P}$ is computably solvable. For a *reductio*, suppose that computable scientist Ψ solves $\mathbf{P}$ stably, and let computable $\psi : SEQ \to N$ underlie Ψ. Since $A \in \mathbf{P}$, it follows from Lemma (67) that:

(69)   There is $\sigma \in SEQ$ such that:

(a) range$(\sigma) \subseteq A$,

(b) $\mathrm{W}_{\psi(\sigma)} = A$, and

(c) for every $\tau \in SEQ$ with range$(\tau) \subseteq A$, $\psi(\sigma * \tau) = \psi(\sigma)$.

Let $\{a_i \mid i \in N\}$ be a recursive enumeration of A, and for all $x \in N$, let e_x be the environment $\sigma * \langle x, a_0, a_1, a_2 \ldots \rangle$. Since range$(e_x) = A$ if $x \in A$, (69) implies:

(70)   For $x \in A$, $\psi(e_x[k]) = \psi(\sigma)$ for all $k > \mathrm{length}(\sigma)$.

On the other hand, if $x \notin A$, then range$(e_x) \neq A = \mathrm{W}_{\psi(\sigma)}$. So, since Ψ solves $A \cup \{x\} \in \mathbf{P}$, it follows that:

(71)   For $x \in \bar{A}$, $\psi(e_x[k]) \neq \psi(\sigma)$ for some $k > \mathrm{length}(\sigma)$.

It is easy to see that (70) and (71) provide a positive test for $\bar{A}$, contradicting the nonrecursivity of A.   ∎

2.5.6   Exercises

For the exercises we fix an acceptable indexing $\{\psi_i \mid i \in N\}$ of the computable functions from SEQ to N. We also rely on the following definition.

(72)   DEFINITION: Let scientist Ψ be given.

(a) The collection of languages that Ψ solves is denoted scope(Ψ).

(b) If Ψ is computable, then the collection of languages that Ψ solves stably is denoted stable-scope(Ψ).

(73)   EXERCISE: Call scientist Ψ "total" just in case for all $\sigma \in SEQ$, $\Psi(\sigma)$ is defined. Exhibit a recursive function $f : N \to N$ with the following property. For all $i \in N$, if ψ_i underlies computable scientist Ψ then $\psi_{f(i)}$ underlies total, computable scientist Ψ' such that scope$(\Psi) \subseteq$ scope(Ψ') and stable-scope$(\Psi) \subseteq$ stable-scope(Ψ').

(74)   EXERCISE:

(a) Exhibit a recursive function $f : N \to N$ with the following property. For all $i \in N$, if ψ_i underlies computable scientist Ψ then $\psi_{f(i)}$ underlies scope(Ψ)-consistent, computable scientist Ψ' such that scope(Ψ) $\subseteq$ scope(Ψ').

(b) Show that for some computable scientist Ψ, if Ψ' is any computable scientist that is stable-scope(Ψ)-consistent, then stable-scope(Ψ) $\not\subseteq$ stable-scope(Ψ').

(75)   EXERCISE: Exhibit a computable scientist Ψ such that:

(a) stable-scope(Ψ) $=$ scope(Ψ), and

(b) if Ψ' is a computable scientist that is scope(Ψ)-conservative, then scope(Ψ) $\not\subseteq$ scope(Ψ').

(76)   EXERCISE: ♣ [Osherson et al., 1988b] Show that there is no total computable $h : N \to N$ such that for all $i \in N$, ψ_i underlies a scientist that fails to solve $W_{h(i)}$.

(77)   EXERCISE: ♣ When defining computable scientists we could have relied on indices for an arbitrary, countable collection of sets, in place of the r.e. sets. Here we limit attention to arithmetical sets.[13] Given $n \in N$, we say that scientist Ψ is Σ_n just in case there is computable $\psi : SEQ \to N$ such that for all $\sigma \in SEQ$, $\psi(\sigma)$ is a Σ_n index for $\Psi(\sigma)$. Given $X \subseteq N$, let $\mathbf{P}_X$ consist of all sets of the form $\{m, 0\}$ for $m \in X$ and $\{m\}$ for $m \in \bar{X}$. Show that for every $n \in N$, if X is a Σ_{n+1} set but not a Σ_n set, then the following holds.

(a) Some computable scientist is Σ_0 and solves $\mathbf{P}_X$ stably.

(b) Some computable scientist is Σ_{n+1} and solves $\mathbf{P}_X$ stably and efficiently.

(c) No Σ_n scientist solves $\mathbf{P}_X$ efficiently.

(78)   EXERCISE: ♦ [Blum & Blum, 1975] Let computable scientist Ψ be given, and suppose that $\psi : SEQ \to N$ underlies Ψ. Call Ψ order independent if for all environments e_0, e_1 for stable-scope(Ψ), if range(e_0) $=$ range(e_1) then ψ converges on e_0 and e_1 to the same index. Exhibit a recursive function $f : N \to N$ such that for every $i \in N$, if ψ_i underlies a computable scientist Ψ then $\psi_{f(i)}$ underlies an order independent, computable scientist Ψ' with stable-scope(Ψ) $\subseteq$ stable-scope(Ψ').

---

13. For background and notation, see [Hinman, 1978, Odifreddi, 1989, Rogers, 1987].

(79) EXERCISE: ♦

(a) Exhibit a recursive function $f : N \times N \to N$ with the following property. For all $i, j \in N$, if $W_i \neq \emptyset$ and $W_j \neq \emptyset$ then $\psi_{f(i,j)}$ underlies a computable scientist Ψ with $\{W_i, W_j\} \subseteq \text{scope}(\Psi)$.

(b) [Osherson *et al.*, 1988b] Let infinite *r.e.* $L \subseteq N$ be given. Show that there is no recursive function $f : N \to N$ with the following property. For all $i \in N$, if $W_i \neq \emptyset$ then $\psi_{f(i)}$ underlies a computable scientist Ψ with $\{L, W_i\} \subseteq$ stable-scope(Ψ).

(80) EXERCISE: ♥ [Fulk, 1990] Scientist Ψ is "prudent" if range$(\Psi) =$ stable-scope(Ψ). Show that for every computable scientist Ψ there is a prudent, computable scientist Ψ' such that stable-scope$(\Psi) \subseteq$ stable-scope (Ψ').

(81) EXERCISE: Scientist Ψ is *nontrivial* just in case for all $\sigma \in SEQ$, $\Psi(\sigma)$ is infinite whenever defined. Show that there is a problem **P** such that:

(a) every language in **P** is infinite;

(b) some computable scientist solves **P** stably and efficiently;

(c) no nontrivial, computable scientist solves **P**.

(82) EXERCISE: ♥ [Fulk *et al.*, 1994]. Let language L and environment e be given. We call e a *noisy environment* for L just in case $L \subseteq$ range(e) and card(range$(e) - L$) is finite. Scientist Ψ solves problem **P** *on noisy environments* just in case for every $L \in$ **P** and noisy environment e for L, Ψ converges on e to L. In this case, **P** is *solvable on noisy environments*. The obvious definitions apply for the criterion of *computable*, and *stable computable* solvability on noisy environments. Exhibit a problem **P** such that:

(a) some computable scientist solves **P** stably on noisy environments, but

(b) no computable scientist solves **P** on incomplete environments.

(Note: Our own proof of the exercise relies on results in Section 2.6, below.)

(83) EXERCISE: ♠ Exhibit a recursive function $f : N^2 \to N$ with the following property. Let problem **P** and $i, j \in N$ be given. Suppose that:

(a) ψ_i underlies a computable scientist that solves **P** efficiently, and

(b) ψ_j underlies a computable scientist that solves **P** stably.

Then $\psi_{f(i,j)}$ underlies a computable scientist that solves **P** stably and efficiently.

(84)   EXERCISE: ◆   Recall that an environment is a total function from N to N. The function may be recursive or nonrecursive.

(a) [Blum & Blum, 1975] Exhibit a recursive function $f : N \to N$ with the following property. For all problems **P** and $i \in N$, if ψ_i underlies a computable scientist that stably solves every recursive environment for **P** then $\psi_{f(i)}$ underlies a computable scientist that stably solves **P**.

(b) [Wiehagen, 1977] Exhibit a recursive function $f : N \to N$ with the following property. For all problems **P** and $i \in N$, if ψ_i underlies a computable scientist that stably solves every nonrecursive environment for **P** then $\psi_{f(i)}$ underlies a computable scientist that stably solves **P**.

2.6   Functional Languages

2.6.1   Numbers as Codes for Phenomena

The numerical paradigm gives an austere impression but it is nonetheless connected to some of the empirical phenomena that scientists must unravel. The connection must be forged, however, by coding the phenomena as natural numbers. Consider, for instance, periodicity present in the microstructure of a certain class of crystals. Not every period may be physically realizable. The realizable ones correspond to a subset of N, hence to a language L. We might wish to know what constraints on L are necessary or sufficient in order for it to be reliably discovered by an agent that is exposed to L's elements. Some information about these constraints is available from facts proved earlier. For example, we learn from Corollary (17)b that it does not suffice for L to be an arbitrary, cofinite subset of a given, infinite set, since no agent can solve every language drawn from such a collection. We learn from Theorem (56) that the constraints might permit reliable discovery of L, but only if the agent is uncomputable. Other results may be interpreted similarly.

Inquiry into one aspect of crystallography is thus formalizable via the numerical paradigm, and the prospects for success can be investigated. The formalization is breathtakingly crude—let it be admitted—but this first step helps clarify many issues, and opens the door to slightly more urbane paradigms, as we hope to illustrate in Chapters 3 and 4.

Sticking with the numerical paradigm for now, the purpose of the present section is to show that there is unsuspected flexibility in the use of languages to represent inquiry. In particular, they can represent the kind of functional relations between variables that often preoccupy scientists, such as the relation between the density and temperature of a given substance. This is achieved by limiting attention to languages that code total functions. We shall now explain and explore this idea.

2.6.2 Definitions and Basic Facts

To get started, we need some notation and terminology. Let us fix a recursive isomorphism $\langle \cdot, \cdot \rangle$ between N^2 and N. A language $L \subseteq N$ is called *functional* if $\{(x, y) \mid \langle x, y \rangle \in L\}$ is the graph of a total function from N to N. The solution of functional languages is simplified by the following fact.

(85)   LEMMA: Let e be an environment, and suppose that both range(e) and L are functional languages. Then, for all $\langle x, y \rangle \in L$, if $\langle x, y \rangle \notin$ range(e), then $\langle x, y' \rangle \in$ range(e) for some $y' \neq y$.

In this sense, environments for a functional language L provide direct evidence for both L and for $N - L$. Information about the latter set is missing in the general, nonfunctional case. Now consider an enumeration of a countable collection **P** of functional languages, and suppose that scientist Ψ faced with data σ conjectures the first language in the enumeration that is consistent with σ (and is undefined if there is none such). In light of Lemma (85), Ψ is easily seen to solve **P**. Moreover, a computable scientist of this kind can be programmed to solve a collection of recursive functional languages for which a set of indices can be recursively enumerated. So we have:

(86)   PROPOSITION:

(a) Every countable collection of functional languages is solvable.

(b) Suppose that S is an *r.e.* set of *r.e.* indexes for functional languages. Then $\{W_i \mid i \in S\}$ is computably solvable.

2.6.3 Computable Solvability of Functional Languages

Despite the richer information carried by environments for functional languages, there remain collections of recursive functional languages that are not computably solvable. This is the content of the following theorem (which, we note, includes no hypothesis of stability in the sense of Section 2.5.5).

(87) THEOREM: [Case & Smith, 1983] The collection of recursive functional languages is not computably solvable.

Proof: Let $\mathbf{F}_0$ be the collection of functional languages L such that for cofinitely many n, $\langle n, 0 \rangle \in L$. Suppose that computable scientist Ψ solves $\mathbf{F}_0$. It suffices to exhibit a recursive, functional language L that Ψ does not solve. Toward this end, call $\sigma \in SEQ$ of length k "orderly" just in case σ has the form $(\langle 0, p_0 \rangle, \ldots, \langle k-1, p_{k-1} \rangle)$. Since Ψ solves $\mathbf{F}_0$, it is easy to verify the existence of computable $f : SEQ \to SEQ$ such that for all orderly $\sigma \in SEQ$:

(88) (a) $f(\sigma)$ is orderly;

     (b) $\sigma \subseteq f(\sigma)$;

     (c) if $k = \text{length}(f(\sigma))$ and $W_m = \Psi(f(\sigma))$, then $\langle k, 0 \rangle \in W_m$.

With the help of f, we construct infinite collection $\{e^i \mid e^i \in SEQ \text{ and } i \in N\}$. It will be the case that:

(89) (a) for all $i, j \in N$, if $j < i$ then $e^j \subset e^i$;

     (b) for all $i \in N$, e^i is orderly;

     (c) for all $i, k, n \in N$, if $\text{length}(e^i) = k + 1$ and $\Psi(e^i[k]) = W_n$, then $\langle k, 0 \rangle \in W_n$ but $e^i(k) = \langle k, 1 \rangle$;

     (d) there is computable $g : N \to SEQ$ such that for all $i \in N$, $e^i = g(i)$.

Construction of the e^i's

Stage 0: $f(\emptyset) = \emptyset$. It follows that conditions (89)a-c are satisfied with respect to e^0.

Stage $i + 1$: Let $k = \text{length}(f(e^i))$, and let $e^{i+1} = f(e^i) * \langle k, 1 \rangle$. By (88)a, e^{i+1} is orderly. By (88)b, $e^i \subset e^{i+1}$. If $\Psi(e^i) = W_n$, then by (88)c, $\langle k, 0 \rangle \in W_n$ whereas we have stipulated that $e^{i+1}(k) = \langle k, 1 \rangle$. It follows that conditions (89)a-c are satisfied with respect to e^{i+1}.

End Construction

Since f is computable, it is easy to see that condition (89)d is satisfied. Let $e = \bigcup_{i \in N} e^i$. By (89)a,b, e is an environment for a functional language L. By (89)d, L is recursive. By (89)c, Ψ does not converge to range(e) on e, hence does not solve e. ∎

2.6.4   Finite Variant Solvability

An approximate form of scientific success consists of discovering a finite variant of the underlying language.[14] The next definition gives substance to this idea.

(90)   DEFINITION: [Case & Smith, 1983, Osherson & Weinstein, 1982a]

(a) We say that scientist Ψ *FV-solves* environment e just in case for cofinitely many k, $\Psi(e[k])$ is a finite variant of range(e). Ψ *FV-solves* problem **P** if Ψ *FV*-solves every environment for **P**.

(b) Suppose that computable $\psi : SEQ \rightarrow N$ underlies scientist Ψ. We say that Ψ *FV*-solves problem **P** *stably* just in case Ψ *FV*-solves **P**, and ψ converges on every environment for **P**.

Notice that to *FV*-solve e, Ψ is allowed to issue infinitely many different hypotheses on e, so long as cofinitely many of them are finite variants of range(e). In contrast, suppose that computable scientist Ψ, with underlying function ψ, *FV*-solves problem **P** stably. Then for every environment e for **P**, ψ converges on e to an index for a finite variant of range(e). Such stability undermines the added flexibility allowed by *FV*-solvability. To see this, let total recursive $\psi' : SEQ \rightarrow N$ be such that for all $\sigma \in SEQ$, $\psi'(\sigma)$ is an index for:

$$(W_{\psi(\sigma)} - \{\langle x, y \rangle \mid \text{ for some } z \neq y, \langle x, z \rangle \in \text{range}(\sigma)\}) \cup \text{range}(\sigma).$$

Then it is easy to see that if ψ' underlies scientist Ψ', then Ψ' solves range(e). So we have proved the following lemma.

(91)   LEMMA: Suppose that computable scientist *FV*-solves stably a collection **P** of recursive functional languages. Then **P** is computably solvable.

The converse is false, that is:

(92)   PROPOSITION: [Barzdin, 1974, Case & Smith, 1983] There is a computably solvable collection **P** of recursive functional languages such that no computable scientist *FV*-solves **P** stably.

For the proof, see [Odifreddi, 1997, Proposition A.1.49]. An immediate corollary to Proposition (92) is the following strengthening of Proposition (68):

---

14. The present subsection may be omitted without loss of continuity. For the intrepid reader, we recall from page 38 that X is a finite variant of Y iff $(X - Y) \cup (Y - X)$ is finite.

There is a collection of recursive functional languages that is computably solvable but not stably.

Without the stability requirement FV-solvability turns out to be very liberal. Indeed:

(93) PROPOSITION: [Case & Smith, 1983, attributed to L. Harrington] There is a computable scientist that FV-solves the entire collection of computable functional languages.

Proof: Let $\{\varphi_i \mid i \in N\}$ be a standard enumeration of the partial recursive functions of one variable. The result of m steps in the computation of $\varphi_i(x)$ is denoted $\varphi_{i,m}(x)$. If $\varphi_i(x)$ is defined, then the "computation time for i on x" is the least m such that $\varphi_{i,m}(x)$ is defined; otherwise, the computation time for i on x is taken to be infinite. Given $n \in N$ and a finite set S of indexes, let race$(S, n) = \varphi_i(n)$, where $i \in S$ and for all $j \in S$ the computation time for i on x is less than the computation time for j on x, or the computation times are equal and $i \leq j$. Intuitively, race(S, n) is the result of applying all the indexes in S to n simultaneously, and considering the output to be the answer of the first index that finishes its computation.

Let $n, i, j \in N$ and $\sigma \in SEQ$ with length$(\sigma) = \ell$ be given. We say that σ *admits* j if for all $\langle x, y \rangle \in$ range(σ) either $\varphi_{j,\ell}(x) = y$ or $\varphi_{j,\ell}(x)$ is infinite. We say that j *agrees with i through n* if for all $x \leq n$, either $\varphi_{j,n}(x) = \varphi_{i,n}(x)$, $\varphi_{j,n}(x)$ is infinite, or $\varphi_{i,n}(x)$ is infinite. Let $D_{n,\sigma}$ consist of all indexes $i \leq$ length(σ) such that σ admits i, and for all $j < i$, if σ admits j then j agrees with i through n. Let total recursive $f : SEQ \to N$ be such that for all $\sigma \in SEQ$, $n \in N$, $\varphi_{f(\sigma)}(n) =$ race$(D_{n,\sigma}, n)$.

Let total recursive $g : N \to N$ be such that for all i, $W_{g(i)} = \{\langle x, \varphi_i(x) \rangle \mid x \in N\}$. (That is, g converts partial recursive indexes into equivalent r.e. indexes.) Suppose that $g \circ f$ underlies scientist Ψ. We claim that Ψ FV-solves every computable functional language. To see this, let e be an environment for computable functional language L. Let i_w be least with $W_{g(i_w)} = L$. For every $j < i_w$ with $\varphi_j \not\subseteq \varphi_{i_w}$ there is k such that $e[k]$ does not admit j. Moreover, for every $j > i_w$ with $\varphi_j \not\subseteq \varphi_{i_w}$ there is $n \in N$ such that i_w does not agree with j through n. It follows easily that:

(94) for all $k \geq i_w$ there are cofinitely many n such that $i_w \in D_{n,e[k]} = \{i \leq k \mid \varphi_i \subseteq \varphi_{i_w}\}$.

Let $k \geq i_w$ be given. Then (94) implies that for cofinitely many n, race$(D_{n,e[k]}, n)$ $= \varphi_{i_w}(n)$. Hence, $\varphi_{f(e[k])}(n) = \varphi_{i_w}(n)$ for cofinitely many n, so $W_{g \circ f(e[k])}$

is a finite variant of L. Since $g \circ f$ underlies Ψ, it follows that Ψ FV-solves
e.   ∎

Many alternative success criteria have been proposed and investigated in the
literature devoted to functional languages. Some are discussed in the exercises
below. For others, see [Case *et al.*, 1993, Jain & Sharma, 1993a, Lange &
Zeugmann, 1993] along with references cited there.

2.6.5   Exercises

(95)   EXERCISE: Environment e is called *orderly* if for all $n \in N$, $e(n) =$
$\langle n, m \rangle$ for some $m \in N$. (Hence, functional languages have exactly one orderly
environment.) Let **P** be a collection of functional languages. Show that some
computable scientist solves **P** [stably] iff some computable scientist solves
[stably] every orderly environment for **P**.

(96)   EXERCISE: [Osherson & Weinstein, 1982a] Exhibit a problem **P** (not
composed of functional languages) with the following properties:

(a) some computable scientist FV-solves **P** stably, but

(b) no scientist solves **P**.

So, Lemma (91) is false if the restriction to functional languages is lifted.

(97)   EXERCISE: ♦ [Case & Ngo-Manguelle, 1979] Scientist Ψ is called
popperian just in case for all $\sigma \in SEQ$, if $\Psi(\sigma)$ is defined then $\Psi(\sigma)$ is a functional language. Such scientists deliver hypotheses that can be tested against
future data (in a functional language), and so conform to the "testability" requirement discussed in [Popper, 1959]. Exhibit a collection **P** of recursive,
functional languages such that **P** is stably solvable but no computable, popperian scientist solves **P**.

(98)   EXERCISE: ♣ Prove the following strengthing of Proposition (46).
There is a collection of recursive functional languages that is solved by no
memory-limited scientist (computable or not).

2.7   Probabilistic Solvability

2.7.1   Allowing Occasional Error

Probability is important to the theory developed within these pages. Its relation
to deterministic induction will occupy Section 3.6 below, and serve to buttress

our claim for the naturalness of the paradigm advanced in Chapter 3. The discussion will then be somewhat technical so it serves a pedagogical purpose to here introduce probabilistic inquiry within the simpler context of the numerical paradigm.[15]

We introduce probability concepts in order to soften the requirements on solvability. At present, a scientist must solve all environments for a given problem in order to be credited with solution of the latter. The following considerations suggest that such a demand is excessive.

Unsolvability never hinges on a single environment. In other words, if some scientist solves every environment for a problem **P** save for just one, then **P** is solvable. The verification of this fact (as well as its extension to any finite number of environments) is easy and left to the reader. It leaves the impression that the all-or-nothing character of our success criterion is needlessly extreme, and that no harm would come of allowing scientists to go astray on a "small subset" of the uncountably many environments for nontrivial problems. Of course, whether anything of substance is conceded by such tolerance depends on the precise meaning accorded to "small subset." In the present section we interpret the phrase probabilistically. Specifically, we shall relax our success criterion to allow scientists to err on a class of environments whose probability is zero. It will be shown that this small concession is not as innocent as it seems. In particular, it yields the solvability of every countable collection of languages.

Let it be noted that our treatment of probabilistic solvability returns us to the ineffective setting introduced at the beginning of the chapter. In particular, scientists are not assumed to be computable.

We begin our discussion by associating probability measures with languages. For background on the measure-theoretic concepts used here, see [Levy, 1979, Sec. VII.3].

2.7.2   Measured Problems

By a *positive probability measure* on nonempty $L \subseteq N$ is meant a function $m_L :$ $N \to [0, 1]$ such that (a) $\sum_{x \in N} m_L(x) = 1$ and (b) for all $x \in N, m_L(x) > 0$ iff $x \in L$. Each such measure determines a unique, complete probability measure M_L on the class of all environments via the following stipulation. For $\sigma \in SEQ$,

---

15. The present section is not essential to understanding Chapters 3 and 4 (although it may help for Section 3.6). It may thus be omitted without loss of continuity.

let E_σ be the class of environments that begin with σ. Then for nonempty σ, $M_L(E_\sigma) = \prod_{j < \text{length}(\sigma)} m_L(\sigma(j))$. Intuitively, for measurable collection E of environments, $M_L(E)$ is the probability that an environment created by infinitely many random draws from L according to m_L belongs to E.

Proof of the following lemma is left as Exercise (110). It tells us that use of m_L to produce a sequence is (probabilistically) guaranteed to create an environment for L, instead of creating, for example, an environment for a proper subset of L.

(99)   LEMMA: Let m_L be a positive probability measure for language L, and let E be the class of environments for L. Then $M_L(E) = 1$.

To introduce our probabilistic conception of success, we need the following definition and lemma.

(100)   DEFINITION: Let scientist Ψ and language L be given. We denote by $E_{solve}(\Psi, L)$ the class of environments for L that Ψ solves.

(101)   LEMMA: Let scientist Ψ, and positive probability measure m_L for language L be given. Then $M_L(E_{solve}(\Psi, L))$ is defined.

Proof: Let Σ consist of just those $\sigma \in SEQ$ such that $\Psi(\sigma) = L$. Then, environment e belongs to $E_{solve}(\Psi, L)$ iff range$(e) = L$ and $e[k] \in \Sigma$ for cofinitely many k. It is easy to verify that the latter conditions exhibit $E_{solve}(\Psi, L)$ as a Borel set.   ∎

Now we are ready to define our probabilistic paradigm.

(102)   DEFINITION: A *measured problem* is a collection of pairs of the form (L, m_L), where L is a language and m_L is a positive probability measure on L. Scientist Ψ *1-solves* measured problem **Q** just in case for every $(L, m_L) \in \mathbf{Q}$, $M_L(E_{solve}(\Psi, L)) = 1$. In this case, **Q** is said to be *1-solvable*.

Thus, solution of a measured problem does not require solving every environment for a language L appearing in it. It is permitted to fail on any class of environments for L whose probability of encounter is zero.[16]

To illustrate, call environment e "segregated" if $\{e(2i) \mid i \in N\}$ contains only even numbers and $\{e(2i + 1) \mid i \in N\}$ contains only odd numbers. There are uncountably many segregated environments for any language with at least

---

16. The definition of **1**-solvability is based on [Wexler & Culicover, 1980, Ch. 3].

three numbers, not all even and not all odd. Segregated environments are nevertheless "rare," for it can be shown with respect to any positive probability measure for any language L that the collection of segregated environments for L has probability 0. Consequently, to **1**-solve a measured problem it suffices to succeed on just its non-segregated environments.

2.7.3   Discovery on Measured Problems

In the foregoing developments, environments are conceived as arising from infinitely many, independent draws from a given language L, where the probability of retrieving $x \in N$ on any given draw is determined by a fixed, positive probability measure m_L. In real inquiry, such assumptions on the availability of data are seldom if ever met since real data are not collected in random fashion. Sets of measure 1 nonetheless provide a compelling interpretation of the idea of "almost all" environments, so it is striking that unsolvability *nearly* disappears for measured problems. We say "nearly" because uncountable problems, measured or not, remain unsolvable [see Lemma (3), page 19].

(103)   THEOREM: [Osherson *et al.*, 1986b] Every countable, measured problem is **1**-solvable.

Proof: Let countable, measured problem **Q** be given. We suppose $\mathbf{Q} \neq \emptyset$ since otherwise there is nothing to prove. Let $\{(L_j, m_{L_j}) \mid j \in N\}$ enumerate **Q**. We say that $\sigma \in SEQ$ "agrees with L_j through n" if range$(\sigma) \cap \{0, 1, \ldots, n\} = L_j \cap \{0, 1, \ldots, n\}$. Given $j, n, k \in N$, define $A_{j,n,k}$ to be the set of environments e for L_j such that $e[k]$ does not agree with L_j through n. It is easy to verify that for fixed j, n, $M_{L_j}(A_{j,n,k})$ is defined and decreases to zero as k increases. We may therefore define a function $d : N \to N$ such that for all n:

$$d(n) = \mu k[M_{L_j}(A_{j,n,k}) < 2^{-n} \text{ for all } j \leq n].$$

Thus, for any language L_j with $j \leq n$, and any environment e for L_j, the probability is at least $1 - 2^{-n}$ that $e[d(n)]$ provides accurate information about $L_j \cap \{0, 1, \ldots, n\}$. By the convergence of $\sum 2^{-n}$:

(104)   For all $j \in N$, $\sum_{n \in N} M_{L_j}(A_{j,n,d(n)})$ is finite.

Let $X_j = \{e \mid e \in A_{j,n,d(n)} \text{ for infinitely many } n\} = \bigcap_{p \in N} \bigcup_{n > p} A_{j,n,d(n)}$. By (104) and the first Borel-Cantelli lemma [Billingsley, 1986, p. 53], it follows that:

(105)   For all $j \in N$, $M_{L_j}(X_j) = 0$.

We infer immediately:

(106)   Let E be the set of environments e for $L_j \in \mathbf{Q}$ such that for all but
finitely many $n \in N$, $e[d(n)]$ agrees with L_j through n. Then $M_{L_j}(E) = 1$.

Now define scientist Ψ as follows. $\Psi(\emptyset) = \emptyset$. For $\sigma \in SEQ$ with $0 <$
length$(\sigma) \notin$ range(d), $\Psi(\sigma) = \Psi(\sigma')$, where σ' is the result of removing
the last element of σ. Otherwise, for $\sigma \in SEQ$ with $0 <$ length$(\sigma) = d(n)$,
$\Psi(\sigma) = L_j$, where j is least such that σ agrees with L_j through n. It is clear
that Ψ converges to L_j on any environment e for L_j with the property: for all
but finitely many $n \in N$, $e[d(n)]$ agrees with L_j through n. By (106), the set
of such environments has M_{L_j}-measure 1.   ∎

2.7.4   Discovery via Coins

Another way of integrating probability into the numerical paradigm is to equip
scientists with a fair coin to be tossed at the arrival of each datum. The scientist's conjecture is allowed to depend on the history of toss outcomes generated
so far. This idea has been richly developed in [Wiehagen et al., 1984, Pitt,
1989]. We limit ourselves to two simple results.[17]

A *coin* is an (infinite) sequence over $\{0, 1\}$, to be thought of as the output
of a random binary generator. The class of all finite initial segments of coins is
denoted *COIN-SEQ*. Let $\Pr(\cdot)$ be the natural probability measure over the class
of all coins. Specifically, $\Pr(\cdot)$ is taken to be the unique, complete measure such
that for all n, and all finite sequences α over $\{0, 1\}$ of length n, $\Pr(C_\alpha) = 2^{-n}$,
where C_α is the set of coins that begin with α.

By a *scientist with coin* is meant a function from *COIN-SEQ* $\times$ *SEQ* to
pow(N). The first argument to such a scientist is interpreted as the outcome
of a finite number of coin tosses; the second argument is the usual kind of
data about the language under examination. Let Θ be a scientist with coin.
In the presence of a coin c, $\lambda\sigma . \Theta(c[\text{length}(\sigma)], \sigma)$ is a scientist in our original sense, namely, a function from *SEQ* to pow(N).[18] Thus, for each lan-

---

17. The present subsection does not presage later developments in Section 3.6 and may be omitted
without loss of continuity.

18. The expression "$\lambda\sigma$" in $\lambda\sigma . \Theta(c[\text{length}(\sigma)], \sigma)$ names the variable taken to be the argument
to Θ. In contrast, the coin c is considered fixed. The λ-notation is used extensively in Chapter 4.
For review, see [Partee et al., 1990, Ch. 13].

guage L and scientist Θ with coin, we can form the set of coins c such that $\lambda\sigma \cdot \Theta(c[\text{length}(\sigma)], \sigma)$ solves L. It is easy to see that this set is measurable under $\text{Pr}(\cdot)$. Its probability quantifies the competence of Θ on L. Officially, we proceed as follows.

(107) DEFINITION: Let $p \in [0, 1]$, language L, and scientist Θ with coin be given. We say that Θ *solves L with probability p* just in case $\text{Pr}(\{\text{coin } c \mid \lambda\sigma \cdot \Theta(c[\text{length}(\sigma)], \sigma)$ solves $L\}) \geq p$. Problem $\mathbf{P}$ is *solvable with probability p* just in case there is scientist Θ with coin that solves every $L \in \mathbf{P}$ with probability p.

The definitions of Section 2.7.2 incorporated probability into environments, whereas scientists remained deterministic. The foregoing definition takes the contrary option. The impact on solvability turns out to be different inasmuch as solvability with probability 1 implies solvability in the original, absolute sense introduced in Section 2.1.2. Indeed, in contrast to Theorem (103), we have the following.

(108) THEOREM: A problem is solvable if and only if it is solvable with probability 1.

Proof: Let nonempty problem $\mathbf{P}$ be given. The left-to-right direction is immediate. For the other direction, let Θ be a scientist with coin that solves every $L \in \mathbf{P}$ with probability 1. Because *COIN-SEQ* $\times$ *SEQ* is countable, so is range(Θ). Hence $\mathbf{P}$ is countable. Define the class C to be:

$$\bigcap_{L \in \mathbf{P}} \{\text{coin } c \mid \lambda\sigma \cdot \Theta(c[\text{length}(\sigma)], \sigma) \text{ solves } L\}.$$

Then, C is a countable intersection of sets of probability 1. Any such class is nonempty, so choose $c_0 \in C$. It follows that $\lambda\sigma \cdot \Theta(c_0[\text{length}(\sigma)], \sigma)$ solves every $L \in \mathbf{P}$. ∎

As the probability required for learning descends from 1 toward 0, more and more problems become solvable [see Exercise (112)]. However, the next proposition shows that no positive probability is low enough to ensure the solvability of all countable problems.

(109) PROPOSITION: The problem consisting of every cofinite subset of N is not solvable with probability p for any $p \in (0, 1]$.

Proof: Exercise (112).   ∎

2.7.5   Exercises

(110)   EXERCISE: Prove Lemma (99).

(111)   EXERCISE: ♣   Show that the problem consisting of N together with every nonempty finite subset of N is not **1**-solved by any memory-limited scientist.

(112)   EXERCISE:

(a)   ♣   Show that each of the problems cited in Corollary (17) is solvable with probability $\frac{1}{2}$ but not with probability greater than $\frac{1}{2}$.

(b)   ♦   Show that for every $n \in N - \{0\}$, $\{N - D \mid D \subseteq N$ and $\mathrm{card}(D) \leq n\}$ is solvable with probability $\frac{1}{n+1}$, but not with probability greater than $\frac{1}{n+1}$. Deduce Proposition (109) as an immediate corollary.

2.8   Transition to Chapter 3

2.8.1   Problems Composed of Propositions

As a last preparation for the work of Chapters 3 and 4, let us briefly examine a numerical model of inquiry that generalizes the one defined in Section 2.1. The generalization offers a revealing analogy to the logical paradigm that occupies us in the sequel, an analogy that is absent from our work so far. The sole innovation is to define a problem not as a class of languages, but as a family of such classes, each disjoint from the others. A family of this kind induces an equivalence relation which we exploit to give meaning to the idea that two languages are "relevantly similar" (namely, by falling into the same equivalence class).

The classes of languages comprising a problem will be called "propositions," thereby using a term whose intuitive significance will become clearer in the next chapter. For a scientist Ψ to solve a problem **P** in the new sense, we require that for any environment e for $L \in \bigcup$ **P**, Ψ converge on e to the proposition that holds L. More precisely, we allow Ψ to converge on e to any *nonempty subset* of the proposition containing L. This added flexibility changes nothing essential here, but will be pivotal when transposed to later paradigms.

Let us now proceed in deliberate fashion.

2.8.2   Definitions and an Example

Formulation of the generalized model requires new definitions of "problem," "scientist," and "solution." The concepts of "language" and "environment" remain the same. Propositions and problems are defined as follows.

(113)   DEFINITION: A nonempty set of languages is a *proposition*. A *problem* is a collection of disjoint propositions.

For example, let E be the collection of languages that contain infinitely many even numbers, and let O be the collection of languages that contain infinitely many odd numbers. Then $E - O$ and $O - E$ are disjoint propositions, and $\{E - O, O - E\}$ is a problem.

Suppose that scientist Ψ is faced with environment e. As explained above, instead of requiring Ψ to converge to range(e), we only require Ψ's conjectures to stabilize to the one proposition containing range(e). Officially:

(114)   DEFINITION: A *scientist* Ψ is any function with domain(Ψ) $\subseteq SEQ$ such that for all $\sigma \in$ domain(Ψ), $\Psi(\sigma) \subseteq$ pow(N). Given scientist Ψ and problem **P**, we say that Ψ *solves* **P** just in case for every $P \in$ **P** and every environment e for a language in P, $\emptyset \neq \Psi(e[k]) \subseteq P$ for cofinitely many k. In this case, **P** is called *solvable*.

As pointed out earlier, scientists may signal their choice of proposition P by announcing any nonempty subset of P. This added flexibility makes no difference to scientists free of computability or other constraints, since they can simply issue all of P instead of a subset. To illustrate the definition:

(115)   PROPOSITION: The problem $\{E - O, O - E\}$ is solvable.

Proof: Let scientist Ψ be defined as follows. Given $\sigma \in SEQ$, $\Psi(\sigma) = E - O$ in case range(σ) contains more distinct even numbers than odd; otherwise, $\Psi(\sigma) = O - E$. It is clear that Ψ solves $\{E - O, O - E\}$.   ∎

Exercise (21) shows that every finite problem is solvable in the numerical paradigm of Section 2.1. In contrast, within the present model we have:

(116)   PROPOSITION: The problem $\{ \{N\}, \{\emptyset \neq D \subseteq N \mid D \text{ finite}\} \}$ is not solvable.

Proof: Let $P_1 = \{N\}$, $P_2 = \{\emptyset \neq D \subseteq N \mid D \text{ finite}\}$. Suppose for a contradiction that scientist Ψ solves $\{P_1, P_2\}$. Define function $\Psi' : SEQ \rightarrow$ pow(N) as follows. For all $\sigma \in SEQ$, if $\emptyset \neq \Psi(\sigma) \subseteq P_1$, then $\Psi'(\sigma) = N$; if $\emptyset \neq \Psi(\sigma) \subseteq P_2$, then $\Psi'(\sigma) = \text{range}(\sigma)$; otherwise, $\Psi'(\sigma)$ is undefined. It is clear that in the paradigm of Section 2.1, Ψ' is a scientist that solves $P_1 \cup P_2$. This contradicts Corollary (17)a.   ∎

In light of our earlier work the reader will have no difficulty elaborating the current paradigm, and analyzing solvability within it. We will not pursue the matter, however, since our sights are set henceforth on a generalization of the numerical paradigm that is more profound. We refer in this way to the logical model of inquiry, evoked on several occasions above. It now takes center stage.

2.8.3 Exercises

(117) EXERCISE: Let $\mathbf{P} = \{\ \{N\}, \{N - \{n\} \mid n \in N\}\ \}$. Show that $\mathbf{P}$ is not solvable.

2.9 Notes

The numerical paradigm derives from the seminal papers of [Solomonoff, 1964, Gold, 1967, Putnam, 1975] and [Blum & Blum, 1975]. Unlike these authors, we have adopted an extensional approach that introduces r.e. indexing and computability at a later stage. For this reason, some of our attributions [e.g., to Dana Angluin in Theorem (16)] are slightly inexact, and reflect the transposition of earlier results to our more abstract framework.

The concept of "reliability" discussed in Exercise (19) was introduced by E. Minicozzi, and exploited by [Blum & Blum, 1975] in a set-up different from ours.

Our definition of "efficiency" in Section 2.3 is a variant of an idea found in [Gold, 1967]. See [Daley & Smith, 1986, Freivalds et al., 1995, Jain & Sharma, 1995] for other approaches. The so-called "PAC" perspective on learning puts efficiency at the top of its agenda, but involves paradigms very different from those considered here (as discussed in Section 1.3.3 above). See [Anthony & Biggs, 1992, Kearns & Vazirani, 1994] for overviews. The efficiency theorem (25) was proved in slightly different form in [Osherson et al., 1986c].

The concept of $\mathbf{P}$-boundedness, introduced in Definition (30), is a variant of "prudence," discussed in [Osherson et al., 1982]. The concept is investigated in [Kurtz & Royer, 1988, Fulk, 1990, Jain & Sharma, 1993b].

The concept of "induction by enumeration" in Exercise (43) is drawn from [Gold, 1967]. The observation embodied in the exercise appears in [Osherson et al., 1986c].

The memory-buffer strategy of Exercise (53) is taken up in [Fulk et al., 1994]. Consistency (Section 2.3.2) is investigated in [Fulk, 1988]. See [Kinber & Stephan, 1995, Jain & Sharma, 1994, Zeugmann et al., 1995] for detailed studies of conservativeness.

The paradigm of Exercise (77), involving arithmetical sets, is inspired by the discussion in [Kugel, 1977].

For a deeper analysis of 1-solvability (Section 2.7), see [Angluin, 1988, Montagna, 1996a, Montagna, 1996b].

Exercise (79)b implies that there is no effective means of combining the competences of computable scientists. Such combination is called "aggregation," and has been investigated from various perspectives, for example: [Jain & Sharma, 1990a, Osherson *et al.*, 1986a, Pitt & Smith, 1988, Baliga *et al.*, 1996]. (See also [Daley, 1986, Jain & Sharma, 1990b, Pitt, 1989] for related work.)

The concepts of incomplete and noisy data [see Exercise (82)] may be combined. The result is environmental "imperfection." See [Fulk *et al.*, 1994] for discussion.

For more on the popperian scientists of Exercise (97), see [Case *et al.*, 1994].

Finite variant solvability, discussed in Section 2.6.4, can be conceived as a crude attempt to model the idea of "approximate inference." More sophisticated models are presented in [Royer, 1986, Fulk & Jain, 1994, Smith & Velauthapillai, 1986].

3 A First-Order Framework for Inquiry

3.1 Fundamentals

3.1.1 The First-Order Perspective

As a model of science the numerical paradigm is limited by its impoverished conception of reality and of data. The latter concepts are respectively represented by subsets of N and by sequences of numbers, and they cast only faint light on the richness of the problems facing real scientists (even allowing for the kind of coding mentioned in Section 2.6.1). For greater realism, potential worlds will henceforth be identified with relational structures for a first-order language, and data will consist of the atomic formulas or their negations made true in whatever structure Nature chooses to be actual. The resulting paradigm will be called *first-order*. The first-order paradigm includes the numerical one as a special case, in a sense to be made clear in Section 3.1.5.

Despite its greater expressiveness, the first-order paradigm remains a crude approximation to genuine scientific activity. It is well to emphasize this point at the outset. Among other things, real theorizing is rarely carried out in the first-order idiom, and the shifting information called "data" cannot be confined to a specific logical form. Its crudeness notwithstanding, we may reasonably hope that the first-order paradigm illuminates aspects of empirical inquiry left out of the numerical paradigm, and that it makes clearer how our idealizations fall short.

The first-order paradigm resembles the model discussed in Section 2.8. The essential difference is the substitution of first-order structures for sets of numbers. Such substitution was first proposed in [Glymour, 1985], and pursued in a variety of articles (for partial reviews, see [Kelly, 1996, Osherson *et al.*, 1996]). A fresh attempt to exploit the same idea will be developed here.

3.1.2 A First-Order Framework

Let us begin by establishing a first-order framework. We fix a countable, decidable set **Sym** consisting of predicates and function symbols of various arities, along with constants. We also give ourselves a countably infinite set *Var* of variables (distinct from all the other symbols). The resulting set of formulas is denoted: $\mathcal{L}_{form}$. Throughout our discussion, **Sym** and *Var* remain fixed. However, many of our results impose special hypotheses on **Sym** (typically, that it include predicates of various arities). Theorems stated without such hypotheses are true for any choice of (countable) **Sym**. The countability of **Sym** is crucial throughout. So we provide a last reminder:

(1)   *Convention:*   We assume that the symbol set **Sym** consists of countably many predicates, function symbols, and constants. We also assume that **Sym** is computably decidable.

Some further notation concerning our language will be used in what follows. The variables of *Var* are enumerated as $\{v_i \mid i \in N\}$. $\mathcal{L}_{sen}$ denotes the set of sentences (no free variables). We use $\mathcal{L}_{basic}$ to denote the set of basic formulas, that is, the subset of $\mathcal{L}_{form}$ consisting of atomic formulas and negations thereof. The set of variables occurring free in $\varphi \in \mathcal{L}_{form}$ is denoted $Var(\varphi)$. An $\exists$ formula is a formula equivalent to a formula in prenex form whose quantifier prefix is limited to existentials. The same terminology applies to $\exists\forall$ formulas and sentences, etc.

We turn now to the semantic side of our first-order framework. It will greatly simplify the discussion to limit attention to countable structures that interpret **Sym**, that is, to structures with finite or denumerable domains. The countability assumption is so pivotal that we enter (once and for all) the following convention.

(2)   *Convention:*   By "structure" will always be meant a *countable* structure that interprets the symbol set **Sym**.

Otherwise, our semantic notions are standard. In particular, structure $\mathcal{S}$ is a model of $\Gamma \subseteq \mathcal{L}_{form}$ just in case there is an assignment $h : Var \to |\mathcal{S}|$ with $\mathcal{S} \models \Gamma[h]$; in this case $\mathcal{S}$ (and also $\mathcal{S}$ with h) are said to "satisfy" Γ.[1] As usual, one set of formulas implies another just in case every structure plus assignment that satisfies the first satisfies the second. (By the Löwenheim-Skolem theorem, limiting ourselves to countable structures has no effect on this definition.) The class of models of $\Gamma \subseteq \mathcal{L}_{form}$ is denoted $MOD(\Gamma)$.[2] For $\varphi \in \mathcal{L}_{form}$, we write $MOD(\varphi)$ in place of $MOD(\{\varphi\})$.

Although we shall rely heavily on Convention (2), it should be noted that our theory can be generalized to structures with domains of arbitrary cardinality. Indeed, such generalization is the topic of the optional Section 3.7. Aside from the latter excursion, however, the remaining discussion bears exclusively on countable models for **Sym**.

---

1. The symbol $|\mathcal{S}|$ denotes the domain of the structure $\mathcal{S}$. For background in logic, see (for example) [Enderton, 1972, Ebbinghaus *et al.*, 1994].

2. We take $MOD(\emptyset)$ to be the class of all structures.

3.1.3 Components of the Paradigm

With the foregoing conventions and notations in hand, the first-order paradigm may now be presented. For this purpose, we step through the items listed in 1.(1).

Worlds. The potential realities of our paradigm are all the (countable) structures that interpret the symbol set **Sym**. Any such structure might turn out to be the scientist's habitat, so the scientist's question is always roughly: "What's true in my structure?"

Problems. It is standard fare in philosophy to construe a proposition as a collection of possible worlds. Relativizing to the present setting, we take a proposition to be a collection of structures. Then, a "problem" is a choice among mutually exclusive propositions. Officially:

(3) DEFINITION: A nonempty class of structures is a *proposition*. A *problem* is a collection of disjoint propositions.

To illustrate, suppose that **Sym** is limited to a single binary predicate. Let proposition P_0 be the collection of strict total orders with a least point, and proposition P_1 be the collection of strict total orders without a least point. Then $\{P_0, P_1\}$ is a problem (similar to those studied in Section 1.2).[3]

Notice the similarity to the Bayesian set-up. A problem is a partition over an event-space, where the events are structures and each cell of the partition is a proposition. A problem thus embodies the question: which of the propositions is true in my world?

Environments. We now consider the information made available to a scientist about a given structure (namely, the structure chosen as the "actual" world). The idea is that the vocabulary of **Sym** is observational, so that the scientist can see of every object which predicates etc. are true of it.[4] If the object bears a name in **Sym**, then the scientist sees that the object bearing name so-and-so has property thus-and-such. If the object is not named in **Sym** then the scientist must rely on some temporary designation, like "the third object shown to me."

---

3. Throughout our discussion we take a total order to be reflexive, connected, transitive, and antisymmetric (like $\leq$ on N). A strict total order is assumed to be irreflexive, connected, transitive, and asymmetric (like $<$ on N).

4. Non-observational, or "theoretical" vocabulary is discussed in Section 3.3.4 below.

As the formal counterpart to such temporary names, we use variables, assigned to the domain of the structure in some arbitrary way. Officially, we proceed as follows.

(4) DEFINITION: Given structure $\mathcal{S}$, a *full assignment* to $\mathcal{S}$ is any mapping of *Var* onto $|\mathcal{S}|$.

Thus, a full assignment h to $\mathcal{S}$ provides temporary names for all the elements of $|\mathcal{S}|$, namely, $s \in |\mathcal{S}|$ is assigned nonempty $h^{-1}(s) \subseteq Var$.

(5) DEFINITION: Let structure $\mathcal{S}$ and full assignment h to $\mathcal{S}$ be given.

(a) An *environment* for $\mathcal{S}$ and h is a sequence e such that range$(e) = \{\beta \in \mathcal{L}_{basic} \mid \mathcal{S} \models \beta[h]\}$.

(b) An *environment* for $\mathcal{S}$ is an environment for $\mathcal{S}$ and h, for some full assignment h to $\mathcal{S}$.

(c) An *environment* is an environment for some structure.

(d) An *environment* for proposition P is an environment for some $\mathcal{S} \in P$.

(e) An *environment* for problem $\mathbf{P}$ is an environment for some $P \in \mathbf{P}$.

Using standard terminology from model theory [Keisler, 1977, Def. 3.1], we may say that an environment for $\mathcal{S}$ and h lists the basic diagram of $\mathcal{S}$, using variables to supply temporary names (via h) for the members of $|\mathcal{S}|$.

(6) *Example:*   Suppose that binary predicate R is the only member of **Sym**, and that structure $\mathcal{S}$ with $|\mathcal{S}| = N$ interprets R as $<$. If full assignment h to $\mathcal{S}$ is $\{(v_i, i) \mid i \in N\}$ then one environment for $\mathcal{S}$ and h begins this way:

$$v_3 \neq v_4 \quad \neg Rv_0v_0 \quad Rv_1v_9 \quad v_9 = v_9 \quad Rv_7v_8 \quad v_0 \neq v_3 \quad v_5 = v_5 \quad \neg Rv_{23}v_8 \quad \ldots$$

If full assignment g to $\mathcal{S}$ is $\{(v_{2i}, i), (v_{2i+1}, i) \mid i \in N\}$ then one environment for $\mathcal{S}$ and g begins this way:

$$v_2 = v_3 \quad \neg Rv_4v_5 \quad Rv_1v_9 \quad v_9 = v_9 \quad Rv_7v_{19} \quad v_0 \neq v_3 \quad \neg Rv_{33}v_2 \quad \neg Rv_{23}v_8 \quad \ldots$$

If P is the proposition containing every strict total order, then this same environment is for P. If $\mathbf{P}$ is a problem that includes P as a component proposition, then the environment is also for $\mathbf{P}$.

As a further aid to intuition, suppose that **Sym** contains the unary predicate H and the binary predicate R. Then, abstracting away from the arbitrary choice of temporary names for domain elements (embodied in the underlying full assignment), an environment provides information like this: "The first object encountered falls under H. The second object does not fall under H, and is related to the first object by R. The third object is identical to the first object. The fourth object is not identical to the second object, and is not related to the third object by R. . . . "

Given environment e and $k \in N$, $e(k)$ denotes the member of e that falls in its kth position, and $e[k]$ is the initial finite segment of e of length k. Thus $e(k)$ comes right after $e[k]$ in e. (The present notation is thus consistent with that for the numerical paradigm, introduced in Section 2.1.1.) Thus, if e is the first environment of Example (6), then $e(2) = Rv_1v_9$, $e[2] = \langle v_3 \neq v_4, \neg Rv_0v_0 \rangle$, $e(0) = v_3 \neq v_4$, and $e[0] = \emptyset$.

The following fact shows that environments offer considerable information about the structures they are for. The proof is easy (and also an immediate consequence of [Keisler, 1977, Prop 3.2(i)]).

(7) LEMMA: Let structures S and T be given.

(a) If S and T are isomorphic then the set of environments for S is identical to the set of environments for T.

(b) If some environment is for both S and T then S and T are isomorphic.

As in earlier paradigms, we let *SEQ* denote the collection of proper initial segments of any environment. Thus, *SEQ* is the countable set of all consistent, finite sequences over $\mathcal{L}_{basic}$, and exhausts the potential data that can become available to scientists. Given nonvoid $\sigma \in SEQ$, we denote by $\bigwedge \sigma$ the conjunction (in order of appearance in σ) of the formulas in range(σ). To cover degenerate cases, we define $\bigwedge \sigma$ to be $\forall v_0(v_0 = v_0)$ in case σ is empty. We denote by $Var(\sigma)$ the set of free variables appearing in σ. Let proposition P and $\sigma \in SEQ$ be given. We say that σ is *for* P just in case $\bigwedge \sigma$ is satisfiable in some member of P. We say that σ is *for* problem $\mathbf{P}$ just in case σ is for some $P \in \mathbf{P}$. Thus, σ is for $\mathbf{P}$ just in case there is $S \in \bigcup \mathbf{P}$ that satisfies $\bigwedge \sigma$.

Scientists.   A scientist in the first-order paradigm is a partial or total mapping of *SEQ* into subclasses of structures. That is, if scientist Ψ is defined on $\sigma \in SEQ$, then $\Psi(\sigma)$ is a collection of structures, thus a proposition.[5] Intuitively, faced with σ, Ψ believes the proposition expressed by $\Psi(\sigma)$.

---

5. More precisely, $\Psi(\sigma)$ is a proposition unless $\Psi(\sigma) = \emptyset$, since by Definition (3) propositions are non-null. We have denied the status "proposition" to the empty set in order to simplify the formulation of subsequent theorems and definitions.

Success. To solve a problem **P**, we require the scientist to reach stable belief in the one true proposition of **P**, namely, the proposition holding the structure presented to the scientist. To express such belief it is sufficient to announce a consistent proposition that implies the true one from **P**; there is no penalty for saying something stronger than the target proposition, provided it is consistent. This added flexibility—perfectly compatible with the goal of discovering the true coset of **P**—will be important to our discussion of computable inquiry, and of belief revision. Before further comment, let us present the official definition of success.

(8)   DEFINITION: Let scientist Ψ be given.

(a) Let environment e for proposition P be given. We say that Ψ *solves P in e* just in case for cofinitely many k, $\emptyset \neq \Psi(e[k]) \subseteq P$. We say that Ψ *solves P* just in case Ψ solves P in every environment for P.

(b) Let problem **P** be given. We say that Ψ *solves* **P** just in case Ψ solves every member of **P**. In this case we say that **P** is *solvable*, and otherwise *unsolvable*.

Hence, solving **P** requires solving every $P \in \mathbf{P}$ in every environment for P. Equivalently: Ψ solves **P** just in case for every $P \in \mathbf{P}$, every $\mathcal{S} \in \mathbf{P}$, and every environment e for $\mathcal{S}$, there are cofinitely many k such that $\emptyset \neq \Psi(e[k]) \subseteq P$. Requiring success on all environments instead of just some eliminates the possibility of communicating the correct answer to a scientist via a coding scheme involving variables (as discussed at length in Section 3.6.1 below).

(9)   *Example:*

(a) Suppose that unary predicate H is the only member of **Sym**. Given $n \in N$, let P_n be the class of all structures $\mathcal{S}$ such that $\mathrm{card}(H^{\mathcal{S}}) = n$. Let $\mathbf{P} = \{P_n \mid n \in N\}$. Then **P** is solvable. To see this, given $\sigma \in SEQ$, let $t(\sigma)$ denote the smallest $n \in N$ such that (a) some structure in P_n satisfies $\bigwedge \sigma$, and (b) for all $m < n$ and all $\mathcal{U} \in P_m$, $\mathcal{U}$ does not satisfy $\bigwedge \sigma$. Define scientist Ψ such that for all $\sigma \in SEQ$, $\Psi(\sigma) = P_{t(\sigma)}$. Then it is easy to see that for all $n \in N$ and all environments e for P_n, $\Psi(e[k]) = P_n$ for cofinitely many k. Thus, Ψ solves **P**.

(b) Suppose that binary predicate R is the only member of **Sym**. Set:

$P_y = \{\langle N, \prec \rangle \mid \prec \text{ is isomorphic to } \omega\}$,

$P_n = \{\langle N, \prec \rangle \mid \prec \text{ is isomorphic to } \omega^*\}$.

Then a straightforward adaptation of the proof of Proposition 1.(7) shows that $\{P_y, P_n\}$ is solvable.[6]

One aspect of Definition (8) needs further discussion. Let P be the proposition consisting of every total order (over a binary predicate R). Let e be an environment for $\langle N, \leq \rangle$, and suppose that scientist Ψ issues the class of *dense* total orders on cofinitely many initial segments of e. Our success criterion credits Ψ with solving P in e, yet Ψ is systematically mistaken in one respect about the structure underlying e, namely, it is not a dense order. Is the criterion too liberal? We think not. If Ψ's behavior is deemed unsatisfactory in this example, the excessive liberality is inherent in P, not in our success criterion. When, for particular purposes, density is an important property of orders, the problem **P** should be defined in such a way that dense and non-dense orders belong to different cosets. In contrast, if P belongs to **P** this can only signify that density is irrelevant to the problem being posed. The criterion introduced in Definition (8) allows the problem itself to determine the desired degree of accuracy.

Before returning to technical issues, we indulge in a philosophical remark. A traditional goal of logic is to assist in the discovery of new truths starting from old. Suppose we conceive of a problem **P** as the true assertion that exactly one proposition in **P** holds the real world, **w**. Then we may ask of logic for help in extending this premise to a new verity, namely, to the true member P of **P**. Once suitably developed, the inductive branch of logic might respond by specifying a formal scientist that stabilizes to P on evidence from **w**—or else by advising us that no such scientist exists, if such is the case. The theory developed in the remainder of the book is intended to help realize an inductive logic of this kind. Notice that such a logic would refrain from giving credence to any particular hypothesis, in light of current data. Instead, the logic would place its faith in particular scientists, conceived as methods for converting data into theories. It is the limiting behavior of the scientist that would be endorsed, not any of the hypotheses issued along the way.[7]

3.1.4   Three Observations

We make three observations about the first-order paradigm.

---

6. For the (standard) notations ω and ω^*, see page 3.

7. An analogous interpretation is often given to statistical confidence intervals. See [Baird, 1992, Sec. 10.5] for discussion.

Countability. Since *SEQ* is countable so is the range of every scientist. We thus have the following counterpart to Lemma 2.(3).

(10) LEMMA: Every solvable problem is countable.

Of course, by a problem being "countable" is meant that it contains countably many propositions. Each of the propositions may have arbitrary cardinality (or even be a proper class, as will often be the case in our examples).

Normal Problems. The next observation employs a notation that will also be useful later.

(11) DEFINITION: For any proposition P, let $\mathfrak{I}(P)$ denote the closure of P under isomorphism. [In other words, $\mathcal{S} \in \mathfrak{I}(P)$ implies that every structure isomorphic to $\mathcal{S}$ is also in $\mathfrak{I}(P)$.] Problem **P** is *normal* just in case for all $P_1, P_2 \in \mathbf{P}, \mathfrak{I}(P_1) \cap \mathfrak{I}(P_2) = \emptyset$.

Thus, in a normal problem there are no isomorphic structures appearing in different propositions. Directly from Lemma (7)a:

(12) LEMMA: Only normal problems are solvable.

In view of the lemma, normal problems will be the focus of our discussion. However, things go more smoothly by keeping non-normal problems within the purview of the first-order paradigm. So it remains the case that a problem is *any* collection of disjoint propositions.

The Necessity of Full Assignments. It might be asked whether recourse to full assignments could be avoided by presenting scientists with all the ∃ sentences made true by the underlying structure. In this set-up, an environment for a structure $\mathcal{S}$ would be a sequence d such that range$(d) = \{\gamma \in \mathcal{L}_{sen} \mid \gamma$ is existential and $\mathcal{S} \models \gamma\}$. No full assignment h appears in the definition.

   In fact, the ∃ sentence approach is not equivalent to the one defined here. This can be seen by considering P_y and P_n of Example (9)b. All of the structures in $P_y \cup P_n$ satisfy the same ∃ sentences, so under the ∃ sentence conception, all would have the same environments. This would render $\{P_y, P_n\}$ unsolvable whereas the example shows it to be solvable under the present conception of environment.

   To make a closely related point, let P_y, P_n be, once again, as specified in Example (9)b. Let e be an environment for $P_y \cup P_n$. Then no formula in $\{\bigwedge e[k] \mid k \in N\}$ "separates" P_y and P_n, in the sense of being satisfiable in just one of them. Indeed, every initial segment of e is satisfiable in every model of $P_y \cup P_n$. The solvability of $\{P_y, P_n\}$ thus illustrates that within our paradigm, separability is not required for successful fixation of belief. The paradigm differs in this respect from the framework established in [Gaifman & Snir, 1982], in which separation plays a major role [see also Exercise (48), below]. Roughly speaking, one motivation for the developments to follow is to understand inquiry without requiring the scientist's data to rule out any theoretically possible world.

It may also be noted here that environments in the first-order paradigm share a feature of environments for functional languages in the numerical paradigm (see Section 2.6). Specifically, if e is an environment and α an atomic formula, then exactly one of α, $\neg\alpha$ belongs to range(e) [compare Lemma 2.(85)]. It is the multiplicity of the potential underlying assignments that prevents the foregoing fact from trivializing the first-order paradigm.

3.1.5 The Numerical Paradigm as a Special Case

We wish now to show that the first-order paradigm is more general than the numerical paradigm of Chapter 2.[8] For this purpose, problems of the numerical variety will be exhibited as special kinds of problems in the present paradigm. For the remainder of the subsection we assume that **Sym** consists of a binary predicate R, a constant $\bar{0}$, and a unary function symbol s. The term that results from n applications of s to $\bar{0}$ is denoted $\bar{n}$. The following definition allows us to represent languages (in the sense of Chapter 2) as propositions.

(13)   DEFINITION: For $L \subseteq N$, denote by P_L the class of structures $\mathcal{S}$ that meet the following conditions, for every $n \in N$:

(a) if $n \in L$ then there is some $p \in N$ with $\mathcal{S} \models R\bar{n}\bar{p}$;

(b) if $n \notin L$, then for all $p \in N$, $\mathcal{S} \models \neg R\bar{n}\bar{p}$.

Now we can formulate the sense in which the numerical paradigm can be embedded into the first-order one.

(14)   THEOREM: Let countable set $\mathbf{P}_\eta$ of nonempty subsets of N be given. Then $\mathbf{P}_\eta$ is solvable in the numerical paradigm iff $\{P_L \mid L \in \mathbf{P}_\eta\}$ is solvable in the present paradigm.

---

8. The present section is not essential to the sequel and may be omitted on a first reading.

Proof: For the purposes of proving Theorem (14), we introduce the follow-ing, temporary convention. A term or notation decorated with η indicates usage in the sense of Chapter 2 (numerical paradigm). Decoration with ℓ indicates usage in the sense of the present chapter (first-order, or "logical" paradigm).

Suppose that $\mathbf{P}_\eta$ is solvable$_\eta$. Proof that $\mathbf{P}_\ell = \{P_L \mid L \in \mathbf{P}_\eta\}$ is solvable$_\ell$ relies on the existence of a total function $f : SEQ_\ell \to SEQ_\eta$ with the following properties.

(15)   (a)   For all $\sigma, \tau \in SEQ_\ell$, if $\tau \subseteq \sigma$ then $f(\tau) \subseteq f(\sigma)$.

     (b)   If $\emptyset \neq L \subseteq N$, and environment$_\ell$ e is for P_L, then $\bigcup_{k \in N} f(e[k])$ is an environment$_\eta$ for L.

A function f satisfying (15) can be defined by induction on SEQ_ℓ as follows:

(16)   (a) $f(\emptyset) = \emptyset$;

     (b)   for all $\sigma \in SEQ_\ell$ and $\beta \in \mathcal{L}_{basic}$, $f(\sigma \star \beta) = f(\sigma) \star n$ if β is of the form $R\bar{n}\bar{p}$ for some $n, p \in N$; otherwise $f(\sigma \star \beta) = f(\sigma)$.

Let scientist$_\eta$ Ψ_η that solves$_\eta$ $\mathbf{P}_\eta$ be given. We define scientist$_\ell$ Ψ_ℓ as fol-lows. For all $\sigma \in SEQ_\ell$, $\Psi_\ell(\sigma)$ is defined just in case $\Psi_\eta(f(\sigma))$ is defined; if $\Psi_\eta(f(\sigma)) = L$ for $L \subseteq N$, then $\Psi_\ell(\sigma) = P_L$. It follows immediately from (15) that Ψ_ℓ solves$_\ell$ $\mathbf{P}_\ell$.

Suppose that $\mathbf{P}_\ell$ is solvable$_\ell$. Proof that $\mathbf{P}_\eta$ is solvable$_\eta$ relies on the existence of a total function $g : SEQ_\eta \to SEQ_\ell$ with the following properties.

(17)   (a)   For all $\sigma, \tau \in SEQ_\eta$, if $\tau \subseteq \sigma$ then $g(\tau) \subseteq g(\sigma)$.

     (b)   For all environments$_\eta$ e for $\emptyset \neq L \subseteq N$, $\bigcup_{k \in N} g(e[k])$ is an environ-ment$_\ell$ for P_L.

The existence of a total function g satisfying (17) is an immediate consequence of the following fact, which is easy to verify.

(18)   FACT: There exists total function $g : SEQ_\eta \to SEQ_\ell$ such that for all $\sigma \in SEQ_\eta$, the following holds:

(a) for every $\tau \in SEQ_\eta$, if $\tau \subseteq \sigma$ then $g(\tau) \subseteq g(\sigma)$;

(b) for every $n \in N$, range$(g(\sigma)) \models R\bar{n}\bar{p}$ for some $p \in N$ iff $n \in$ range(σ);

(c) range$(g(\sigma))$ is consistent with $\{v_n = \bar{n} \mid n \in N\}$;

(d) if $\beta \in \mathcal{L}_{basic}$ contains variables and terms included in $\{v_0 \ldots v_n, \bar{0} \ldots \bar{n}\}$, where $n = $ length(σ), then either β or $\neg\beta$ occurs in $g(\sigma)$.

Suppose that scientist$_\ell$ Ψ_ℓ solves$_\ell$ $\mathbf{P}_\ell$. We define scientist$_\eta$ Ψ_η as follows. For all $\sigma \in SEQ_\eta$, $\Psi_\eta(\sigma)$ is defined just in case $\emptyset \neq \Psi_\ell(g(\sigma)) \subseteq P_L$ for some $L \subseteq N$, in which case $\Psi_\eta(\sigma) = L$. It follows immediately from (17) that Ψ_η solves$_\eta$ $\mathbf{P}_\eta$.   ∎

3.1.6   Exercises

(19)   EXERCISE: Let Q be the rationals. Suppose that **Sym** contains a unary function symbol F along with a constant c_q for every $q \in Q$. Given $f : Q \to Q$, let S_f be the structure with domain Q that interprets F as f and each c_q as q. Show that $\{\{S_f\} \mid f : Q \to Q \text{ a polynomial}\}$ is solvable.

(20)   EXERCISE: Fix an enumeration $\{\alpha_i \mid i \in N\}$ of all atomic formulas. We say that an environment e is *standard* just in case for all $i \in N$, either $e(i) = \alpha_i$ or $e(i) = \neg\alpha_i$. We say that scientist Ψ solves problem $\mathbf{P}$ *on standard environments* just in case for every standard environment e for $P \in \mathbf{P}$, $\emptyset \neq \Psi(e[k]) \subseteq P$ for cofinitely many k. Let problem $\mathbf{P}$ be given. Show that $\mathbf{P}$ is solvable iff some scientist solves $\mathbf{P}$ on standard environments.

(21)   EXERCISE: Suppose that environment e is for structure S and full assignment h. We say that e is *bijective* just in case h is a bijection between *Var* and $|S|$. We say that scientist Ψ solves problem $\mathbf{P}$ *on bijective environments* just in case for every bijective environment e for $P \in \mathbf{P}$, $\emptyset \neq \Psi(e[k]) \subseteq P$ for cofinitely many k. Let problem $\mathbf{P}$ be such that all $S \in \bigcup \mathbf{P}$ is infinite. Show that $\mathbf{P}$ is solvable iff some scientist solves $\mathbf{P}$ on bijective environments.

(22)   EXERCISE: ♣   Suppose that **Sym** consists of a binary predicate R, a constant $\bar{0}$, and a unary function symbol s. The term that results from n applications of s to $\bar{0}$ is denoted $\bar{n}$. Say that a structure S is *standard* if $|S| = N$, and for all $n \in N$ the interpretation of $\bar{n}$ in S is n. Let collection $\mathbf{K}$ of strict total orders over N be given. Denote by $\mathbf{P_K}$ the collection of all standard structures S such that the interpretation of R in S belongs to $\mathbf{K}$. Let $\theta = \exists x \forall y (x \neq y \to Rxy)$ ("least point"). Show that $\mathbf{K}$ is solvable in the order paradigm of Section 1.2 iff $\{\mathbf{P_K} \cap MOD(\theta), \mathbf{P_K} \cap MOD(\neg\theta)\}$ is solvable in the first-order paradigm.

3.2   Solvability Characterized

A fundamental question to be answered about any paradigm is what problems can be solved within it. The present section offers a characterization of solvability within the first-order paradigm. The necessary and sufficient condition that we develop resembles the one offered in Section 2.2 for the numerical paradigm. Just as in Chapter 2, we begin with the idea of data that "locks" a scientist onto the correct conjecture.

3.2.1   Locking Pairs

In place of the locking sequences discussed in Section 2.2.1, we have the following concept.

(23)   DEFINITION: Let scientist Ψ, proposition P, $\mathcal{S} \in P$, $\sigma \in SEQ$, and finite assignment $a : Var \to |\mathcal{S}|$ be given. We say that (σ, a) is a *locking pair* for Ψ, $\mathcal{S}$, and P just in case the following conditions hold.

(a) domain$(a) \supseteq Var(\sigma)$.

(b) $\mathcal{S} \models \bigwedge \sigma[a]$.

(c) For every $\tau \in SEQ$, if $\mathcal{S} \models \exists \bar{x} \bigwedge (\sigma * \tau)[a]$, where $\bar{x}$ contains the variables in $Var(\tau) - \text{domain}(a)$, then $\emptyset \neq \Psi(\sigma * \tau) \subseteq P$.

The following fact is analogous to Lemma 2.(8). It will be exploited in providing a necessary condition for solvability.

(24)   LEMMA: Let scientist Ψ, proposition P, and $\mathcal{S} \in P$ be given. Suppose that scientist Ψ solves P in every environment for $\mathcal{S}$. Then there is a locking pair for Ψ, $\mathcal{S}$, and P.

Proof: Suppose there is no locking pair for Ψ, $\mathcal{S}$, and P. Then:

(25)   For every $\sigma \in SEQ$ and finite assignment $a : Var \to |\mathcal{S}|$, if domain$(a) \supseteq Var(\sigma)$ and $\mathcal{S} \models \bigwedge \sigma[a]$, then there is $\tau \in SEQ$ and finite extension $a' : Var \to |\mathcal{S}|$ of a such that:

(a) $Var(\tau) \subseteq \text{domain}(a')$,

(b) $\mathcal{S} \models \bigwedge \tau[a']$, and

(c) either $\Psi(\sigma * \tau)$ is not defined or $\Psi(\sigma * \tau) = \emptyset$, or $\Psi(\sigma * \tau) \not\subseteq P$.

We shall use (25) to construct an environment e for $\mathcal{S}$ such that Ψ does not solve P in e. This suffices to finish the proof.

Recall our enumeration $\{v_i \mid i \in N\}$ of Var. Let $\{s_i \mid i \in N\}$ enumerate $|\mathcal{S}|$. Let $\{\alpha_i \mid i \in N\}$ enumerate the atomic formulas of $\mathcal{L}_{form}$. Set $a^{-1} = \emptyset$ and $\sigma^{-1} = \emptyset$. The construction of e proceeds by defining $\sigma^k \in SEQ$ and finite assignment $a^k : Var \to |\mathcal{S}|$ for each $k \in N$ such that:

(26) (a) a^k extends a^{k-1};

(b) $v_k \in \text{domain}(a^k)$ and $s_k \in \text{range}(a^k)$;

(c) σ^k extends σ^{k-1};

(d) $Var(\sigma^k) \subseteq \text{domain}(a^k)$ and $\mathcal{S} \models \bigwedge \sigma^k[a^k]$;

(e) $\alpha_k \in \text{range}(\sigma^k)$ or $\neg\alpha_k \in \text{range}(\sigma^k)$;

(f) either $\Psi(\sigma^k)$ is not defined or $\Psi(\sigma^k) = \emptyset$, or $\Psi(\sigma^k) \not\subseteq P$.

By (26)a,b, $\bigcup_{k \in N} a^k$ is a full assignment to $\mathcal{S}$. By (26)c-e, $\bigcup_{k \in N} \sigma^k$ is an environment e for $\mathcal{S}$ and $\bigcup_{k \in N} a^k$. By (26)f, Ψ does not solve P in e.

Let $k \in N$ be given, and suppose that $a^{-1} \ldots a^{k-1}, \sigma^{-1} \ldots, \sigma^{k-1}$ have been defined. We define a^k and σ^k. Let $n \geq k$ be least such that every variable appearing in α_k has index less than or equal to n. Let a be any finite extension of a^{k-1} such that $\{v_i \mid i \leq n\} \subseteq \text{domain}(a)$ and $s_k \in \text{range}(a) \subseteq |\mathcal{S}|$. Define $\sigma \in SEQ$ to be $\sigma^{k-1} * \alpha_k$ if $\mathcal{S} \models \alpha_k[a]$; otherwise, define σ to be $\sigma^{k-1} * \neg\alpha_k$. By (25), let $\tau \in SEQ$ and finite extension $a' : Var \to |\mathcal{S}|$ of a be such that $Var(\tau) \subseteq \text{domain}(a')$, $\mathcal{S} \models \bigwedge \tau[a']$, and either $\Psi(\sigma * \tau)$ is not defined or $\Psi(\sigma * \tau) = \emptyset$, or $\Psi(\sigma * \tau) \not\subseteq P$. Define $a^k = a'$ and $\sigma^k = \sigma * \tau$. It is easy to verify that a^k and σ^k satisfy (26). ∎

3.2.2 Tip-offs in the First-Order Paradigm

With Lemma (24) in hand, we are ready to characterize solvability. Analogously to the numerical paradigm, the solvability of a problem requires that each of its elements be associated with a faithful signal. The signals are once again called "tip-offs," and defined in the present subsection. After tip-offs are defined, we show that their existence is both sufficient and necessary for solvability.

(27) DEFINITION: By a π-set is meant any collection of $\forall$ formulas all of whose free variables are drawn from the same finite set.

(28) DEFINITION: Let problem $\mathbf{P}$ and $P \in \mathbf{P}$ be given. A *tip-off* for P in $\mathbf{P}$ is a countable collection t of π-sets such that:

(a) for every $\mathcal{S} \in P$ and full assignment h to $\mathcal{S}$, there is $\pi \in t$ with $\mathcal{S} \models \pi[h]$;

(b) for all $\mathcal{U} \in P' \in \mathbf{P}$ with $P' \neq P$, all full assignments g to $\mathcal{U}$, and all $\pi \in t$, $\mathcal{U} \not\models \pi[g]$.

If every member of $\mathbf{P}$ has a tip-off in $\mathbf{P}$, then we say that $\mathbf{P}$ *has tip-offs*.

In contrast to the numerical tip-offs of Section 2.2.2, tip-offs in the sense of Definition (28) are infinite sets. The difference arises from the role of structures within the first-order paradigm, and the need to choose a full assignment to present them to scientists. Languages, in comparison, are simpler objects. Here is an example of a problem with tip-offs, and a problem without.

(29) *Example:* Suppose that binary predicate R is the only symbol of **Sym**. Let T be the theory of total orders (with respect to R) with either a least or a greatest point. Let $\theta = \exists x \forall y R x y$. Then $\mathbf{P} = \{MOD(T \cup \{\theta\}), MOD(T \cup \{\neg\theta\})\}$ has tip-offs.

Proof: A tip-off for $MOD(T \cup \{\theta\})$ in $\mathbf{P}$ is $\{\{\forall x R v_i x \mid i \in N\}\}$ and a tip-off for $MOD(T \cup \{\neg\theta\})$ in $\mathbf{P}$ is $\{\{\forall x R x v_i \mid i \in N\}\}$. ∎

(30) *Example:* Suppose that binary predicate R is the only symbol of **Sym**. Let $\mathcal{S} = \langle N, \preceq \rangle$ be isomorphic to ω, let $\mathcal{T} = \langle Z, \preceq^* \rangle$, with Z the set of integers, be isomorphic to $\omega^* \omega$, and let disjoint propositions P_1, P_2 be such that $\mathcal{S} \in P_1$ and $\mathcal{T} \in P_2$. Then $\{P_1, P_2\}$ does not have tip-offs.

Proof: We show that there is no tip-off for P_2 in $\{P_1, P_2\}$. Let π-set π be satisfiable in $\mathcal{T}$. Let $n \in N$ and variables $x_0 \ldots x_n$ be such that for all $\varphi \in \pi$, $Var(\varphi) \subseteq \{x_0 \ldots x_n\}$. Choose $a_o \ldots a_n \in Z$ such that for all $\varphi \in \pi$, $\mathcal{T} \models \varphi[a_0/x_0 \ldots a_n/x_n]$. Choose $k \geq 0$ such that $\{a_i + k \mid i \leq n\} \subseteq N$. It is easy to verify that for all $\varphi \in \pi$, $\mathcal{T} \models \varphi[a_0 + k/x_0 \ldots a_n + k/x_n]$. Since $\forall$ formulas are preserved in substructures (see [Hodges, 1993, Cor. 2.4.2]), it follows that for all $\varphi \in \pi$, $\mathcal{S} \models \varphi[a_0 + k/x_0 \ldots a_n + k/x_n]$. Hence π is satisfiable in $\mathcal{S}$. Hence every π-set of formulas is satisfiable in $\mathcal{S}$ whenever it is satisfiable in $\mathcal{T}$. Since $\mathcal{S} \in P_1$ and $\mathcal{T} \in P_2$, this implies that there is no tip-off for P_2 in $\{P_1, P_2\}$. ∎

3.2.3 Tip-offs Are Sufficient for Solvability

Countable problems are solvable if and only if they have tip-offs. This fact will be proved in the present subsection and the next. We start with sufficiency.

(31) PROPOSITION: If problem $\mathbf{P}$ is countable and has tip-offs, then $\mathbf{P}$ is solvable.

Proof: The proposition is trivial if $\mathbf{P} = \emptyset$, so suppose otherwise. Let $\{P_j \mid j < \kappa\}$ be a repetition-free enumeration of the countably many propositions in $\mathbf{P}$, where $\kappa = \mathrm{card}(\mathbf{P})$. Let $\{t_j \mid j < \kappa\}$ enumerate tip-offs corresponding to the

P_j. Since **P** is nonempty and countable, and since each tip-off is countable, we may enumerate the π-sets in $\bigcup_{j < \kappa} t_j$ as $\{\pi_i \mid i \in N\}$. By Definition (28) we have:

(32)   (a) For every $S \in \bigcup \mathbf{P}$ and full assignment h to S, there is $i \in N$ such that $S \models \pi_i[h]$.

   (b) For every i such that π_i is satisfiable in some $S \in \bigcup \mathbf{P}$, there is exactly one j such that $\pi_i \in t_j$.

We specify scientist Ψ that solves **P**. Let $\sigma \in SEQ$ be given. If for all i, range$(\sigma) \cup \pi_i$ is not satisfiable in any $S \in \bigcup \mathbf{P}$, then $\Psi(\sigma)$ is undefined. Otherwise, let i be least such that range$(\sigma) \cup \pi_i$ is satisfiable in some $S \in \bigcup \mathbf{P}$. Then $\Psi(\sigma) = P_j$, where by (32)b, j is unique with $\pi_i \in t_j$. To see that Ψ solves **P**, let $j < \kappa$, environment e for $S \in P_j$, and full assignment h to S be given. By (32)a, let i_0 be least with $S \models \pi_{i_0}[h]$. Then by Definition (28), j is unique with $\pi_{i_0} \in t_j$. We need to show that $\Psi(e[k]) = P_j$ for cofinitely many k. By the definition of Ψ, it suffices for this purpose to show:

(33)   For all $i < i_0$ there are cofinitely many initial segments σ of e such that range$(\sigma) \cup \pi_i$ is inconsistent.

To prove (33), let $i < i_0$ be given. Then $S \not\models \pi_i[h]$. So there is $\varphi \in \pi_i$ such that $S \not\models \varphi[h]$. Since h is onto $|S|$ and φ is a $\forall$ formula, there is $k \in N$ such that range$(e[k]) \cup \{\varphi\}$ is inconsistent. So range$(e[k']) \cup \{\varphi\}$ is inconsistent for all $k' \geq k$, implying (33).   ∎

As an application, from Example (29) we obtain the following.

(34)   PROPOSITION: Suppose that binary predicate R is the only symbol of **Sym**. Let T be the theory of total orders (with respect to R) with either a least or a greatest point. Let $\theta = \exists x \forall y\, Rxy$. Then $\{MOD(T \cup \{\theta\}), MOD(T \cup \{\neg\theta\})\}$ is solvable.

3.2.4   Tip-offs Are Necessary for Solvability

(35)   PROPOSITION: Every solvable problem has tip-offs.

Proof: Let problem **P** and scientist Ψ be such that Ψ solves **P**. Let $P \in \mathbf{P}$ be given. We exhibit a tip-off for P in **P**. Given $\sigma \in SEQ$, define X_σ to be the collection of pairs (S, a) such that $S \in P$, a is a finite assignment to S, and (σ, a) is a locking pair for Ψ, S and P. Define Π_σ to be the collection of $\forall$

formulas φ such that $Var(\varphi) \subseteq Var(\sigma)$ and for all $(\mathcal{S}, a) \in X_\sigma$, $\mathcal{S} \models \varphi[a]$. Given $\sigma \in SEQ$ and $f : Var(\sigma) \to Var$, set $\Pi_{\sigma,f} = \{\varphi[f(x)/x, x \in Var(\sigma)] \mid \varphi \in \Pi_\sigma\}$. We claim that $t = \{\Pi_{\sigma,f} \mid \sigma \in SEQ, f : Var(\sigma) \to Var\}$ is a tip-off for P in $\mathbf{P}$. It is evident that t is a countable collection of π-sets. To prove clause (a) of Definition (28), let $\mathcal{S} \in P$ and full assignment h to $\mathcal{S}$ be given. By Lemma (24), let $\sigma \in SEQ$ and finite assignment $a : Var \to |\mathcal{S}|$ be such that (σ, a) is a locking pair for Ψ, $\mathcal{S}$, and P. Then $(\mathcal{S}, a) \in X_\sigma$, so $\mathcal{S} \models \Pi_\sigma[a]$. Since h is onto $|\mathcal{S}|$, we can choose $f : Var(\sigma) \to Var$ such that $\mathcal{S} \models \Pi_{\sigma,f}[h]$. Hence, $\Pi_{\sigma,f}$ is the π-set in t that witnesses (28)a.

To finish the proof, we show that (28)b is satisfied. Let $\sigma \in SEQ$, $f : Var(\sigma) \to Var$, $\mathcal{U} \in P' \in \mathbf{P}$ with $P' \neq P$, and full assignment g to $\mathcal{U}$ be given. Suppose for a contradiction that $\mathcal{U} \models \Pi_{\sigma,f}[g]$. Then we can choose full assignment g' to $\mathcal{U}$ such that:

(36) $\mathcal{U} \models \Pi_\sigma[g']$.

It is clear that $\bigwedge \sigma \in \Pi_\sigma$, so there is an environment e for $\mathcal{U}$ and g' such that $\sigma \subset e$. We will show that for all $k \geq length(\sigma)$, $\emptyset \neq \Psi(e[k]) \subseteq P$. Since e is for $\mathcal{U} \in P' \in \mathbf{P}$ with $P' \neq P$, this implies that Ψ does not solve P', contradicting our choice of Ψ. So let $k \geq length(\sigma)$ be given. Choose $\tau \in SEQ$ such that $e[k] = \sigma * \tau$. We must show:

(37) $\emptyset \neq \Psi(\sigma * \tau) \subseteq P$.

For this purpose, we establish:

(38)   CLAIM: There is $(\mathcal{S}, a) \in X_\sigma$ such that $\mathcal{S} \models \exists \bar{x} \bigwedge(\sigma * \tau)[a]$, where $\bar{x}$ contains the variables in $Var(\tau) - domain(a)$.

Proof of Claim (38):   For a contradiction, suppose that for all $(\mathcal{S}, a) \in X_\sigma$, $\mathcal{S} \models \forall \bar{x} \neg \bigwedge(\sigma * \tau)[a]$. Then $\forall \bar{x} \neg \bigwedge(\sigma * \tau) \in \Pi_\sigma$, so by (36), $\mathcal{U} \models \forall \bar{x} \neg \bigwedge(\sigma * \tau)[g']$. However, this is impossible since e was chosen to be for $\mathcal{U}$ and g', and $\sigma * \tau \subset e$.   ∎

We now use (38) to prove (37). For any $(\mathcal{S}, a) \in X_\sigma$, (σ, a) is a locking pair for Ψ, $\mathcal{S}$, and P. Hence, by Claim (38) and clause (c) of Definition (23), $\emptyset \neq \Psi(\sigma * \tau) \subseteq P$.   ∎

As an application, from Example (30) we obtain the following.

(39)   PROPOSITION: Suppose that binary predicate R is the only symbol of **Sym**. Let $\mathcal{S} = \langle N, \preceq \rangle$ be isomorphic to ω, let $\mathcal{T} = \langle Z, \preceq^* \rangle$, with Z the set of

integers, be isomorphic to $\omega^*\omega$, and let P_1, P_2 be two disjoint propositions such that $\mathcal{S} \in P_1$ and $\mathcal{T} \in P_2$. Then $\{P_1, P_2\}$ is not solvable.

Along with Lemma (10), Propositions (31) and (35) yield the following counterpart to Theorem 2.(16).

(40) THEOREM: A problem is solvable if and only if it is countable and has tip-offs.

3.2.5 Exercises

(41) EXERCISE: ♣ Suppose that **Sym** is limited to the vocabulary of Boolean algebra. Let $T \subseteq \mathcal{L}_{sen}$ be such that $MOD(T)$ is the class of Boolean algebras, and let $\theta \in \mathcal{L}_{sen}$ be true of a Boolean algebra iff it is atomic. Show that $\{MOD(T \cup \{\theta\}), MOD(T \cup \{\neg\theta\})\}$. is not solvable. (For Boolean algebra, see [Levy, 1979, Ch. VIII].)

(42) EXERCISE: ♣ Suppose that **Sym** is limited to a unary constant $\overline{0}$, a unary function symbol s, and two binary function symbols $\oplus$, $\otimes$. Let Q be the conjunction of the seven axioms of "Robinson's arithmetic" (see [Boolos & Jeffrey, 1989, Ch. 14]). Show that the problem $\{MOD(Q), MOD(\neg Q)\}$ is not solvable.

(43) EXERCISE: Suppose that **Sym** $= \emptyset$. Let $P_1 = \{N\}$, $P_2 = \{\emptyset \neq D \subseteq N \mid D$ finite$\}$. Show that $\{P_1, P_2\}$ is not solvable. [Compare Proposition 2.(116).]

(44) EXERCISE: Suppose that **Sym** is limited to the binary predicate R. Let T be the theory of strict total orders (with respect to R). Denote by θ the sentence $\forall xz[Rxz \rightarrow \exists y(Rxy \wedge Ryz)]$ ("R is dense"). Show that **P** $= \{MOD(T \cup \{\theta\}), MOD(T \cup \{\neg\theta\})\}$ is not solvable.

(45) EXERCISE: Suppose that **Sym** is limited to a binary predicate R. Say that structure $\mathcal{S}$ is *standard* if $|\mathcal{S}| = N$. We specify a countable collection $\{\mathcal{S}_j \mid j \in N\}$ of standard structures by specifying the extension $R^{\mathcal{S}_j}$ of R for all $j \in N$. $R^{\mathcal{S}_0}$ is the relation $\{(i, i + 1) \mid i \in N\}$. For $j > 0$, $R^{\mathcal{S}_j}$ is the finite relation $\{(i, i + 1) \mid i < j\}$. Let $P_1 = \{\mathcal{S}_0\}$ and $P_2 = \{\mathcal{S}_j \mid j > 0\}$. Show that $\{P_1, P_2\}$ is not solvable.

(46) EXERCISE: Given a well-ordering $\prec$ over collection **K** of structures, we define an associated scientist $\Psi_\prec$ as follows. For all $\sigma \in SEQ$, $\Psi_\prec(\sigma)$ is the $\prec$-least member $\mathcal{S}$ of **K** such that $\bigwedge \sigma$ is satisfiable in $\mathcal{S}$. $\Psi_\prec(\sigma)$ is undefined if there is no such. Problem **P** is *solvable by enumeration* just in case there is a

well-ordering $\prec$ of $\bigcup \mathbf{P}$ such that $\Psi_\prec$ solves $\mathbf{P}$. Exhibit structures $\mathcal{S}$ and $\mathcal{T}$ such that $\{\{\mathcal{S}\}, \{\mathcal{T}\}\}$ is solvable, but not by enumeration. [Compare Exercise 2.(43).]

(47)   EXERCISE: Let scientist Ψ and problem $\mathbf{P}$ be given. We say that Ψ is *P-conservative* just in case for all $P \in \mathbf{P}$, σ for $\mathbf{P}$ and $\beta \in \mathcal{L}_{basic}$, if $\emptyset \neq \Psi(\sigma) \subseteq P$ and every structure in $\Psi(\sigma)$ satisfies range$(\sigma) \cup \{\beta\}$, then $\emptyset \neq \Psi(\sigma * \beta) \subseteq P$. Exhibit structures $\mathcal{S}$ and $\mathcal{T}$ such that $\mathbf{P} = \{\{\mathcal{S}\}, \{\mathcal{T}\}\}$ is solvable, but no $\mathbf{P}$-conservative scientist solves $\{\{\mathcal{S}\}, \{\mathcal{T}\}\}$.

(48)   EXERCISE: ♣ Call problem $\mathbf{P}$ "separated" (respectively "elementary separated") just in case for all $P_1 \neq P_2 \in \mathbf{P}$, $\mathcal{S} \in P_1$, and $\mathcal{T} \in P_2$, there is no isomorphic (respectively elementary) embedding from $\mathcal{S}$ into $\mathcal{T}$. (For background on embeddings, see [Hodges, 1993, Secs. 1.2, 2.5].)

(a) Exhibit a separated, unsolvable problem.

(b) Show that every solvable problem is elementary separated.

The present exercise thus strengthens Lemma (12).

(49)   EXERCISE: ♣ Call problem $\mathbf{P}$ *elementary* just in case for all $\mathcal{S}, \mathcal{U} \in \bigcup \mathbf{P}$, $\mathcal{S} \equiv \mathcal{U}$. Exhibit an elementary, solvable problem consisting of more than one proposition.

(50)   EXERCISE: ♦ Scientist Ψ *weakly* solves problem $\mathbf{P}$ just in case for every environment e for $P \in \mathbf{P}$, the following conditions are met:

(a) There are infinitely many k such that $\emptyset \neq \Psi(e[k]) \subseteq P$.

(b) For every $P' \in \mathbf{P}$ with $P' \neq P$, there are only finitely many k such that $\emptyset \neq \Psi(e[k]) \subseteq P'$.

In this case $\mathbf{P}$ is said to be *weakly solvable*.

(a) Exhibit a problem that is weakly solvable but not solvable.

(b) Show that no unsolvable problem consisting of finitely many propositions is weakly solvable.

3.3   Four Applications

The preceding section was devoted to characterizing solvability in terms of tip-offs. In the present section we exploit the characterization in the service of four ideas. First we show how the necessary and sufficient condition for solvability provided by Theorem (40) can be neatly expressed in the infinitary language

$\mathcal{L}_{\omega_1\omega}$. Next we introduce notation for problems of a specific and natural form that will occupy much of Chapter 4. The solvable subclass of these restricted problems is then characterized via a simple, first-order condition. Next we introduce a yet broader class of problems and highlight one of their appealing properties. Finally, we discuss a modification of the first-order paradigm that represents (very roughly, to be sure) the distinction between "observational" and "theoretical" vocabulary.

3.3.1   Solvability and the Infinitary Language $\mathcal{L}_{\omega_1\omega}$

Theorem (40) reveals an intimate association between solvability and express-ibility in the infinitary language $\mathcal{L}_{\omega_1\omega}$.[9] To formulate the matter, we rely on the following definition.

(51)   DEFINITION: A sentence in the language $\mathcal{L}_{\omega_1\omega}$ is called *extended-∃∀* just case it has the form:

$$\bigvee_{i<\kappa_1} \exists \bar{x}_i \bigwedge_{j<\kappa_2} \forall \bar{y}_j \varphi_{i,j},$$

where $\kappa_1, \kappa_2 \leq \omega$, and $\varphi_{i,j} \in \mathcal{L}_{form}$ is quantifier-free, for all $i < \kappa_1$, $j < \kappa_2$.

As an immediate consequence of Theorem (40), we have:

(52)   COROLLARY: A problem **P** is solvable iff it consists of countably many propositions, each of which is equivalent in $\bigcup \mathbf{P}$ to an extended-∃∀ sentence of $\mathcal{L}_{\omega_1\omega}$.

At first encounter, Corollary (52) seems to be no improvement over Theo-rem (40). Whereas the theorem evokes countably infinite sets of finite formulas (namely, tip-offs), the corollary evokes a countably infinite formula. So neither characterization has a finite character. Nonetheless, the corollary turns out to facilitate proofs of solvability, as the reader will appreciate by working Exer-cise (68).

3.3.2   Problems of Form $(T, \{\theta_0, \ldots, \theta_n\})$

We now introduce notation for a special kind of problem. Such problems become increasingly important as our theory develops.

---

9. For $\mathcal{L}_{\omega_1\omega}$, see [Ebbinghaus *et al.*, 1994, Section IX.2].

(53)   DEFINITION: Let problem **P**, $T \subseteq \mathcal{L}_{sen}$, and $\theta_0 \ldots \theta_n \in \mathcal{L}_{sen}$ be given. We say that **P** has the form $(T, \{\theta_0, \ldots, \theta_n\})$ just in case:

(a) for every model $\mathcal{S}$ of T there is exactly one $i \in \{0 \ldots n\}$ such that $\mathcal{S} \models \theta_i$;

(b) $\mathbf{P} = \{MOD(T \cup \{\theta_i\}) \mid 0 \leq i \leq n\}$.

By Definition (3) of "problem," **P** can have the form $(T, \{\theta_0, \ldots, \theta_n\})$ only if $MOD(T \cup \{\theta_i\})$ is nonempty for all $i \leq n$. Hence $T \cup \{\theta_i\}$ must be consistent for all $i \leq n$, which implies that T is consistent. Intuitively, given problem $(T, \{\theta_0, \ldots, \theta_n\})$, we may think of T as an accepted background theory that leaves open which among the θ_i's is true. The θ_i's thus partition the models of T. A special case arises when the partition has just two cosets. The problem then has the form $(T, \{\theta, \neg\theta\})$. In contrast, the more general form $(T, \{\theta_0, \theta_1, \ldots\})$ is of no use since an easy compactness argument shows that there is no infinite partition of $MOD(T)$ of the form $\{MOD(T \cup \{\theta_i\}) \mid i \in N\}$.

To make sure that our notation is clear, let us consider the following example. Suppose that **Sym** is limited to a binary function symbol $\circ$, and let T be the theory of groups. Let θ be the sentence $\forall xy(x \circ y = y \circ x)$. Then $(T, \{\theta, \neg\theta\})$ is the problem whose propositions are:

- the class of abelian groups, namely, $MOD(T \cup \{\theta\})$;

- the class of non-abelian groups, namely, $MOD(T \cup \{\neg\theta\})$.

Of course, $(T, \{\theta, \neg\theta\})$ is trivially solvable. For a more substantial illustration, Propositions (34) and (39) yield the following.

(54)   PROPOSITION: Suppose that **Sym** is limited to the binary predicate R. Let $\theta = \exists x \forall y Rxy$, let $T_0 \subseteq \mathcal{L}_{sen}$ be the theory of total orders over R, and let $T_1 \subseteq \mathcal{L}_{sen}$ be the theory of total orders with either a greatest point or a least point, but not both. Then the problem $(T_1, \{\theta, \neg\theta\})$ is solvable but $(T_0, \{\theta, \neg\theta\})$ is not.

We can use Proposition (54) to make an important point about inquiry within the first-order paradigm. Let a problem of form $(T, \{\theta_0, \ldots, \theta_n\})$ be given, and suppose that e is one of its environments. If there is $k \in N$ such that $T \cup \{\bigwedge e[k]\}$ implies the right choice among $\theta_0 \ldots \theta_n$, then the problem can be solved in e by simply issuing $MOD(T \cup \{\bigwedge e[k]\})$ at each stage k of inquiry. If the same is true of all the environments, then it may be said that $(T, \{\theta_0, \ldots, \theta_n\})$ is solved by "waiting for deduction to work." For, it suffices to wait for the background theory and current data to imply the right answer.

The first-order paradigm would have a trivial character if the solvability of $(T, \{\theta_0, \ldots, \theta_n\})$ implied solvability via such a strategy. But in fact, awaiting deduction is not sufficient. This can be seen from the fact that $(T_1, \{\theta, \neg\theta\})$ of Proposition (54) is solvable even though for all $\sigma \in SEQ$, $\bigwedge \sigma$ implies neither θ nor $\neg\theta$ in the models of T_1. So, some kind of genuinely inductive strategy is needed to solve $(T_1, \{\theta, \neg\theta\})$.

Recall that Theorem (40) offers an infinitary condition on the solvability of arbitrary problems since the tip-offs it evokes are infinite sets of π-sets. The restricted form of problems $(T, \{\theta_0, \ldots, \theta_n\})$, however, leads us to hope for deeper characterization of their solvability. In fact, for problems of form $(T, \{\theta_0, \ldots, \theta_n\})$ we have the following, finitary condition.

(55) THEOREM: A problem of form $(T, \{\theta_0, \ldots, \theta_n\})$ is solvable if and only if for every $i \leq n$, θ_i is equivalent in T to an $\exists\forall$ sentence.

Proof: The right-to-left direction of the theorem follows immediately from Corollary (52). For the left-to-right direction, let solvable problem of form $(T, \{\theta_0, \ldots, \theta_n\})$ be given. Let $j \leq n$ also be given. By Proposition (35), let t be a tip-off for $MOD(T \cup \{\theta_j\})$ in $(T, \{\theta_0, \ldots, \theta_n\})$. By Definition (28)b, for all $i \leq n$ with $i \neq j$, every member of t is unsatisfiable in every $\mathcal{U} \in MOD(T \cup \{\theta_j\})$. Since the θ_i's partition the models of T, it follows that $T \cup \pi \models \theta_j$ for all $\pi \in t$. By compactness, for all $\pi \in t$ there is finite $Y \subseteq \pi$ with $T \cup Y \models \theta_j$. From this and Definition (28)a, we infer:

(56) For every $\mathcal{S} \in MOD(T \cup \{\theta_j\})$ there is $\exists\forall$ $\chi_\mathcal{S} \in \mathcal{L}_{sen}$ such that:

(a) $\mathcal{S} \models \chi_\mathcal{S}$.

(b) $T \cup \{\chi_\mathcal{S}\} \models \theta_j$.

Let $\Sigma = \{\chi_\mathcal{S} \mid \mathcal{S} \in MOD(T \cup \{\theta_j\})\}$. We show that θ_j is equivalent over T to some finite disjunction of members of Σ. This suffices to complete the proof. By (56)b, for every disjunction ρ of members of Σ, $T \cup \{\rho\} \models \theta_j$. Hence, it suffices to show that for some such ρ, $T \cup \{\theta_j\} \models \rho$. For a reductio suppose that for every disjunction ρ of members of Σ, $T \cup \{\theta_j\} \not\models \rho$. Then for every finite $\Delta \subseteq \Sigma$, $T \cup \{\theta_j\} \cup \{\neg\varphi \mid \varphi \in \Delta\}$ is satisfiable. Hence by the compactness and Löwenheim-Skolem theorems there is (countable) $\mathcal{S} \in MOD(T \cup \{\theta_j\})$ such that $\mathcal{S} \models \{\neg\varphi \mid \varphi \in \Sigma\}$. But this implies that there is $\mathcal{S} \in MOD(T \cup \{\theta_j\})$ such that $\mathcal{S} \not\models \chi_\mathcal{S}$, contradicting (56)a. ∎

Theorem (55) will be central to the discussion that follows. It would be equally possible to rely on the following corollary.

(57)   COROLLARY: A problem of form $(T, \{\theta_0, \ldots, \theta_n\})$ is solvable if and only if for all $i \leq n$, θ_i is equivalent in T to a boolean combination of $\exists$ sentences.

Proof: An easy adaptation of [Chang & Keisler, 1977, Theorem 3.1.16] shows the following.

(58)   Let $T \subseteq \mathcal{L}_{sen}$ and $\theta \in \mathcal{L}_{sen}$ be given. Suppose that θ is equivalent in T both to an $\exists\forall$ sentence and to a $\forall\exists$ sentence. Then θ is equivalent in T to a boolean combination of $\exists$ sentences.

The corollary follows easily from (58) and Theorem (55).   ∎

Like the theorem, Corollary (57) offers a finitary characterization of solvability for problems of form $(T, \{\theta_0, \ldots, \theta_n\})$. Since just one such characterization suffices for the sequel, we will rely on the theorem rather than its corollary.

3.3.3   Problems of Form $(T, \{P_0, P_1, \ldots\})$

We now define another special class of problems.

(59)   DEFINITION: Let $T \subseteq \mathcal{L}_{sen}$ be given. We say that problem **P** has the form $(T, \{P_0, P_1, \ldots\})$ just in case $\mathbf{P} = \{P_0, P_1, \ldots\}$ and $\bigcup \mathbf{P} = MOD(T)$.

Of course, a problem can have the form $(T, \{P_0, P_1, \ldots\})$ only if T is a consistent theory. In this case, a problem of form $(T, \{P_0, P_1, \ldots\})$ is a partition of the models of T. Plainly, any problem of form $(T, \{\theta_0, \ldots, \theta_n\})$ also has the more general form $(T, \{P_0, P_1, \ldots\})$. To get a grip on Definition (59) it is nice to have an example of a solvable problem of form $(T, \{P_0, P_1, \ldots\})$ that does not have the form $(T, \{\theta_0, \ldots, \theta_n\})$. Here is one.

(60)   *Example:*   Suppose that **Sym** consists of a unary function symbol s and a constant $\bar{0}$. The term that results from n applications of s to $\bar{0}$ is denoted $\bar{n}$. Let P_0 be the class of models of $\{\bar{n} = \bar{0} \mid n \in N\}$, and for $i > 0$, let P_i be the class of models of $\{\bar{n} = \bar{0} \mid n < i\} \cup \{\bar{i} \neq \bar{0}\}$. Then it is easy to verify that $(\emptyset, \{P_0, P_1, \ldots\})$ is solvable.

The solvable problems of form $(T, \{P_0, P_1, \ldots\})$ share a feature that is described in the following proposition.

(61)   PROPOSITION: Let solvable problem **P** of form $(T, \{P_0, P_1, \ldots\})$ be given. Then for all $P \in \mathbf{P}$, there is an enumeration $\{X_i \mid i \in N\}$ of sets of $\exists\forall$ sentences such that $P = \bigcup_{i \in N} MOD(T \cup X_i)$.

The matter might be put this way: first-order sentences are sufficient to distinguish among any collection **P** of propositions that partition a Δ-elementary class, provided only that **P** is solvable. It is striking that no further constraint need be placed upon the propositions.

Proof of Proposition (61): Let $P \in \mathbf{P}$ be given. By Proposition (35), let enumeration $\{\pi_i \mid i \in N\}$ of π-sets be a tip-off for P in **P**. For all $i \in N$ denote by X_i the set of existential closures of all formulas of form $\bigwedge D$ with D a nonempty, finite subset of π_i. We claim that $P = \bigcup_{i \in N} MOD(T \cup X_i)$. To show the left-to-right inclusion, let $S \in P$ be given. By Proposition (35) let $i \in N$ be such that π_i is satisfiable in S. Hence, $S \models T \cup X_i$. So $P \subseteq \bigcup_{i \in N} MOD(T \cup X_i)$. For the other direction, suppose for a contradiction that $\bigcup_{i \in N} MOD(T \cup X_i) - P \neq \emptyset$. Let structure S and $i \in N$ be such that $S \in MOD(T \cup X_i) - P$. Since $\bigcup \mathbf{P} = MOD(T)$, S belongs to some $P' \in \mathbf{P}$ with $P' \neq P$. By Proposition (35) let t be a tip-off for $P' \in \mathbf{P}$, and let $\pi \in t$ be such that π is satisfiable in S. Without loss of generality we may suppose that π_i and π share no free variables. It is clear that $T \cup \pi_i \cup \pi$ is unsatisfiable, since otherwise some model of T satisfies π-sets from distinct tip-offs, contradicting Definition (28). Summarizing:

(62) (a) all variables that occur free in π_i do not occur free in π;

   (b) π is satisfiable in S;

   (c) $T \cup \pi_i \cup \pi$ is unsatisfiable.

By (62)c and compactness, let finite $D \subseteq \pi_i$ and finite $D' \subseteq \pi$ be such that:

(63) $T \cup D \cup D'$ is unsatisfiable.

Denote by φ the existential closure of $\bigwedge D$ and by φ' the universal closure of $\neg \bigwedge D'$. It follows from (62)a and (63) that:

(64) $T \models \varphi \rightarrow \varphi'$.

We deduce from (62)b that $S \not\models \varphi'$, which with (64) and $S \models T$ implies that $S \models \neg\varphi$. Since $\varphi \in X_i$ this contradicts the hypothesis that $S \in MOD(X_i)$. ∎

   The converse of the proposition is false. To see this, suppose **Sym** $= \emptyset$, and for $i > 0$ let χ_i be an $\exists\forall$ sentence true in just the structures of cardinality i. Let $X = \{\exists x_0 \ldots x_n \bigwedge_{0 \leq i < j \leq n} x_i \neq x_j \mid n > 0\}$. Then $\mathbf{P} = \{MOD(X), MOD(\chi_1), MOD(\chi_2) \ldots\}$ partitions the class of all structures, hence has the form $(T, \{P_0, P_1, \ldots\})$. But an easy consequence of Exercise (43) shows **P** to be unsolvable.

3.3.4   Theoretical Terms

Our fourth application of tip-offs bears on the distinction between observational and theoretical terms. Our discussion of this matter will not be followed up in the remainder of the book, so the present subsection may be skipped without loss of continuity.

The vocabulary employed by real scientists is partly theoretical in the sense that certain predicates can be attributed to objects only in the presence of a background theory; the "raw" data do not suffice to determine their applicability.[10] As a first attempt to incorporate this distinction between kinds of predicates, we divide **Sym** into observational and theoretical parts. Let us admit that such a paradigm remains a very rough approximation to scientific reality since the observational/theoretical distinction is not fixed and sharp, but shifts with the character and stage of inquiry. The cautionary note notwithstanding, for the remainder of this subsection we let symbol sets **Sym** and $\mathbf{Sym}_o$ be such that $\mathbf{Sym}_o \subseteq \mathbf{Sym}$. The idea is that $\mathbf{Sym}_o$ contains the observational vocabulary whereas $\mathbf{Sym} - \mathbf{Sym}_o$ contains the theoretical vocabulary. The next definition generalizes the idea of solvability in a straightforward way.

(65)   DEFINITION: Let scientist Ψ and proposition P be given. We say that Ψ solves P *on* $\mathbf{Sym}_o$ *environments* just in case for every $S \in P$, if S^o is the restriction of S to $\mathbf{Sym}_o$, and if e is an environment for S^o, then for cofinitely many k, $\emptyset \neq \Psi(e[k]) \subseteq P$. A problem is *solvable on* $\mathbf{Sym}_o$ *environments* just in case some scientist solves each of its members on $\mathbf{Sym}_o$ environments.

From the definition it follows immediately that if disjoint propositions P, P' share a $\mathbf{Sym}_o$ reduct, then $\{P, P'\}$ is not solvable on $\mathbf{Sym}_o$ environments. This is a rather trivial form of unsolvability, of course, since it rests on radical underdetermination by data. An example of unsolvability without this trivializing feature is the topic of Exercise (71), below.

An easy adaptation of the proofs of Propositions (31) and (35) yields the following.

(66)   PROPOSITION: A problem **P** is solvable on $\mathbf{Sym}_o$ environments iff **P** consists of countably many propositions, each of which is equivalent in $\bigcup \mathbf{P}$ to an extended-$\exists\forall$ sentence of $\mathcal{L}_{\omega_1\omega}$ in the vocabulary $\mathbf{Sym}_o$.

---

10. To illustrate, a theoretical predicate might be: "is a rock that fell from Mars to Earth." See A. Treiman, "To See a World in 80 Kilograms of Rock," *Science* 272 (7 June 1996), pp. 1447 - 1448. The article begins with the mesmerizing assertion: "Of the more than 7500 known meteorites, only 12 are certain to have come from Mars."

Similarly, adaptating the proof of Theorem (55) yields:

(67) PROPOSITION: A problem of form $(T, \{\theta_0, \ldots, \theta_n\})$ is solvable on **Sym**$_o$ environments if and only if for every $i \leq n$, θ_i is equivalent in T to an $\exists\forall$ sentence in the vocabulary **Sym**$_o$.

3.3.5 Exercises

(68) EXERCISE: Suppose that **Sym** contains just two binary function symbols. For $i \in N$ either 0 or prime, let proposition P_i be the collection of all fields of characteristic i in this vocabulary. Show that $\{P_i \mid i$ is either 0 or prime$\}$ is solvable.

(69) EXERCISE: ♦ Suppose that **Sym** is limited to a binary function symbol. Let **G** be all groups, **T** all torsion groups, and **F** all torsion-free groups. Show that $\{\mathbf{F}, \mathbf{G} - \mathbf{F}\}$ is solvable but that $\{\mathbf{T}, \mathbf{G} - \mathbf{T}\}$ is not.

(70) EXERCISE: Show that any problem of form $(T, \{\theta_0, \ldots, \theta_n\})$ is solvable if T is model-complete. (For model-completeness, see [Hodges, 1993, Sec. 8.3].)

(71) EXERCISE: Exhibit finite symbols sets **Sym**$_o \subseteq$ **Sym**, theory T in the vocabulary of **Sym**, and sentence θ in the vocabulary of **Sym**$_o$ such that $(T, \{\theta, \neg\theta\})$ is solvable but not on **Sym**$_o$ environments.

3.4 Efficient Discovery

The solvability criterion of the first-order paradigm imposes no requirement of speedy convergence, so is open to the objection that tardy success is often no better than failure. We are therefore led to reinforce our conception of success by including a standard of efficiency. Similarly to the numerical paradigm (see Section 2.3), we formulate the matter in terms of data use that cannot be uniformly improved. The relevant definitions are introduced in the next subsection. Then we look at some results.

3.4.1 Success Points and Efficiency

Given proposition P and environment e for P, scientist Ψ solves P in e just in case Ψ's successive conjectures ultimately become non-void subsets of P. The particular propositions that Ψ announces, however, may vary indefinitely as Ψ progresses through e. So, success does not imply convergence to a single proposition. To define efficiency in the first-order paradigm it is thus necessary to define a convergence-concept that is more general than the "convergence points" of Section 2.3.1.

(72) DEFINITION: Let proposition P, environment e for P, and scientist Ψ be given. We say that Ψ is P-correct on e at $k \in N$ just in case $\emptyset \neq \Psi(e[k]) \subseteq P$. If Ψ solves P in e, we write SP(Ψ, e, P) to denote the least $k_0 \in N$ such that Ψ is P-correct on e at k for all $k \geq k_0$.

Such a k_0 may be called the "success point" of Ψ on e, relative to P. Efficient discovery may now be defined in terms of success points.

(73) DEFINITION: Suppose that scientist Ψ solves problem **P**. We say that Ψ solves **P** *efficiently* just in case for every scientist Ψ' that solves **P**, if there is an environment e_0 for $P_0 \in \mathbf{P}$ with $k_0 = \mathrm{SP}(\Psi', e_0, P_0) < \mathrm{SP}(\Psi, e_0, P_0)$ then there is $P_1 \in \mathbf{P}$, $S \in P_1$, and full assignment h to S such that:

(a) $S \models \bigwedge e_0[k_0][h]$ and

(b) SP(Ψ, e, P_1) $\leq k_0 <$ SP(Ψ', e, P_1) for every environment e for S and h that extends $e_0[k_0]$.

In this case, **P** is solvable *efficiently*.

Intuitively, Ψ is efficient if it succeeds faster than any rival on many environments—where a "rival" is a scientist that succeeds faster than Ψ on some environment.

In parallel with the numerical paradigm [see Definition 2.(28)], it will prove useful to define a strong form of *in*efficiency.

(74) DEFINITION: Suppose that scientist Ψ solves problem **P**. We say that Ψ is *dominated* on **P** just in case there is scientist Ψ' such that:

(a) Ψ' solves **P**;

(b) for all $P \in \mathbf{P}$ and environments e for P, SP(Ψ', e, P) $\leq$ SP(Ψ, e, P);

(c) for some $P \in \mathbf{P}$ and environment e for P, SP(Ψ', e, P) $<$ SP(Ψ, e, P).

As an easy consequence of Definitions (73) and (74), we have:

(75) LEMMA: Suppose that scientist Ψ solves problem **P**. If Ψ is dominated on **P** then Ψ does not solve **P** efficiently.

3.4.2 Solvability Implies Efficient Solvability

Theorem 2.(25) shows that solvability and efficient solvability are coextensive properties of problems in the numerical paradigm. We shall now see that the same is true of the first-order paradigm. Specifically, requiring scientists to be efficient as well as successful does not reduce the class of solvable problems. Such is the content of the following theorem.

(76) THEOREM: Every solvable problem is efficiently solvable.

Proof: Let problem $\mathbf{P}$ be solvable, and let scientist Ψ be as specified in the proof of Proposition (31). It follows easily that for every $\sigma \in SEQ$ for $\mathbf{P}$, we may choose $P \in \mathbf{P}$, $\mathcal{S} \in P$, and full assignment h to $\mathcal{S}$ with the following properties:

(77) (a) $\mathcal{S} \models \bigwedge \sigma[h]$;

(b) $\Psi(e[k]) = P$, for every $k \geq \text{length}(\sigma)$, and every environment e for $\mathcal{S}$ and h that extends σ.

Relying on Exercise (87) we conclude from (77) that Ψ solves $\mathbf{P}$ efficiently. ∎

3.4.3 Discovery from Oracles

Our concept of environment (Section 3.1.3) is a passive affair inasmuch as the scientist has no control over the order in which it examines basic formulas. In this subsection—which may be omitted without loss of continuity—we consider more active exploration of the environment. This is achieved via an "oracle" that the scientist uses to determine the truth-value of any atomic formula according to the environment it faces. For example, the scientist may request the truth-value of Hv_3, and then use the answer to determine its next query. The answers to the queries are drawn from a single environment, hence are consistent with each other. Officially, an oracle is defined as follows.

(78) DEFINITION: An *oracle* is a mapping of SEQ into the set of atomic formulas of $\mathcal{L}_{form}$. The *data-sequence generated by oracle $\boldsymbol{o}$ for environment e* is the sequence d defined inductively as follows.

Suppose that $d[k]$ is defined for some $k \geq 0$. Let $o(d[k]) = \alpha$. Then $d(k) = \alpha$ if $\alpha \in \text{range}(e)$; otherwise $d(k) = \neg\alpha$. (Recall that $d(k)$ comes just after $d[k]$.)

The data sequence generated by $\boldsymbol{o}$ for environment e is denoted $\boldsymbol{o}(e)$.

Thus, an oracle $\boldsymbol{o}$ progressively interrogates whatever environment e it faces, and chooses its next query as a function of the information acquired so far. Note that $\text{range}(o(e)) \subseteq \text{range}(e)$, with the inclusion possibly proper. We conceive an oracle as mediating the interaction of a scientist with its environment. The definition of success using an oracle gives formal meaning to this idea.

(79) DEFINITION: Let proposition P, oracle o, and scientist Ψ be given. We say that Ψ solves P *using* o just in case for all environments e for P, $\emptyset \neq \Psi(o(e)[k]) \subseteq P$ for cofinitely many k. Ψ solves problem $\mathbf{P}$ *using* o just in case Ψ solves every member of $\mathbf{P}$ using o. In this case, we say that $\mathbf{P}$ is solvable *using* o.

Although there are pathological oracles that interfere with discovery, it is easy to see that a problem is solvable if and only if it is solvable using some oracle. Moreover, it will be seen in Exercise (88) that using an oracle sometimes allows more efficient use of data than when an environment is received passively.

We wish now to further explore the efficient use of oracles. It will be shown that for some problems there is no best oracle. For such problems, no matter what oracle is selected there is another one that allows uniform improvement in time to success. To state the matter precisely, we must generalize the idea of "success points" in Definition (72) to data sequences generated by oracles. This is achieved (once again) by allowing the oracle to stand between the scientist and its environment.

(80) DEFINITION: Let proposition P, environment e for P, oracle o, and scientist Ψ be given. We say that Ψ is P-*correct* on $o(e)$ at $k \in N$ just in case $\emptyset \neq \Psi(o(e)[k]) \subseteq P$. If Ψ solves P on e using o, we write $\mathrm{SP}(\Psi, o(e), P)$ to denote the least $k_0 \in N$ such that Ψ is P-correct on $o(e)$ at k for all $k \geq k_0$.

Inefficient use of an oracle may be defined similarly to Definition (74).

(81) DEFINITION: Suppose that scientist Ψ solves problem $\mathbf{P}$ using oracle o. We say that Ψ is *dominated* on $\mathbf{P}$ using o just in case there is scientist Ψ' and oracle o' such that:

(a) for all $P \in \mathbf{P}$ and environments e for P, $\mathrm{SP}(\Psi', o'(e), P) \leq \mathrm{SP}(\Psi, o(e), P)$, and

(b) for some $P \in \mathbf{P}$ and environment e for P, $\mathrm{SP}(\Psi', o'(e), P) < \mathrm{SP}(\Psi, o(e), P)$.

Now we can formulate the phenomenon alluded to above. In contrast to Theorem (76), we have the following fact about use of an oracle.

(82) PROPOSITION: Suppose that **Sym** consists of a unary function symbol and a unary predicate. Then there is $T \subseteq \mathcal{L}_{sen}$ and $\theta \in \mathcal{L}_{sen}$ with the following properties.

(a) $(T, \{\theta, \neg\theta\})$ is solvable.

(b) Suppose that scientist Ψ solves $(T, \{\theta, \neg\theta\})$ using oracle o. Then Ψ is dominated on $(T, \{\theta, \neg\theta\})$ using oracle o.

Proof: Let H be the unary predicate, s the unary function symbol of **Sym**. Set $T = \{\forall xy[(Hy \wedge y = sx) \rightarrow Hx]\}$, $\theta = \forall x Hx$. We claim that $\mathbf{P} = (T, \{\theta, \neg\theta\})$ witnesses the proposition. By Theorem (55), $\mathbf{P}$ is solvable. Let scientist Ψ and oracle o be such that Ψ solves $\mathbf{P}$ using o. Let $P_1 = MOD(T \cup \{\theta\})$, and $P_2 = MOD(T \cup \{\neg\theta\})$. To finish the proof it suffices to exhibit scientist Ψ' and oracle o' such that the following holds.

(83)   (a) Ψ' solves $\mathbf{P}$ using o'.

    (b) For $P \in \{P_1, P_2\}$ and all environments e for P, $SP(\Psi', o'(e), P) \leq SP(\Psi, o(e), P)$.

    (c) For some environment e_0 for P_2, $SP(\Psi', o'(e_0), P_2) < SP(\Psi, o(e_0), P_2)$.

We define such an o', followed by Ψ'.

For every atomic formula β of form $Hs^m v_n$ denote by β^+ the formula $Hs^{m+1} v_n$; for all other basic formulas β, β^+ denotes the same formula as β. For all $\sigma \in SEQ$ denote by σ^- the sequence that results from σ by replacing every formula of form $Hs^{m+1} v_n$ by the formula $Hs^m v_n$. Let oracle o' be such that for all $\sigma \in SEQ$, $o'(\sigma) = (o(\sigma^-))^+$. It follows that for all $k \in N$ and environments e, $o'(e)[k]$ is the result of replacing in $o(e)[k]$ each formula of form $Hs^m v_n$ by $Hs^{m+1} v_n$.

We define Ψ' as follows. Let $\sigma \in SEQ$ be given. If $\Psi(\sigma)$ is undefined then $\Psi'(\sigma)$ is undefined. Otherwise, either $\bigwedge \sigma \models \neg\theta$ and $\Psi'(\sigma) = P_2$, or $\bigwedge \sigma \not\models \neg\theta$ and $\Psi'(\sigma) = \Psi(\sigma^-)$. Conditions (83)a,b follow easily from the definition of o', the hypothesis that Ψ solves $\mathbf{P}$ using o, and the downward closure guaranteed by T. Regarding (83)c, since Ψ solves $\mathbf{P}$ using o, there are $k_0, n, m \in N$ and environment e for P_1 such that:

(84)   (a) $\emptyset \neq \Psi(o(e)[k_0]) \subseteq P_1$,

    (b) $\mathrm{range}(o(e)[k_0]) \models Hs^m v_n$, and

    (c) $\mathrm{range}(o(e)[k_0]) \not\models Hs^{m+1} v_n$.

By (84)c, choose environment e_0 for P_2 such that:

(85)   (a) $\mathrm{range}(o(e)[k_0]) \subseteq \mathrm{range}(e_0)$;

    (b) $\neg Hs^{m+1} v_n \in \mathrm{range}(e_0)$.

By (84)a and (85)a, $\emptyset \neq \Psi(o(e_0)[k_0]) \subseteq P_1$. By (84)b, (85)b, and the definition of o', $\bigwedge o'(e_0)[k_0] \models \neg H s^{m+1} v_n$, hence $\bigwedge o'(e_0)[k_0] \models \neg\theta$. So $\Psi'(o(e_0)[k])$ $= P_2$ for all $k \geq k_0$. Since e_0 is for P_2, it follows that $k_0 \leq SP(\Psi', o'(e_0), P_2)$ $< SP(\Psi, o(e_0), P_2)$, verifying (83)c.   ∎

Proposition (82) reveals that efficiency in the context of oracles is a more complex affair than efficiency in the original sense of passively received environments. For this reason, as our model of efficient inquiry in the sequel we shall rely on the environment based Definition (73) rather than on concepts involving oracles.

3.4.4 Exercises

(86) EXERCISE: Let solvable problem **P** and scientist Ψ that solves **P** be given. Show that Ψ is not dominated on **P** if and only if for all $\sigma \in SEQ$, if σ is for **P** then there exists $P \in \mathbf{P}$ and environment e for **P** such that:

(a) e extends σ, and

(b) for every $k \geq \text{length}(\sigma)$, $\emptyset \neq \Psi(e[k]) \subseteq P$.

(87) EXERCISE: Let solvable problem **P** and scientist Ψ that solves **P** be given. Show that Ψ solves **P** efficiently if and only if for every $\sigma \in SEQ$, if σ is for **P** then there exists $P \in \mathbf{P}$, $S \in P$, and full assignment h to S such that:

(a) $S \models \bigwedge \sigma[h]$, and

(b) for every environment e for S and h that extends σ, for every $k \geq \text{length}(\sigma)$, $\emptyset \neq \Psi(e[k]) \subseteq P$.

(88) EXERCISE: Exhibit scientist Ψ, oracle o and problem **P** with the following properties:

(a) Ψ solves **P** using o.

(b) For every scientist Ψ' that solves **P**,

(i) $SP(\Psi, o(e), P) \leq SP(\Psi', e, P)$ for all $P \in \mathbf{P}$ and environments e for P,

(ii) $SP(\Psi, o(e), P) < SP(\Psi', e, P)$ for some $P \in \mathbf{P}$ and environment e for P.

This provides a sense in which data can be used more efficiently with an oracle than without.

3.5   Computable Solvability

Parallel to similar concerns in the numerical context (see Section 2.5), the goal
of the present section is to isolate and study the computable scientists. Crucial
to this enterprise is finding a way to represent the scientist's hypotheses, which
are more complex objects in the first-order paradigm than in the numerical one.
The matter is resolved in the following subsection, after which we define the
computable scientists and then investigate their competence.[11]

3.5.1   Finitization of Propositions

It is an engaging (but far from proven) hypothesis that human scientific behavior is Turing simulable. If the hypothesis is true, then the first-order paradigm
would benefit from greater realism by isolating the computable scientists for
separate study. For this to be possible, we must find a way to represent the inputs and outputs of formal scientists as finite objects. Inputs pose no problem
since SEQ is already a countable and recursive set of finite objects. The difficulty is the scientist's outputs, namely, propositions. Propositions, it will be
recalled, are vast objects that comprehend an arbitrary class of structures. To
represent them finitely it seems necessary to limit attention to *definable* classes
of structures, that is, to classes composed of every structure that satisfies some
given set of formulas. However, if the set of formulas is not recursively enumerable (*r.e.*), then such a proposition remains ineffable (at least, in one important
sense) by computable scientists. So we must further restrict propositions to
classes of structures that are definable by an *r.e.* set S of formulas. The *r.e.* index of S can then be used to express the proposition defined by S. Finitizing
outputs in this way makes it possible for some scientists to be simulated by
computable functions from SEQ to the set of indices. These will be called the
"computable scientists."

   To keep things tidy, it is necessary to distinguish the indices used for *r.e.*
sets of formulas from those used for *r.e.* sets of numbers. So we fix for the
remainder of our discussion an acceptable indexing $\{W_i^{form} \mid i \in N\}$ of the *r.e.*
subsets of $\mathcal{L}_{form}$, along with an acceptable indexing $\{W_i^{num} \mid i \in N\}$ of the *r.e.*
subsets of N. Observe that for $i \in N$ with W_i^{form} consistent, $MOD(W_i^{form})$ is

---

11. As before, our discussion presupposes elementary concepts of computability theory, as presented, for example, in [Rogers, 1987, Machtey & Young, 1978, Davis & Weyuker, 1983, Boolos
& Jeffrey, 1989].

a proposition; namely, it is the class of structures that satisfy all the formulas enumerated by the ith Turing Machine.[12] Hence, given mapping $\psi : SEQ \to N$ and $\sigma \in \text{domain}(\psi)$, $MOD(W^{form}_{\psi(\sigma)})$ is also a proposition (when nonempty), namely, $MOD(W^{form}_i)$ for $i = \psi(\sigma)$. In this way, ψ uses i to express its view that a certain proposition is true.

3.5.2   Computable Scientists

Pursuant to the preceding discussion, the class of computable scientists is defined as follows.

(89)   DEFINITION: Scientist Ψ is *computable* just in case there is computable $\psi : SEQ \to N$ such that for all $\sigma \in SEQ$, $\Psi(\sigma)$ is defined iff $\psi(\sigma)$ is defined, and when both are defined $\Psi(\sigma) = MOD(W^{form}_{\psi(\sigma)})$. In this case, we say that ψ *underlies* Ψ. A problem that is solved by a computable scientist is *computably solvable*.

So, a computable scientist Ψ is associated with a computable function $\psi : SEQ \to N$ that simulates Ψ's behavior. If Ψ issues proposition P on input σ, then ψ issues an index i on σ with $P = MOD(W^{form}_i)$. Computability thus imposes a double constraint on scientists. On the one hand, conjectures must be expressed via (indices for) recursively axiomatizable theories.[13] On the other hand, the conversion of data into theories must be Turing simulable. The two conditions are necessary and jointly sufficient for computer implementation of a scientist, at least in principle.

3.5.3   Competence of Computable Scientists

It is evident that the computable scientists are restricted in their competence. For, let T be a complete, nonrecursive theory. Then the degenerate problem $\{MOD(T)\}$ is trivially solvable but not computably. This is because there is no consistent, r.e. extension of T, hence no means whereby a computable scientist can express a nonempty subset of $MOD(T)$. It is a more trenchant fact that computable scientists have limited competence even when there is no computability obstacle to their announcing an adequate theory. To prove the point, we use the following definition.

---

12. By footnote 2 on page 62, $MOD(W^{form}_i)$ is the universal proposition (all structures) if $W^{form}_i = \emptyset$.

13. Let us recall that a set S of formulas is *r.e.* if and only if S is recursively axiomatizable, i.e., if and only if S is a equivalent to a recursive subset of $\mathcal{L}_{form}$ [Craig, 1953]. Therefore, S has a recursive set of axioms if and only if S has an *r.e.* set of axioms.

(90) DEFINITION: A proposition is *elementary* just in case it has the form $MOD(\theta)$ for some $\theta \in \mathcal{L}_{sen}$.

So, elementary propositions can be expressed by just a single sentence. A fortiori, for every elementary proposition P there is an index i with $P = MOD(W_i^{form})$. Indeed, it suffices to take i such that $W_i^{form} = \{\theta\}$, where θ defines P. The upshot is that problems consisting of elementary propositions pose no difficulty specifically linked to expressing the propositions. This leaves open the possibility that a *collection* of elementary propositions is difficult for a computable scientist to manipulate, and indeed we shall see that such collections can be solvable but not computably. This fact is an easy corollary of the following result, analogous to Theorem 2.(56).

(91) THEOREM: Suppose that **Sym** is limited to a binary predicate, two constants, and a unary function symbol. Then for every countable collection Σ of scientists there is a problem **P** with the following properties.

(a) Every member of **P** is elementary.

(b) **P** is solvable.

(c) No member of Σ solves **P**.

Proof: Let R be the predicate, $\bar{0}$, a be the constants, and s be the unary function symbol of **Sym**. We denote n applications of s to $\bar{0}$ by $\bar{n}$. Let $\delta \in \mathcal{L}_{sen}$ be true in a structure iff (a) R is interpreted as a discrete total ordering of the entire domain, and (b) s is interpreted as sending each element into the R-next element [that is, the structure must satisfy $\forall x(\, s(x) \neq x \wedge Rxs(x) \wedge \neg \exists y(y \neq x \wedge y \neq s(x) \wedge Rxy \wedge Rys(x))\,)$]. Given $i \in N$, let θ_i^+ be $\bar{i} = a \wedge \delta \wedge \exists x \forall y Rxy$, and let θ_i^- be $\bar{i} = a \wedge \delta \wedge \neg \exists x \forall y Rxy$.

Let $X \subseteq N$ be given. We denote by $\mathbf{P}_X$ the problem $\{MOD(\theta_i^+) \mid i \in X\} \cup \{MOD(\theta_i^-) \mid i \notin X\}$. It is obvious that $\mathbf{P}_X$ is a solvable problem whose members are elementary. So let countable collection Σ of scientists be given. To finish the proof it suffices to show that there is $X \subseteq N$ such that no member of Σ solves $\mathbf{P}_X$. Because there are uncountably many subsets of N, it thus suffices to show:

(92) If X and Y are distinct subsets of N then $\mathbf{P}_X \cup \mathbf{P}_Y$ is not solvable.

To demonstrate (92), let $X, Y \subseteq N$ and $i \in N$ be such that $i \in X$ iff $i \notin Y$. Set $P = MOD(\theta_i^+)$ and $P' = MOD(\theta_i^-)$. Observe that $\{P, P'\} \subseteq \mathbf{P}_X \cup \mathbf{P}_Y$. Hence it suffices to show that $\{P, P'\}$ is not solvable. However, this is easily shown by applying Proposition (35). ∎

Of course, the computable scientists form a countable set, since their number is bounded by the number of Turing Machines. So we obtain the immediate corollary:

(93)   COROLLARY: Let **Sym** be as in Theorem (91). Then there is a solvable problem whose members are elementary but which is not solvable computably.

3.5.4   Efficiency and Computability

We have so far conceived efficiency in terms of the number of data examined prior to convergence (see Section 3.4.1). In the context of computable scientists, however, a variety of additional efficiency concepts may be defined. The new concepts are based on the time that a given scientists spends examining individual $\sigma \in SEQ$. For example, given a function $f : SEQ \to N$, it may be said that scientist Ψ is "f-fast" just in case the running time of Ψ on $\sigma \in SEQ$ is bounded by $f(\sigma)$. The running-time conception of efficiency may be studied in conjunction with the data-use conception, to provide an overall picture of resource consumption during inquiry. For our part, however, we shall continue to focus just on data-use, leaving running-time to one side. The reason for our choice is the shape of the ensuing theory of revision-based inquiry, to be developed in the next chapter. Only data-use will be relevant (see Section 4.4 below).

Recall that Theorem (76) shows every solvable problem to be solvable efficiently. This reassuring fact does not survive the transition to computable solvability. Instead, the following proposition shows that there are computably solvable problems that cannot be solved efficiently by computable scientists.

(94)   PROPOSITION: Suppose that **Sym** is limited to the vocabulary of arithmetic (including $\bar{0}$ and a unary function symbol s) plus the additional constant a. Then there is a problem **P** with the following properties.

(a)   Every member of **P** is elementary.

(b)   **P** is solvable computably.

(c)   Every computable scientist that solves **P** is dominated on **P**.

Proof: Let n applications of s to $\bar{0}$ be denoted $\bar{n}$. Let Q be the finite set of axioms of Robinson's Arithmetic. By standard results [Boolos & Jeffrey, 1989, Ch. 14], let $\phi(x, y) \in \mathcal{L}_{form}$ have two free variables x, y, exclude a, and be such that for all $i \in N$, $Q \models \exists x \phi(x, \bar{i})$ iff $i \in W_i^{num}$. Given $i \in N$, set $P_i^+ = MOD(Q \cup \{a = \bar{i}, \exists x \phi(x, \bar{i})\})$, and $P_i^- = MOD(Q \cup \{a = \bar{i}, \neg \exists x \phi(x, \bar{i})\})$. We claim that $\mathbf{P} = \{P_i^+ \mid i \in W_i^{num}\} \cup \{P_i^- \mid i \notin W_i^{num}\}$ witnesses the proposition.

It is immediate that the propositions of **P** are elementary. To show that
P is computably solvable, define computable $\psi : SEQ \to N$ as follows. Let
$\sigma \in SEQ$ be given. Suppose that $i \in N$ is unique with $a = \bar{i} \in \text{range}(\sigma)$. Then if
i has not appeared in W_i^{num}, within length(σ) steps of its standard enumeration,
$\psi(\sigma)$ equals an index for $Q \cup \{a = \bar{i}, \neg \exists x \phi(x, \bar{i})\}$; and if i has appeared in
W_i^{num} within length(σ) steps of its standard enumeration, $\psi(\sigma)$ is an index for
$Q \cup \{a = \bar{i}, \exists x \phi(x, \bar{i})\}$. For all other $\sigma \in SEQ$, $\psi(\sigma)$ is undefined. It is easy to
verify that if ψ underlies Ψ, then Ψ solves **P**.

Finally, for a contradiction, suppose that computable $\psi : SEQ \to N$ underlies a scientist that solves **P** and is not dominated on **P**. Then Definition (74) is
easily seen to imply the following.

(95)   For all $i \in N$,

(a) if $i \in W_i^{num}$ then $\psi(a = \bar{i})$ is an *r.e.* index for $X \subseteq \mathcal{L}_{form}$ with $X \models$
$\exists x \phi(x, \bar{i})$,

(b) if $i \notin W_i^{num}$ then $\psi(a = \bar{i})$ is an *r.e.* index for $X \subseteq \mathcal{L}_{form}$ with $X \models$
$\neg \exists x \phi(x, \bar{i})$.

But (95) yields a decision procedure for the halting problem.   ∎

3.5.5   *R.e.* Problems of Form $(T, \{\theta_0, \dots, \theta_n\})$

Problems of form $(T, \{\theta_0, \dots, \theta_n\})$ are central to our theory so it is interesting
to consider their computable solvability. Of course, if T is not recursively
axiomatizable then $(T, \{\theta_0, \dots, \theta_n\})$ may not be computably solvable for want
of the ability to name its propositions. To set aside this kind of inexpressible
case, we rely on the following definition.

(96)   DEFINITION: A problem of form $(T, \{\theta_0, \dots, \theta_n\})$ is called *r.e.* just in
case T is a recursively enumerable set of sentences.

All the propositions of an *r.e.* problem of form $(T, \{\theta_0, \dots, \theta_n\})$ are recursively axiomatizable, so computable scientists can express any of them. Do
there remain further obstacles to solving such problems, obstacles that are specific to computable scientists? Surprisingly, this question receives a negative
answer. In other words, the computable solvability of an *r.e.* problem of form
$(T, \{\theta_0, \dots, \theta_n\})$ follows from its solvability *tout court*. The matter may be
stated as follows.

(97)   THEOREM: Every solvable, *r.e.* problem of form $(T, \{\theta_0, \dots, \theta_n\})$ is
computably solvable.

Proof: Let solvable, *r.e.* problem of form $(T, \{\theta_0, \ldots, \theta_n\})$ be given. Let $\{\varphi_j(\bar{x}_j) \mid j \in N\}$ be a recursive enumeration of all $\forall$ formulas $\varphi(\bar{x})$ such that for some $0 \leq i \leq n$, $T \models \exists \bar{x} \varphi(\bar{x}) \to \theta_i$. Define computable $\psi : SEQ \to N$ as follows. Let $\sigma \in SEQ$ be given. Then $\psi(\sigma)$ is undefined if $\bigwedge \sigma \models \neg \varphi_j$ for all $j \in N$. Otherwise, $\psi(\sigma)$ is an index for $T \cup \{\varphi_j\}$, where j is least with $\bigwedge \sigma \not\models \neg \varphi_j$.

Suppose that ψ underlies Ψ. To see that Ψ solves $(T, \{\theta_0, \ldots, \theta_n\})$, let $i_0 \leq n$ and environment e for $MOD(T \cup \{\theta_{i_0}\})$ be given. By Theorem (55), let j_0 be least such that range$(e) \cup \{\varphi_{j_0}\}$ is consistent. Then by compactness, for all $j < j_0$, $\bigwedge e[k] \models \neg \varphi_j$ for cofinitely many k. Hence $\psi(e[k])$ is an index for $T \cup \{\varphi_{j_0}\}$ for cofinitely many k. Since the θ_i's partition the models of T, it follows that $T \models \varphi_{j_0} \to \theta_{i_0}$. So $T \cup \{\varphi_{j_0}\}$ is consistent, and $T \cup \{\varphi_{j_0}\} \models T \cup \{\theta_{i_0}\}$. ∎

A uniform version of the theorem is given in Exercise (104).

3.5.6 The Numerical Paradigm as a Special Case

The results of Section 3.1.5 exhibited the numerical paradigm as a special case of the logical one. It will now be shown that this is also true with respect to computable solvability.[14] For the remainder of the section we assume that **Sym** consists of a binary predicate R, a constant $\bar{0}$, and a unary function symbol s. The term that results from n applications of s to $\bar{0}$ is denoted $\bar{n}$. We rely again on Definition (13), which we recall here for convenience.

(98) DEFINITION: For $L \subseteq N$, denote by P_L the class of structures $\mathcal{S}$ that meet the following conditions, for every $n \in N$:

(a) if $n \in L$ then there is some $p \in N$ with $\mathcal{S} \models R\bar{n}\bar{p}$;

(b) if $n \notin L$, then for all $p \in N$, $\mathcal{S} \models \neg R\bar{n}\bar{p}$.

The following theorem formulates the sense in which the computable version of the numerical paradigm can be embedded in the first-order paradigm. It is the computable version of Theorem (14).

(99) THEOREM: Let countable set $\mathbf{P}_\eta$ of nonempty *r.e.* subsets of N be given. Then $\mathbf{P}_\eta$ is computably solvable in the numerical paradigm if and only if $\mathbf{P}_\ell = \{P_L \mid L \in \mathbf{P}_\eta\}$ is computably solvable in the present paradigm.

Proof: We introduce a temporary convention, familiar from the proof of Theorem (14). A term or notation decorated with η indicates usage in the sense

---

14. The present subsection may be omitted without loss of continuity.

of Chapter 2 (numerical paradigm). Decoration with ℓ indicates usage in the sense of the present chapter (logical paradigm).

For the left-to-right direction, let computable $\psi_\eta : SEQ_\eta \to N$ underlie a scientist$_\eta$ that solves$_\eta$ $\mathbf{P}_\eta$. Examination of (16) in the proof of Theorem (14) shows that there is total recursive $f : SEQ_\ell \to SEQ_\eta$ such that:

(100) (a) For all $\sigma, \tau \in SEQ_\ell$, if $\tau \subseteq \sigma$ then $f(\tau) \subseteq f(\sigma)$.

   (b) If $\emptyset \neq L \subseteq N$, and environment$_\ell$ e is for P_L, then $\bigcup_{k \in N} f(e[k])$ is an environment$_\eta$ for L.

It is easy to see that there is computable $\psi_\ell : SEQ_\ell \to N$ which satisfies the following properties, for all $\sigma \in SEQ_\ell$. If $\psi_\eta(f(\sigma))$ is undefined then $\psi_\ell(\sigma)$ is undefined. If $\psi_\eta(f(\sigma)) = i$ then $\psi_\ell(\sigma) = j$, where $j \in N$ is such that:

(101) (a) W_j^{form} is consistent;

   (b) for all $n \in N$, if $n \in W_i^{num}$ then for some $p \in N$, $R\bar{n}\bar{p} \in W_j^{form}$;

   (c) for all $n \in N$, if $n \notin W_i^{num}$ then for all $p \in N$, $\neg R\bar{n}\bar{p} \in W_j^{form}$.

Let Ψ_ℓ be the (computable) scientist$_\ell$ that ψ_ℓ underlies. From (101) we infer that for all $\sigma \in SEQ_\ell$ and $L \subseteq N$, if $\Psi_\eta(f(\sigma))$ is defined and $\Psi_\eta(f(\sigma)) = L$, then $\emptyset \neq \Psi_\ell(\sigma) \subseteq P_L$. From (100) it follows immediately that Ψ_ℓ solves$_\ell$ $\mathbf{P}_\ell$.

For the right-to-left direction, let computable $\psi_\ell : SEQ_\ell \to N$ underlie a scientist$_\ell$ that solves$_\ell$ $\mathbf{P}_\ell$. Examination of Fact (18) in the proof of Theorem (14) shows that there is total recursive $g : SEQ_\eta \to SEQ_\ell$ such that:

(102) (a) For all $\sigma, \tau \in SEQ_\eta$, if $\tau \subseteq \sigma$ then $g(\tau) \subseteq g(\sigma)$.

   (b) For all environments$_\eta$ e for $\emptyset \neq L \subseteq N$, $\bigcup_{k \in N} g(e[k])$ is an environment$_\ell$ for P_L.

It is easy to see that there is computable $\psi_\eta : SEQ_\eta \to N$ which satisfies the following properties, for all $\sigma \in SEQ_\eta$. If $\psi_\ell(g(\sigma))$ is undefined then $\psi_\eta(\sigma)$ is undefined. If $\psi_\ell(g(\sigma)) = i$ then $\psi_\eta(\sigma) = j$, where $j \in N$ is such that for all $n \in N$:

(103) $n \in W_j^{num}$ iff there is $p \in N$ such that $W_i^{form} \models R\bar{n}\bar{p}$.

Let Ψ_η be the (computable) scientist$_\eta$ that ψ_η underlies. From (103) we infer that for all $\sigma \in SEQ_\eta$ and $L \subseteq N$, if $\Psi_\ell(g(\sigma))$ is defined and $\emptyset \neq \Psi_\ell(g(\sigma)) \subseteq P_L$, then $\Psi_\eta(\sigma) = L$. From (102) it follows immediately that Ψ_η solves$_\eta$ $\mathbf{P}_\eta$.   ∎

The construction in the proofs of Theorems (14) and (99) allows us to carry over much of the fine-grained theory of computable scientists developed in Chapter 2 to the present chapter. We give an example in Exercise (106).

3.5.7   Exercises

(104)   EXERCISE: Show that there is total recursive $F : \mathcal{L}_{sen}^{<\omega} \times SEQ \to \mathcal{L}_{sen}$ with the following property. Let solvable problem **P** be of form $(T, \{\theta_0, \ldots, \theta_n\})$, with $T = \{\delta_i \mid i \in N\}$. Then $\lambda\sigma \, . \, MOD(T \cup \{F(\delta_0 \ldots \delta_{\text{length}(\sigma)}, \theta_0 \ldots \theta_n, \sigma)\})$ solves **P**.

(105)   EXERCISE: [Osherson *et al.*, 1996] Let $T, X \subseteq \mathcal{L}_{sen}$ be given. We say that scientist Ψ *X-solves* T just in case for every environment e for $\mathcal{S} \in MOD(T)$, there is $\theta \in X$ such that $\mathcal{S} \models \theta$, and $\Psi(e[k]) = MOD(\theta)$ for cofinitely many k. Show that there is total recursive $F : \mathcal{L}_{sen}^{<\omega} \times SEQ \to \mathcal{L}_{sen}$ with the following property. Let $T = \{\delta_i \mid i \in N\} \subseteq \mathcal{L}_{sen}$ and $X = \{\theta_i \mid i \in N\} \subseteq \mathcal{L}_{sen}$ be such that T is X-solvable. Then $\lambda\sigma \, . \, MOD(F(\delta_0 \ldots \delta_{\text{length}(\sigma)}, \theta_0 \ldots, \theta_{\text{length}(\sigma)}, \sigma))$ X-solves T.

(106)   EXERCISE: Suppose that scientist Ψ solves problem **P**. We say that Ψ *respects* **P** just in case for all $\sigma \in SEQ$ for **P**, $\Psi(\sigma)$ is a nonempty subset of some $P \in \mathbf{P}$. Show that there is a computably solvable problem **P** such that no computable scientist that solves **P** respects **P**.

3.6   Probabilistic Solvability

3.6.1   Success on Enough Environments

It is time to address the concern that our requirements for success are excessive. Specifically, the refusal to allow failure on even a single environment has the ring of hysteria, especially since the odds of running into any one environment seem low. So why not relax a little, and settle for success on "almost" every environment? We now use the tools of probability to make this idea precise and explore it. Since probabilistic solvability is not pursued beyond the present section, it can be skipped on a first reading.

As a preliminary to the technicalities that follow, let us recall the motive for the present, "absolute" conception of success [Definition (8)]. By requiring scientists to succeed on every environment for a given problem, we prevent Nature from passing along illicit hints. To appreciate the issue, consider the problem $\mathbf{P} = \{P_\ell, P_n\}$, where P_ℓ is the class of total orders with a least point, and P_n is the class of total orders with no least point. Proposition (54) shows **P** to be unsolvable. It is nonetheless easy to specify scientist Ψ and uncountable collections $\mathcal{E}_\ell$, $\mathcal{E}_n$ of environments such that:

- $\mathcal{E}_\ell$ is for P_ℓ and $\mathcal{E}_n$ is for P_n;
- Ψ solves P_ℓ in every $e \in \mathcal{E}_\ell$, and solves P_n in every $e \in \mathcal{E}_n$.

One way to achieve this is to distinguish which of the two variables v_0, v_1 appears first in an environment, and to define:

$\mathcal{E}_\ell = \{e \mid e$ is for P_ℓ, and v_0 appears before v_1 in $e\}$;

$\mathcal{E}_n = \{e \mid e$ is for P_n, and v_1 appears before v_0 in $e\}$.

The right answer is thus coded into the environments of $\mathcal{E}_\ell \cup \mathcal{E}_n$, and may be used in underhanded fashion by a scientist designed just for them. All such coding schemes are foiled in a single stroke by asking scientists to succeed on every environment for a given problem.

Still, such a categorical policy might be an over-reaction to the danger of illicit hints. In particular, it seems sufficient to construct environments according to some random process that respects the underlying structure. The randomization can then be counted on to block collusion while nonetheless allowing scientists to fail in the rare case. Let us now give substance to this idea.

3.6.2   Framework and Conventions

To guide our conception of a randomly constructed environment, we shall picture environments as created by a stochastic process, similarly to Section 2.7. In particular, the elements of a model's domain are assumed to be delivered for inspection via independent, identically distributed sampling according to a probability law which may be unknown to the scientist. Just as before, the scientist's mission is to discover the proposition from which her environment is drawn. However, this need only be achieved on "enough" environments. The latter expression is interpreted probabilistically, in terms of the measure of the class of environments on which the scientist succeeds. The resulting paradigm classifies performance in probabilistic rather than all-or-none fashion, and thus differs from the absolute conception of scientific success considered up to now. The facts to be discussed below indicate the robustness of our theory of inquiry. This is because weak conditions suffice to render solvability in the absolute sense equivalent to solvability in the probabilistic sense.

Just for this section, we rely on some conventions that will ease our notation. Here is the first one.

(107)   *Convention:*   The domain of every structure is taken to be N.

It is clear that Convention (107) has no impact on the generality of our results except insofar as finite models are excluded from the discussion. In light of this convention, in the present section we use the expression "full assignment" in place of "full assignment to S." The class of full assignments is denoted H. Probabilities will be defined over subsets of H, and to make this possible, we abide by the following convention.

(108)   *Convention:*   Members of H will be conceived as drawn from the Baire space N^ω, by identifying each $v_i \in Var$ with the integer i.

For example, $\langle 1, 0, 3, 2, 5, 4, 7, 6, 9, 8 \cdots \rangle \in N^\omega$ represents the full assignment that sends v_0 to 1, v_1 to 0, v_2 to 3, v_3 to 2, etc. Convention (108) is essential to what follows. Nevertheless, for $h \in H$ and $i \in N$, we continue to write $h(v_i)$, not $h(i)$.

3.6.3   Measures over Assignments and Probabilistic Solvability Defined

The class of all positive probability distributions over N is denoted M. (A distribution μ over N is positive if for all $n \in N$, $\mu(n) > 0$.) Given $\mu \in M$, we extend μ to the product measure over N^ω in the usual way (as reviewed, for example, in [Levy, 1979, Section VII.3]). The collection of such measures is also denoted by M. Context will make it clear whether $\mu \in M$ refers to a positive distribution over N, or to its product-extension to N^ω.

By our identification of assignments with members of N^ω, each $\mu \in M$ induces probabilities over the measurable subsets of H. Observe that H does not exhaust N^ω since the latter includes sequences whose range is not all of N whereas H is limited to full assignments. However, in what follows we will be able to ignore $N^\omega - H$ because of the following fact, proved in [Billingsley, 1986, Ch. 4].

(109)   LEMMA: Let $\mu \in M$ be given. Then $\mu(H) = 1$.

For a given scientist and structure we now wish to measure the class of full assignments that lead the scientist to success. For this purpose we rely on the following definition.

(110)   DEFINITION: Let scientist Ψ, proposition P, and $S \in P$ be given. The class of $h \in H$ such that Ψ solves P in every environment for S and h is denoted: $H_{solve}(\Psi, S, P)$.

Thus, the definition of $H_{solve}(\Psi, S, P)$ abstracts from the order in which basic formulas appear in an environment, focussing instead on the full assignments that give rise to them. Success is defined as follows.

(111)   DEFINITION: Let scientist Ψ, problem $\mathbf{P}$, $M_0 \subseteq M$, and $p \in [0, 1]$ be given. We say that Ψ *solves* $\mathbf{P}$ *on* M_0 *with probability* p just in case for all $\mu \in M_0$, all $P \in \mathbf{P}$, and all $S \in P$, $\mu(H_{solve}(\Psi, S, P)) \geq p$. In this case $\mathbf{P}$ is *solvable on* M_0 *with probability* p.

The foregoing definition is specialized to individual propositions P and individual distributions μ in the obvious way. Thus, Ψ *solves* P *on* μ *with probability* p just in case Ψ solves $\{P\}$ on $\{\mu\}$ with probability p.

   We remarked above that for measuring probability it is the full assignments that count, not the environments. Indeed, it would be possible to limit attention to just a single environment for each full assignment. To make this clear, fix for the remainder of the section an enumeration $\{\alpha_i \mid i \in N\}$ of all atomic formulas.

(112)   DEFINITION: An environment e is *standard* just in case for all $i \in N$, either $e(i) = \alpha_i$ or $e(i) = \neg\alpha_i$. For structure S and full assignment h we denote by $[S, h]$ the standard environment for S and h.

As an immediate consequence of Exercise (20) we have the following.

(113)   LEMMA: Let problem $\mathbf{P}$, $M_0 \subseteq M$, and $p \in [0, 1]$ be given. Then $\mathbf{P}$ is solvable on M_0 with probability p if and only if there is scientist Ψ such that for every $\mu \in M_0$, every $P \in \mathbf{P}$, and every $S \in P$, $\mu(\{h \in H \mid \Psi$ solves P in $[S, h]\}) \geq p$.

The lemma will be useful in several proofs below since it allows us to concentrate on just the standard environments when demonstrating probabilistic solvability.

3.6.4   Probabilistic Solvability under Isomorphic Closure

We now show that under a mild condition on problems, the probabilistic conception of solvability is identical to the absolute one.

(114) THEOREM: Suppose that problem **P** consists of propositions that are closed under isomorphism. Let $\mu \in M$ be given. Then **P** is solvable if and only if **P** is solvable on μ with probability greater than .5.

Proof of the theorem relies on the following lemma.

(115) LEMMA: Suppose that scientist Ψ solves problem **P** on $\mu \in M$ with probability greater than .5. Then for every $P \in \mathbf{P}$ and every $\mathcal{S} \in P$, there are cofinitely many k such that:

$$\mu(\{h \in H \mid \emptyset \neq \Psi([\mathcal{S}, h][k]) \subseteq P\}) > .5.$$

Proof: Suppose that the conclusion to Lemma (115) is false. Then for infinitely many k, $\mu(Z_k) \geq .5$, where Z_k is the complement of $\{h \in H \mid \emptyset \neq \Psi([\mathcal{S}, h][k]) \subseteq P\}$. From the Bolzano-Weierstrass theorem we infer $\limsup_k \mu(Z_k) \geq .5$. By [Billingsley, 1986, Theorem 4.1(i)], this implies $\mu(\limsup_k Z_k) \geq .5$. Now, for every $h \in \limsup_k Z_k$ there are infinitely many k such that either $\Psi([\mathcal{S}, h][k])$ is undefined, equal to $\emptyset$, or not a subset of P. Hence Ψ fails to solve P on $[\mathcal{S}, h]$ for every $h \in \limsup_k Z_k$. By Definition (111), this contradicts our choice of Ψ. ∎

For the sequel we rely on the following notation.

(116) DEFINITION: Given finite sequence $\zeta \in N^{<\omega}$ we denote by O_ζ the basic open set of N^ω consisting of all $h \in H$ that begin with ζ.

Proof of Theorem (114): The left-to-right direction of Theorem (114) is immediate. For the other direction, suppose that scientist Ψ solves **P** on μ with probability greater than .5. Towards specifying a scientist that solves **P**, we introduce some terminology and notation.

We say that environment e is *bijective* just in case there is structure $\mathcal{S}$ and bijective full assignment h such that e is for $\mathcal{S}$ and h. Denote by h_0 the bijective full assignment $\{(v_i, i) \mid i \in N\}$. Recall our enumeration $\{\alpha_i \mid i \in N\}$ of atomic formulas, defining standard environments, fixed earlier. For all $n \in N$ denote by $f(n)$ the least integer greater than all the indexes of the members of $\bigcup\{Var(\alpha_i) \mid i < n\}$. For all $n \in N$ denote by SEQ_n the set of all $\tau \in SEQ$ such that $\text{length}(\tau) = n$, and some standard environment extends τ.

We can now define scientist Ψ' that solves every $P \in \mathbf{P}$ in every bijective environment for P. By Exercice (21) this suffices to show that **P** is solvable.

Let $\sigma \in SEQ$ be given. Suppose there is $n \leq \text{length}(\sigma)$ and (unique) $P \in \mathbf{P}$ such that the following holds.

(117)   There is finite $A \subset N^{f(n)}$ such that:

(a) $\mu(\bigcup\{O_\zeta \mid \zeta \in A\}) > .5$;

(b) for all $\zeta \in A$ there is $\tau \in SEQ_n$ such that $\emptyset \neq \Psi(\tau) \subseteq P$ and for all structures $\mathcal{U}$, $\mathcal{U} \models \bigwedge \sigma[h_0]$ implies $\mathcal{U} \models \bigwedge \tau[\zeta]$.

Then choose greatest $n \leq \text{length}(\sigma)$ and (unique) $P \in \mathbf{P}$ such that (117) holds, and set $\Psi'(\sigma) = P$. Otherwise, $\Psi'(\sigma)$ is undefined.

Let $P \in \mathbf{P}$, $\mathcal{S} \in P$, and bijective environment e for $\mathcal{S}$ be given. It suffices to show that for cofinitely many k, $\Psi'(e[k]) = P$. Observe that there is a unique structure $\mathcal{S}_e$ such that e is for $\mathcal{S}_e$ and h_0. By Lemma (7)b, $\mathcal{S}_e$ is isomorphic to $\mathcal{S}$. Hence, since P is closed under isomorphism, $\mathcal{S}_e \in P$. So Lemma (115) implies that there is $k_0 \in N$ such that:

(118)   for all $k \geq k_0$, $\mu(\{h \in H \mid \emptyset \neq \Psi([\mathcal{S}_e, h][k]) \subseteq P\}) > .5$.

Given $k \in N$ and $\zeta \in N^{f(k)}$, let $[\mathcal{S}_e, \zeta][k] \in SEQ$ be the initial segment of length k of $[\mathcal{S}_e, h]$, for any h that extends ζ. From (118) we infer the existence of finite $A \subset N^{f(k_0)}$ with:

(119) (a) $\mu(\bigcup\{O_\zeta \mid \zeta \in A\}) > .5$, and

(b) for all $\zeta \in A$, $\emptyset \neq \Psi([\mathcal{S}_e, \zeta][k_0]) \subseteq P$.

Since A is finite it is easily verified that there is $k_1 \geq k_0$ such that:

(120)   for all $\zeta \in A$, and all structures $\mathcal{U}$ such that $\mathcal{U}$ and h_0 satisfy $\bigwedge[\mathcal{S}_e, h_0][k_1]$, $\mathcal{U}$ and ζ satisfy $\bigwedge[\mathcal{S}_e, \zeta][k_0]$.

Now let $k_2 \geq k_1$ be given. Then (119) and (120) imply the existence of $k \leq k_2$ (for example: k_0), finite $A' \subseteq N^{f(k)}$ (for example: A), and $P' \in \mathbf{P}$ (for example: P) such that:

(a) $\mu(\bigcup\{O_\zeta \mid \zeta \in A'\}) > .5$;

(b) for all $\zeta \in A'$ there is $\tau \in SEQ_k$ (for example: $[\mathcal{S}_e, \zeta][k_0]$) such that:

(i) $\emptyset \neq \Psi(\tau) \subseteq P'$, and

(ii) for all structures $\mathcal{U}$, if $\mathcal{U}$ and h_0 satisfy $\bigwedge[\mathcal{S}_e, h_0][k_2]$, then $\mathcal{U} \models \bigwedge \tau[\zeta]$.

Since $[\mathcal{S}_e, h_0] = e$, this implies the existence of greatest $k_3 \leq k_2$, finite $A' \subseteq N^{f(k_3)}$, and $P' \in \mathbf{P}$ such that:

(a) $\mu(\bigcup\{O_\zeta \mid \zeta \in A'\}) > .5$;

(b) for all $\zeta \in A'$ there is $\tau \in SEQ_{k_3}$ such that:

(i) $\emptyset \neq \Psi(\tau) \subseteq P'$, and

(ii) for all structures $\mathcal{U}$, $\mathcal{U} \models \bigwedge e[k_2][h_0]$ implies $\mathcal{U} \models \bigwedge \tau[\zeta]$.

From the definition of Ψ' we infer immediately that $\Psi'(e[k_2]) = P'$. It thus remains to show that $P' = P$. Note that $k_3 \geq k_0$, and suppose for a contradiction that $P' \neq P$. Then the latter inequality implies $\mu(\{h \in \mathbf{H} \mid \Psi(\[\mathbb{S}, h\][k_3]) \nsubseteq P\}) > .5$. Since $\mathbb{S}$ and $\mathbb{S}_e$ are isomorphic, this with Lemma (7)a yields $\mu(\{h \in \mathbf{H} \mid \Psi(\[\mathbb{S}_e, h\][k_3]) \nsubseteq P\}) > .5$. In view of the fact that $k_3 \geq k_0$, the latter claim contradicts (118). ∎

We will see in Proposition (128), below, that the condition of isomorphic closure cannot be lifted in Theorem (114). The theorem nonetheless demonstrates the close proximity of the absolute and probabilistic criteria of solvability.

3.6.5 Probabilistic Solvability under Permutation

There is a dual to Theorem (114). To state it, we rely on the following definition.

(121) DEFINITION: Let $\mu \in M$ be given. The class of $\mu' \in M$ such that for some permutation $\pi : N \to N$, $\mu' = \{(i, \mu(\pi(i))) \mid i \in N\}$ is denoted $PERM(\mu)$.

So, $PERM(\mu)$ is the class of distributions that apply the same probabilities as μ to N, but not necessarily in the same order. The next theorem shows that no matter what distribution μ we start with, the class $PERM(\mu)$ is diverse enough to represent all of M. In other words, we will see that a problem is solvable on M with probability greater than .5 if and only if it is solvable on $PERM(\mu)$ with probability greater than .5. Indeed, it turns out that solvability in the latter sense implies solvability *tout court*, that is, with respect to *all* environments.

(122) THEOREM: Let problem $\mathbf{P}$ and $\mu \in M$ be given. Then $\mathbf{P}$ is solvable if and only if $\mathbf{P}$ solvable on $PERM(\mu)$ with probability greater than .5.

Proof: The left-to-right direction is immediate. For the other direction, recall Definition (11), and note the following immediate consequence of Definition (28) and Theorem (40).

(123) A normal problem $\mathbf{P}$ is solvable if and only if $\{\mathcal{I}(P) \mid P \in \mathbf{P}\}$ is solvable.

Suppose that scientist Ψ solves $\mathbf{P}$ on $PERM(\mu)$ with probability greater than .5. Then, by Lemma (7)a, it is easy to see that $\mathbf{P}$ is normal. We will show that Ψ solves $\mathfrak{I}(\mathbf{P}) = \{\mathfrak{I}(P) \mid P \in \mathbf{P}\}$ on μ. Theorem (114) will then imply that $\mathfrak{I}(\mathbf{P})$ is solvable, from which the solvability of $\mathbf{P}$ follows by (123).

The following notation will be used. Let permutation $\pi : N \to N$ be given. We denote by μ_π the element $\{(i, \mu(\pi(i))) \mid i \in N\}$ of $PERM(\mu)$. Given $h \in H$, let $h_\pi = \{(v_i, \pi(h(v_i))) \mid i \in N\}$, and given $H \subseteq H$, let $H_\pi = \{h_\pi \mid h \in H\}$. Now let $P \in \mathbf{P}$ and $\mathcal{S} \in \mathfrak{I}(P)$ be given. It suffices to show that:

(124) $\mu(H_{solve}(\Psi, \mathcal{S}, \mathfrak{I}(P))) > .5$.

Since $\mathcal{S} \in \mathfrak{I}(P)$ there is permutation $\pi : N \to N$ and $\mathcal{S}_\pi \in P$ such that for all $\alpha \in \mathcal{L}_{basic}$ and full assignments h, $\mathcal{S} \models \alpha[h]$ if and only if $\mathcal{S}_\pi \models \alpha[h_\pi]$. Observe:

(125)   (a) Every environment for $\mathcal{S}$ and h is an environment for $\mathcal{S}_\pi$ and h_π.

        (b) For all μ-measurable $H \subseteq H$, $\mu_\pi(H) = \mu(H_\pi)$.

By our assumption on Ψ, $\mu_\pi(H_{solve}(\Psi, \mathcal{S}_\pi, P)) > .5$. Since $P \subseteq \mathfrak{I}(P)$, this immediately implies:

(126) $\mu_\pi(H_{solve}(\Psi, \mathcal{S}_\pi, \mathfrak{I}(P))) > .5$.

From (125)b and (126) we obtain:

(127) $\mu(H_{solve}(\Psi, \mathcal{S}_\pi, \mathfrak{I}(P))) > .5$.

Finally, (127) and (125)a imply (124).   ∎

3.6.6   Separating Probabilistic and Absolute Solvability

Our results so far raise the possibility that probabilistic solvability collapses into absolute solvability. So the question arises whether there are problems $\mathbf{P}$ and classes M_0 of probabilities such that $\mathbf{P}$ is solvable on M_0 but unsolvable absolutely. From Theorems (114) and (122), for this to be possible at least one member of $\mathbf{P}$ must not be closed under isomorphism, and M_0 must not include $PERM(\mu)$ for any $\mu \in M$. Under these constraints the following fact provides an affirmative answer to our question. Indeed, it suffices to consider a "small" problem and a "large" class of distributions.

(128)   PROPOSITION: Suppose that **Sym** consists of a binary predicate. Then there exist disjoint propositions P_1, P_2 and uncountable $M_0 \subseteq M$ such that $\{P_1, P_2\}$ is unsolvable, but $\{P_1, P_2\}$ is solvable on M_0 with probability 1.

Proof: Let the binary predicate of **Sym** be R. We specify a countable collection $\{S_j \mid j \in N\}$ of structures by specifying the extension R^{S_j} of R for all $j \in N$. R^{S_0} is the successor function $\{(i, i + 1) \mid i \in N\}$. For $j > 0$, R^{S_j} is the finite relation $\{(i, i + 1) \mid i < j\}$. Let $P_1 = \{S_0\}$ and $P_2 = \{S_j \mid j > 0\}$. To prove the proposition it suffices to show:

(129) (a) $\{P_1, P_2\}$ is not solvable.

  (b) Let $M_0 \subset M$ be any class of measures such that for all $i \in N$, $\mathrm{glb}\{\mu(i) \mid \mu \in M_0\} > 0$. Then $\{P_1, P_2\}$ is solvable on M_0 with probability 1.

Exercise (45) yields (129)a, so we now prove (129)b. Towards specifying a scientist that solves **P** on M_0 with probability 1, we introduce some notation and facts.

Given $i, j \in N$, let $B_{i,j}$ be the collection of $h \in H$ such that not all of $0, 1, \ldots, i$ occur in $\{h(v_0) \ldots h(v_j)\}$. By the assumption on M_0, let strictly increasing $f : N \to N$ be such that for all $i \in N$ and $\mu \in M_0$, $\mu(B_{i,f(i)}) < \frac{1}{2^i}$. So for each $\mu \in M_0$, $\Sigma_i \mu(B_{i,f(i)})$ converges. Hence, by the first Borel-Cantelli lemma ([Billingsley, 1986, Theorem 4.3]):

(130) $\mu(\limsup_i B_{i,f(i)}) = 0$ for every $\mu \in M_0$.

Via the definition of $B_{i,j}$, (130) yields:

(131)   For every $\mu \in M_0$ the class of $h \in H$ such that $\{0 \ldots i\} \nsubseteq \{h(v_0) \ldots h(v_{f(j)})\}$ for infinitely many $i \in N$ has μ-measure 0.

Call $\sigma \in SEQ$ "complete through m" if m is the greatest number such for all $i, j \leq m$, either $\bigwedge \sigma \models Rv_i v_j$ or $\bigwedge \sigma \models \neg Rv_i v_j$. We now define scientist Ψ that solves **P** on M_0 with probability 1. Let $\sigma \in SEQ$ be given. If σ is not complete through $f(0)$, then $\Psi(\sigma)$ is undefined. Otherwise, suppose that n is greatest such that σ is complete through $f(n)$. If $\bigwedge \sigma$ implies the existence of an R-chain of length at least n, then $\Psi(\sigma) = P_1$; otherwise, $\Psi(\sigma) = P_2$. Because f is strictly increasing, Ψ is well defined.

To prove that Ψ solves P_2 on M_0 with probability 1, let environment e for P_2 be given. Then for all but finitely many $n \in N$, range(e) does not imply the existence of an R-chain of length at least n. So $\Psi(e[k]) = P_2$ for cofinitely many k. Hence Ψ solves P_2. Hence Ψ solves P_2 on M_0 with probability 1. To prove that Ψ solves P_1 on M_0 with probability 1, call $h \in H$ "bad" just in case there is an environment e for S_0 and h with the following property. For infinitely many k, $e[k]$ is complete through $f(k)$ but $\bigwedge e[k]$ does not imply the existence of an R-chain of length at least k. Let $\mu \in M_0$ be given. It follows directly from (131) that the class of bad $h \in H$ has μ-measure 0. Hence, there is a class of $h \in H$ of μ-measure 1 such that for every environment e for S_0 and h, $\Psi(e[k]) = P_1$ for cofinitely many k. That is, $\mu(H_{solve}(\Psi, S_0, P_1)) = 1$. Hence Ψ solves P_1 on M_0 with probability 1.   ∎

3.6.7   Reliable Solvability with Specified Probability

To this point our results have concerned probability of success greater than .5. We wish now to consider the competence of scientists whose probability of success is small. To formulate interesting results requires reflection about the case in which success is not achieved. A scientist who solves a given structure with small probability is worse than useless if she exhibits high probability of misleading an external observer. In particular, it is misleading to converge to a false theory; for in this case the mistaken theory appears to be held with confidence, and risks being accredited. If the probability that the scientist misleads us this way is high, and the probability of genuine success low, it might be better to show her no data at all. So we are led to the following definition.

(132)   DEFINITION: Let scientist Ψ, proposition P, structure $S \notin P$, and full assignment h be given. We say that Ψ is *unreliable on S, h and P* if there is an environment e for S and h such that $\emptyset \neq \Psi(e[k]) \subseteq P$ for cofinitely many k. We denote by $H_{unrel}(\Psi, S, P)$ the class of $h \in H$ for which Ψ is unreliable on S, h and P.

The ensuing definition of reliable, probable success includes the requirement that the chance of being permanently misled is zero.

(133)   DEFINITION: Let scientist Ψ, problem $\mathbf{P}$, $M_0 \subseteq M$, and $p \in [0, 1]$ be given. We say that Ψ *reliably solves $\mathbf{P}$ on M_0 with probability p* just in case for every $\mu \in M_0$ and for every $P \in \mathbf{P}$, the following holds:

(a) for all $S \in P$, $\mu(H_{solve}(\Psi, S, P)) \geq p$;

(b) for all $S \in \bigcup \mathbf{P} - P$, $\mu(H_{unrel}(\Psi, S, P)) = 0$.

In this case $\mathbf{P}$ is *reliably solvable on M_0 with probability p*.

In particular, Ψ reliably solves $\mathbf{P}$ on M_0 with positive probability just in case there is no chance that Ψ will mislead us in an environment for $\mathbf{P}$ and some chance that Ψ will succeed. We now show that reliable solvability in the foregoing sense embodies a surprising degree of scientific competence.

(134)   THEOREM: Let problem **P** and $M_0 \subseteq M$ be given. Then **P** is solvable on M_0 with probability 1 if and only if **P** is reliably solvable on M_0 with probability greater than 0.

Proof: The left-to-right direction is immediate. To prove the other direction, let scientist Ψ reliably solve **P** on M_0 with probability greater than 0. Without loss of generality, we may assume that:

(135) Ψ is total and for all $\sigma \in SEQ$, $\Psi(\sigma) \in$ **P**.

Given $\sigma \in SEQ$ and $k \leq \text{length}(\sigma)$, let $\sigma[k]$ be the initial sequence of length k in σ. By the "score" of $\sigma \in SEQ$ will be meant $\text{length}(\sigma) - k_0$ for the smallest $k_0 \leq \text{length}(\sigma)$ such that $\Psi(\sigma[k]) = \Psi(\sigma)$ for all k with $k_0 \leq k \leq \text{length}(\sigma)$. The score of any $\sigma \in SEQ$ is well defined since Ψ is total, and it is denoted by score σ.

We shall exhibit scientist Ψ' that uses Ψ as a subroutine in order to solve **P** on M_0 with probability 1. Specifically, faced with environment $[S, h]$, Ψ' will examine Ψ's score on initial segments of environments generated from the information in the "tails" of $[S, h]$. It will turn out that Ψ is guaranteed to succeed on one such tail, and be unreliable on none. Hence, Ψ's scores will grow fast on at least one tail, and such growth will be guaranteed to signal correct choice of $P \in$ **P**. This will be enough for Ψ' to eventually spot the stable and correct response that Ψ makes on one of the tails. We now introduce definitions and facts that make this idea precise.

Call $h \in N$ "saturated" iff for all $n \in N$, $h^{-1}(n)$ is infinite. By positivity:

(136)   Let $\mu \in M$ be given. Then $\mu(\{h \in H \mid h \text{ is saturated}\}) = 1$.

Given $n \in N$ and $h \in H$, we denote by h^n the assignment $\{(v_i, h(v_{i+n})) \mid i \in N\}$. Thus, h^n is the assignment generated by "starting over" at n. If h is saturated, then h^n is a full assignment. It is easy to verify the following fact.

(137)   For every $n \in N$ there is a (computable) function $F_n : SEQ \to SEQ$ such that:

(a) for all $\sigma, \tau \in SEQ$ with $\sigma \subseteq \tau$, $F_n(\sigma) \subseteq F_n(\tau)$;

(b) for all structures S and saturated $h \in H$, $\bigcup_{k \in N} F_n([S, h][k]) = [S, h^n]$.

The following lemma is the key to our construction.

(138)   LEMMA: Let scientist Ψ reliably solve **P** on M_0 with probability greater than 0. Then for all $P, Q \in$ **P** with $P \neq Q$, all $S \in P$, and all $\mu \in M_0$ the following holds.

(a) Let X be the class of all saturated $h \in H$ such that for some $n \in N$, Ψ solves P on $[S, h^n]$. Then $\mu(X) = 1$.

(b) Let Z be the class of all saturated $h \in H$ such that for some $n \in N$, Ψ is unreliable on S, h^n, and Q. Then $\mu(Z) = 0$.

Proof of Lemma (138): Let $P, Q \in \mathbf{P}$ with $P \neq Q$, $S \in P$, and $\mu \in M_0$ be given, and let X, Z be as described. By the assumption on Ψ and taking $n = 0$ in the description of X, we infer that $\mu(X) > 0$. In conjunction with (136), Kolmogorov's zero-one law for tail events [Billingsley, 1986, Theorem 4.5] then implies that $\mu(X) = 1$. This proves part (a) of the Lemma. We now prove part (b). Let Y be the class of all $h \in H$ such that Ψ is unreliable on S, h, and Q. It is immediate that $Z \subseteq \bigcup_{\alpha \in N^{<\omega}} \alpha * Y$. Hence:

$$\mu(Z) \leq \sum_{\alpha \in N^{<\omega}} \mu(\alpha * Y) = \sum_{\alpha \in N^{<\omega}} \mu(Y) \times \mu(O_\alpha).$$

By the assumption on Ψ, $\mu(Y) = 0$. This shows that $\mu(Z) = 0$, as required. ∎

We now describe a scientist Ψ' such that for every $P \in \mathbf{P}$, every $S \in P$, and every $\mu \in M$, $\mu(\{h \in H \mid \Psi' \text{ solves } P \text{ in } [S, h]\}) = 1$. By Lemma (113) this suffices to prove the proposition. Let $\sigma \in SEQ$ be given. Suppose there is $n \leq$ length(σ) such that for all $n' \leq$ length(σ), score $F_n(\sigma) - n \geq$ score $F_{n'}(\sigma) - n'$. Then $\Psi'(\sigma) = \Psi(F_{n_0}(\sigma))$, where n_0 is the least such n. Otherwise $\Psi'(\sigma)$ is undefined. Let $\mu \in M_0$, $P \in \mathbf{P}$, and $S \in P$ be given. We show that the set of all $h \in H$ such that $\Psi'([S, h][k]) = P$ for cofinitely many k has μ-measure 1. By Lemma (138)a, there is a set X with μ-measure 1 of saturated $h \in H$ each with the property that for some $n, k_0 \in N$, the score of $[S, h^n][k]$ is $k - k_0$ for cofinitely many k. So in view of (135), for all $h \in X$ there are $n \in N$ and $P' \in \mathbf{P}$ such that for cofinitely many k, $\Psi'([S, h][k]) = \lim_{k' \to \infty} \Psi([S, h^n][k']) = P'$. Let Z be the set of $h \in X$ such that for some $n \in N$, $\lim_{k' \to \infty} \Psi([S, h^n][k']) = P'$ and $P' \neq P$. To finish the proof, it suffices to show that $\mu(Z) = 0$. But this follows immediately from Lemma (138)b and the choice of Ψ. ∎

3.6.8 Partial Summary

The collection of problems of form $(T, \{\theta_0, \dots, \theta_n\})$, where T has no finite models, is a rich and natural class for the theory of empirical inquiry. The results in the present chapter provide alternative characterizations of its solvable subclass. To recapitulate, we state the characterizations in the following corollary, which is an immediate consequence of Theorems (55), (114), and (134), along with Corollary (57).

(139)   COROLLARY: Let $\mu \in M$ and problem **P** of form $(T, \{\theta_0, \ldots, \theta_n\})$ be given. Suppose that T has no finite models. Then the following conditions are equivalent.

(a) **P** is solvable.

(b) For all $i \leq n$, θ_i is equivalent in T to an $\exists\forall$ sentence.

(c) For all $i \leq n$, θ_i is equivalent in T to a boolean combination of $\exists$ sentences.

(d) **P** is solvable on μ with probability greater than .5.

(e) **P** is reliably solvable on μ with probability greater than 0.

3.6.9   Bayesian Scientists

The definitions introduced in the present section may be employed to contrast the first-order paradigm with the Bayesian approach to inquiry. For this purpose we shall isolate the class of scientists that may fairly be said to exhibit "Bayesian behavior." Then it will be asked whether every solvable problem is solved by one of them. The answer will be negative.[15] Before turning to the details, we indicate the importance of the question to our enterprise.

The idea of "Bayesian behavior" will be defined in probabilistic terms, yet the success criterion at issue is the absolute one proper to the first-order paradigm: all environments must lead the scientist to success. Under these conditions it will be possible to exhibit a simple, solvable problem that is not solved by any scientist (computable or otherwise) that behaves in Bayesian fashion. However, the reader might object that such a finding is irrelevant to the Bayesian since she is willing to settle for success with probability 1 instead of success with certainty (as required in the first-order paradigm). In response, we note that our goal is only to show that within the first-order paradigm, Bayesian scientists do not constitute a canonical form for inquiry, even with respect to some simple classes of problems. In this way we hope to motivate the developments of Chapter IV, in which a variety of schemes for inquiry will be presented that are in fact canonical.

For the remainder of the section we take **Sym** to be limited to the binary predicate R. We let $\mathcal{S}_\ell$ be the structure $(N, \leq)$, and $\mathcal{S}_g$ be the structure $(N, \geq)$. (The subscript ℓ stands for "least point for R" whereas g refers to "greatest point for R.") We let $\mathbf{P}^* = \{\{\mathcal{S}_\ell\}, \{\mathcal{S}_g\}\}$. It follows immediately from Proposition (34) that $\mathbf{P}^*$ is solvable (indeed, it is computably solvable).

---

15. For an alternative perspective on Bayesianism within a paradigm somewhat different from ours, see [Juhl, 1993].

To define "Bayesian behavior" we rely on the following definition, whose intuitive significance is explained subsequently.

(140)   DEFINITION: A *Bayesian set-up* is a pair (p, μ), where $p \in (0, 1)$ and $\mu \in M$.

The interpretation of p in a Bayesian set-up (p, μ) is the prior probability of Nature choosing S_ℓ as the actual world; her probability of choosing S_g is $1 - p$. As before, for $n \in N$, $\mu(n)$ is the (positive) probability that Nature chooses n when assigning a value to a variable. We will shortly be able to extend μ to the conditional probability $\mu(\sigma \mid S_\ell)$ of observing some data σ given that the world is S_ℓ, and also to the conditional probability $\mu(\sigma \mid S_g)$ of observing σ given S_g. Once these probabilities are well defined, it will be possible to provide a precise sense in which a scientist behaves like a Bayesian.

Given $\mu \in M$, $\mathcal{U} \in \{S_\ell, S_g\}$, and $\sigma \in SEQ$, the μ-measure of the collection of full assignments h with $\mathcal{U} \models \bigwedge \sigma[h]$ is well defined since it is equal to the countable sum of $\mu(O_\zeta)$, for all $\zeta \in N^{<\omega}$ such that: (a) $Var(\sigma) \subseteq \text{domain}(\zeta)$, (b) $\mathcal{U} \models \bigwedge \sigma[\zeta]$, and (c) for no proper initial segment ζ' of ζ, $Var(\sigma) \subseteq \text{domain}(\zeta')$. This number is the conditional probability $\mu(\sigma \mid \mathcal{U})$ according to μ of observing σ given that Nature chooses $\mathcal{U}$ as the world. Now recall Bayes' Theorem in the present context: $\mu(S_\ell \mid \sigma) \sim \mu(\sigma \mid S_\ell) \times \mu(S_\ell)$, and similarly for $\mu(S_g \mid \sigma)$. It follows that:

$$\mu(S_\ell \mid \sigma) > \mu(S_g \mid \sigma) \quad \text{iff} \quad \mu(\sigma \mid S_\ell) \times \mu(S_\ell) > \mu(\sigma \mid S_g) \times \mu(S_g).$$

We thus have the elements necessary to define the Bayesian strategy.

(141)   DEFINITION: Let (p, μ) be a Bayesian set-up.

(a) Given $\sigma \in SEQ$, we define $\mu(\sigma \mid S_\ell)$ to be $\mu(h \in H \mid S_\ell \models \bigwedge \sigma[h])$. Similarly, we define $\mu(\sigma \mid S_g)$ to be $\mu(h \in H \mid S_g \models \bigwedge \sigma[h])$.

(b) Scientist Ψ is *Bayesian with respect to* (p, μ) just in case for all $\sigma \in SEQ$ for $\mathbf{P}^*$, $\Psi(\sigma) = \{S_\ell\}$ if $\mu(\sigma \mid S_\ell) \times p > \mu(\sigma \mid S_g) \times (1 - p)$, and $\Psi(\sigma) = \{S_g\}$ if $\mu(\sigma \mid S_\ell) \times p < \mu(\sigma \mid S_g) \times (1 - p)$.

(c) Scientist Ψ is *Bayesian* just in case Ψ is Bayesian with respect to some Bayesian set-up.

Note that in case $\mu(\sigma \mid S_\ell) \times p = \mu(\sigma \mid S_g) \times (1 - p)$, we place no constraint on the behavior of a scientist that is Bayesian with respect to (p, μ). We can finally make the point indicated at the beginning of this subsection.

(142)   PROPOSITION: No Bayesian scientist solves $\mathbf{P}^*$.

Proof: Suppose that scientist Ψ is Bayesian with respect to Bayesian set-up (p, μ). Let $\sigma \in SEQ$, and finite assignment $a : Var \to N$ be given with domain$(a) \supseteq Var(\sigma)$ and $S_\ell \models \bigwedge \sigma[a]$. We shall exhibit $\tau \in SEQ$ such that:

(143)   (a) $S_\ell \models \exists \bar{x} \bigwedge (\sigma * \tau)[a]$, where $\bar{x}$ contains the variables in $Var(\tau) -$ domain(a), and

    (b) $\Psi(\sigma * \tau) = \{S_g\}$.

This suffices to show that there is no locking pair for Ψ, S_ℓ, and $\{S_\ell\}$, in the sense of Definition (23). It then follows from Proposition (24) that Ψ does not solve $\{S_\ell\}$ in every environment for S_ℓ, hence does not solve $\mathbf{P}^*$.

Choose i to be greater than any index for a variable in domain(a). Let $G(n, m) \in SEQ$ be a uniformly chosen finite sequence with range $\{Rv_{i+j}v_{i+j+1} \mid j \leq n\} \cup \{v_{i+n+1} = v_{i+n+j} \mid 1 < j \leq m\}$. Intuitively, S_ℓ interprets $G(n, m)$ as: "there is a long ascending sequence whose last member must be far from 0, and is sampled many times," whereas S_g interprets $G(n, m)$ as: "there is a long descending sequence whose last member may be close to 0 and is sampled many times." Since μ must ultimately attach smaller probabilities to large numbers than to 0, as n, m get large, $G(n, m)$ becomes much less likely under S_ℓ than under S_g. Using the fact that an arbitrarily large fraction of the probability assigned by μ is concentrated in a finite subset of N, it is easy to verify that:

(144)   For all $\delta \in (0, 1)$, there is $n, m \in N$ such that $\mu(G(n, m) \mid S_\ell) < \mu(G(n, m) \mid S_g) \times \delta$.

By the choice of i, $\mu(\sigma * G(n, m) \mid S_\ell) = \mu(\sigma \mid S_\ell) \times \mu(G(n, m) \mid S_\ell)$, and $\mu(\sigma * G(n, m) \mid S_g) = \mu(\sigma \mid S_g) \times \mu(G(n, m) \mid S_g)$. Since S_ℓ and S_g satisfy the same members of SEQ, and $S_\ell \models \bigwedge \sigma[a]$, both $\mu(\sigma \mid S_\ell)$ and $\mu(\sigma \mid S_g)$ are fixed numbers greater than 0. The same is true of p, $(1 - p)$ by Definition (140). So from (144) we infer:

(145)   There are $n, m \in N$ such that $p \times \mu(\sigma * G(n, m) \mid S_\ell) < (1 - p) \times \mu(\sigma * G(n, m) \mid S_g)$.

Let $\tau = G(n, m)$ for n, m that satisfy (145). Since any finite chain is satisfiable in both S_ℓ and S_g, the choice of i implies (143)a. Definition (141)b and (145) imply (143)b.   ∎

To exploit again our favorite example, let $\theta = \exists x \forall y Rxy$, and let $T_1 \subseteq \mathcal{L}_{sen}$ be the theory of total orders with either a greatest point or a least point, but not both. By Proposition (54), $(T_1, \{\theta, \neg\theta\})$ is solvable. In contrast, since no Bayesian scientist solves $\mathbf{P}^*$, none solves $(T_1, \{\theta, \neg\theta\})$ either. Therefore Bayesian scientists do not offer a cannonical means of solving problems of the form $(T, \{\theta_0, \ldots, \theta_n\})$, at least, not within the first-order paradigm. There is something disappointing in this outcome, inasmuch as Bayesian behavior is often considered to be irreproachably rational. A challenge is posed thereby, namely, to exhibit other forms of rational hypothesis selection that succeed where Bayesianism fails. We pick up the challenge in the next and final chapter.

3.7   Generalization to Arbitrary Structures

3.7.1   Inquiry in an Uncountable Setting

The first-order paradigm introduced in Section 3.1 makes essential use of the limitation to countable structures, particularly in its simple conception of "environment" [see Definition (4)]. It was for this reason that we introduced Convention (2), which binds the concept "structure" to its countable case. It must be admitted, however, that legitimate complaints may arise about the exclusion of structures with uncountable domains. If the universe has uncountably many "things" in it, then real scientists must deal with a domain that fits into none of the structures for which the first-order paradigm is currently adapted.[16] Moreover, even without speculating about the numerical size of the universe, it is interesting to see how the first-order paradigm can be extended to structures with arbitrary domains. Such is the purpose of the present section, which is entirely optional, since we return in the sequel to the first-order paradigm defined in Section 3.1. Why retreat to the original paradigm after having extended it to arbitrary structures? The reason is the greater simplicity of the countable case, compared to the sophisticated tools that will be needed to assimilate uncountable structures. The simplicity of the original paradigm favors clarity in discussing other issues, notably, belief revision in the next chapter.

The present section is organized as follows. Section 3.7.2 introduces the technical machinery to be used in generalizing the first-order paradigm. The new paradigm is then presented and motivated in Section 3.7.3. The relation

---

16. The cardinality of "things" in the physical universe need not be identified with the set of elementary particles, or even with the set of its subsets. For example, it might be held that every potential wavelength of light is "something" in the physical universe. In this case the numerical size of the universe has at least the power of the continuum, should it be true that light can assume every wavelength in some non-void interval.

between the original and generalized paradigms is explored in Sections 3.7.4–3.7.6. The following convention is central to the discussion. Whenever we wish to denote an object in its general sense (without limitation to the countable case), we attach the subscript g (for "general"). Thus, in particular:

(146)   *Convention:*   By a *structure*$_g$ is meant a model for **Sym** with an arbitrary domain (either countable or uncountable). We continue to use the term "structure" (without the subscript g) in the sense of Convention (2), namely, as involving a countable domain.

Similarly, the class of models$_g$ (arbitrary domain!) for $T \subseteq \mathcal{L}_{sen}$ is denoted $MOD(T)_g$, etc. The convention extends even to our name for the new paradigm; it will be qualified as: first-order$_g$.

3.7.2   Partial Isomorphisms

To define scientific success within the first-order$_g$ paradigm, we rely on the follow idea. Every uncountable structure$_g$ $\mathcal{S}$ has many countable substructures. For the scientist to succeed on $\mathcal{S}$ we require that she succeed on a collection of its countable substructures that are "sufficiently similar" to $\mathcal{S}$. To cash in the latter idea, we use a standard concept from model theory, namely "partial isomorphism." Except for a few results to be cited from other sources, our presentation of the relevant concepts is self-contained. For background, see [Ebbinghaus *et al.*, 1994, Ch. XII].

(147)   DEFINITION: Let structures$_g$ $\mathcal{S}$ and $\mathcal{T}$ be given. Let p be an injective map from $|\mathcal{S}|$ into $|\mathcal{T}|$. (Note: p need be neither total nor onto.) Then p is a *partial isomorphism* from $\mathcal{S}$ to $\mathcal{T}$ just in case the following conditions are satisfied.

(a) For every n-place predicate R in **Sym**, and every n elements $x_1 \ldots x_n$ in domain(p), $(x_1 \ldots x_n) \in R^{\mathcal{S}}$ iff $(p(x_1) \ldots p(x_n)) \in R^{\mathcal{T}}$.

(b) For every n-ary function symbol f in **Sym**, and every $n + 1$ elements $x_1 \ldots x_n, x$ in domain(p), $f^{\mathcal{S}}(x_1 \ldots x_n) = x$ iff $f^{\mathcal{T}}(p(x_1) \ldots p(x_n)) = p(x)$.

(c) For every individual constant c in **Sym**, and every x in domain(p), $c^{\mathcal{S}} = x$ iff $c^{\mathcal{T}} = p(x)$.

The class of partial isomorphisms between two structures$_g$ can be used to assess their similarity, where ordinal numbers serve as the similarity measure. The next definition shows how such an ordinal conception of similarity arises.

(148)   DEFINITION: Let structures$_g$ S and T be given. Denote by $PI(S, T)$ the class of partial isomorphisms from S to T. For each ordinal α we define a class $I_\alpha(S, T)$ of partial isomorphisms from S to T. This is accomplished inductively, as follows.

$I_0(S, T) = PI(S, T)$.

Let ordinal α be given. Suppose that $I_\beta(S, T)$ has been defined for all $\beta < \alpha$. We distinguish two cases.

Case 1: $\alpha = \beta + 1$. Then $I_\alpha(S, T)$ is the class of all $f \in PI(S, T)$ such that: for all $a \in |S|$ there is $g \in I_\beta(S, T)$ with $g \supseteq f$ and domain$(g) =$ domain$(f) \cup \{a\}$; for all $a \in |T|$ there is $g \in I_\beta(S, T)$ such that $g \supseteq f$ and range$(g) =$ range$(f) \cup \{a\}$.

Case 2: α is a limit ordinal. Then $I_\alpha(S, T) = \bigcap_{\beta < \alpha} I_\beta(S, T)$.

We say that S and T are *finitely isomorphic* just in case $I_\omega(S, T) \neq \emptyset$.
We say that S and T are *partially isomorphic* just in case $I_\alpha(S, T) \neq \emptyset$ for all ordinals α.

Our definitions of "finitely isomorphic" and "partially isomorphic" are not standard, but they are equivalent to those usually given, for example, in [Ebbinghaus *et al.*, 1994, Sec. XII.1]; see Exercises (179), (180), below. The next example should help in understanding the hierarchy introduced in Definition (148).

(149)   *Example:*   Suppose that **Sym** is limited to a binary predicate. Let $S = (Z, <)$ and $T = (Z + Z, <)$, where Z is the ring of integers, and $Z + Z$ is two copies of the ring, one above the other. Then it is easy to verify that $I_{\omega+1}(S, T) \neq \emptyset$, but $I_{\omega+2}(S, T) = \emptyset$.

The following lemma draws together some facts to be used later.

(150)   LEMMA: Let structures$_g$ S and T be given.

(a) The empty partial isomorphism belongs to $I_0(S, T)$.

(b) For all ordinals α, β, if $\alpha \leq \beta$ then $I_\beta(S, T) \subseteq I_\alpha(S, T)$.

(c) Suppose that **Sym** is finite. Then S and T are finitely isomorphic iff S and T are elementarily equivalent.

(d) Suppose that $|\mathcal{S}|$ and $|\mathcal{T}|$ are countable. Then $\mathcal{S}$ and $\mathcal{T}$ are partially isomorphic iff $\mathcal{S}$ and $\mathcal{T}$ are isomorphic.

Proof: The proofs of (150)a,b are trivial. We obtain (150)c directly from Fraïssé's theorem, [Ebbinghaus *et al.*, 1994, Thm XII.2.1]. Similarly, (150)d is a generalization of a famous result of Cantor, proved in [Ebbinghaus *et al.*, 1994, Thm XII.1.5d]. ∎

Thus, Lemma (150)c and Example (149) imply that $(Z, <)$ and $(Z + Z, <)$ are elementarily equivalent (although not isomorphic).

3.7.3 The Paradigm

Now we are ready to specify the first-order$_g$ paradigm. As usual, this is achieved by stepping through its components, as listed in 1.(1). The subscript g continues to signal usage in the general sense.

Worlds. The potential realities of our paradigm are all the structures$_g$ that interpret the symbol set **Sym**.

Problems. The concepts of *proposition$_g$* and *problem$_g$* are straightforward generalizations of the countable case. Officially:

(151) DEFINITION: A nonempty class of structures$_g$ is a *proposition$_g$*. A *problem$_g$* is a collection of disjoint propositions$_g$.

Environments. For the first-order$_g$ paradigm there is no need to extend the concept of "environment" to structures$_g$. Instead, we continue to rely on Definition (4), which introduced full assignments and environments just for structures. As before, we let *SEQ* denote the collection of proper initial segments of any environment.

Scientists. Again, the generalization from the countable case is straightforward. Officially:

(152) DEFINITION: A *scientist$_g$* is any partial or total mapping of *SEQ* into subclasses of structures$_g$.

Success. The key definition is now given. Subsequent to its presentation we provide some motivating remarks.

(153) DEFINITION: Let scientist$_g$ Ψ and proposition$_g$ P be given. We say that Ψ *solves$_g$* P just in case for all $\mathcal{T} \in P$, there is ordinal α such that:

(a) there is countable substructure S of T with $I_\alpha(S, T) \neq \emptyset$;

(b) for all countable substructures S of T with $I_\alpha(S, T) \neq \emptyset$ and for all environments e for S, $\emptyset \neq \Psi(e[k]) \subseteq P$ for cofinitely many k.

Let problem$_g$ **P** be given. We say that Ψ *solves$_g$* **P** just in case Ψ solves$_g$ every $P \in$ **P**. In this case, **P** is *solvable$_g$*.

Lemma (150)a implies that condition (153)a can always be met by choosing ordinal α small enough.

Intuitively, for Ψ to solve$_g$ P in structure$_g$ T, Ψ must recognize that $T \in P$ through the examination of any countable substructure of T that is sufficiently similar to T. Moreover, the degree of similarity to T can be chosen arbitrarily, provided that there are countable substructures of T with the similarity in question. From Lemma (150)b we see that Ψ is better off choosing a high degree of similarity to T if possible, since Ψ will then be responsible for a smaller class of substructures. It is also immediate that solvability$_g$ is a nonvacuous concept, admitting both the solvable$_g$ and unsolvable$_g$ case. This is because all singleton problems$_g$ are solvable$_g$, whereas any problem$_g$ that separates an uncountable structure$_g$ from all of its countable substructures is unsolvable$_g$.

In the remainder of this subsection, we attempt to motivate Definition (153) by considering some of its special cases. For this purpose, let uncountable structure$_g$ T be given, and suppose that some countable substructure of T is partially isomorphic to T. That this case is possible is shown by the following example.

(154) *Example:*   Suppose that **Sym** consists of a sole, binary predicate. Let R be the reals, and let Q be the rationals, both under their natural order. Then it is easy to see that Q is a countable substructure of R that is partially isomorphic to R.

Among its countable substructures, a partially isomorphic substructure of T might be considered to be "as similar as possible" to T.[17] So it is reasonable to require the scientist to succeed just on such substructures, when they exist.

---

17. In discussing partially isomorphic structures M and N, Barwise [Barwise, 1975, p. 292] remarks: "Some authors prefer the more picturesque terminology *potentially isomorphic*, to suggest that M and N would become isomorphic if only they were to become countable, say in some larger universe of set theory."

Fortunately, the following lemma [in conjunction with Lemma (150)b] shows this to be exactly the requirement imposed by Definition (153).

(155) LEMMA: Let structure$_g$ $\mathcal{T}$ be given. Then there is an ordinal α such that for all countable substructures $\mathcal{S}$ of $\mathcal{T}$, $I_\alpha(\mathcal{S}, \mathcal{T}) \neq \emptyset$ iff $\mathcal{S}$ is partially isomorphic to $\mathcal{T}$.

To explain the point of the lemma, suppose that structure$_g$ $\mathcal{T}$ in proposition$_g$ P has a countable, partially isomorphic substructure, and that scientist$_g$ Ψ solves$_g$ P. Then to behave successfully with respect to $\mathcal{T}$ it is necessary and sufficient that Ψ issue a nonempty subset of P cofinitely often on any environment for a countable, partially isomorphic substructure of $\mathcal{T}$.

Proof of Lemma (155): For all countable substructures $\mathcal{S}$ of $\mathcal{T}$, $PI(\mathcal{S}, \mathcal{T})$ is a set (rather than a proper class). Hence, $\bigcup \{PI(\mathcal{S}, \mathcal{T}) \mid \mathcal{S}$ is a countable substructure of $\mathcal{T}\}$ is a set. It follows that the ordinals β satisfying

$$I_{\beta+1}(\mathcal{S}, \mathcal{T}) \subset I_\beta(\mathcal{S}, \mathcal{T}) \text{ for some countable substructure } \mathcal{S} \text{ of } \mathcal{T}$$

also form a set. We may therefore choose an ordinal α_0 such that for all $\beta \geq \alpha_0$ and all countable substructures $\mathcal{S}$ of $\mathcal{T}$, $I_\beta(\mathcal{S}, \mathcal{T}) \not\subset I_{\alpha_0}(\mathcal{S}, \mathcal{T})$. It follows directly from Lemma (150)b that:

(156)   for all $\beta \geq \alpha_0$ and countable substructures $\mathcal{S}$ of $\mathcal{T}$, $I_\beta(\mathcal{S}, \mathcal{T}) = I_{\alpha_0}(\mathcal{S}, \mathcal{T})$.

For all countable substructures $\mathcal{S}$ of $\mathcal{T}$, if $\mathcal{S}$ is partially isomorphic to $\mathcal{T}$ then $I_{\alpha_0}(\mathcal{S}, \mathcal{T}) \neq \emptyset$ by Definition (148). Conversely, suppose that $\mathcal{S}$ is a countable substructure of $\mathcal{T}$ with $I_{\alpha_0}(\mathcal{S}, \mathcal{T}) \neq \emptyset$. Then by (156) and Definition (148), $\mathcal{S}$ and $\mathcal{T}$ are partially isomorphic.   ∎

If there were guaranteed to be partially isomorphic, countable substructures of $\mathcal{T}$, we would gladly restrict our definition of solvability$_g$ to this case. There is, however, no such guarantee, as shown by the next lemma.

(157) LEMMA: Suppose that **Sym** contains countably many unary predicates. Then there is a structure$_g$ $\mathcal{T}$ with cardinality of the continuum such that $I_1(\mathcal{S}, \mathcal{T}) = \emptyset$, for all countable substructures $\mathcal{S}$ of $\mathcal{T}$.

Proof: Let the unary predicates be enumerated as P_i, $i \in N$. Let structure$_g$ $\mathcal{T}$ be such that $|\mathcal{T}| = \mathrm{pow}(N)$, and for all $i \in N$, $P_i^{\mathcal{T}} = \{X \subseteq N \mid i \in X\}$. To prove the lemma, let countable substructure $\mathcal{S}$ of $\mathcal{T}$ be given. Since there are uncountably many subsets of N, we may choose $X_0 \subseteq N$ such that:

(158) for all $X \in |\mathcal{S}|$ there is $i \in N$ such that $X \in P_i^\mathcal{S}$ iff $X_0 \notin P_i^\mathcal{T}$.

Directly from Definition (148), (158) is enough to show that $I_1(\mathcal{S}, \mathcal{T}) = \emptyset$. ∎

In the absence of partially isomorphic, countable substructures of $\mathcal{T}$ there may be a greatest ordinal α that satisfies (153)a. This situation is illustrated in extreme form by the preceding lemma: for all countable substructures $\mathcal{S}$ of $\mathcal{T}$, the maximal ordinal α for which $I_\alpha(\mathcal{S}, \mathcal{T}) \neq \emptyset$ is $\alpha = 0$. When such a maximal α exists (be it 0 or a greater ordinal), the class of countable substructures $\mathcal{S}$ of $\mathcal{T}$ with $I_\alpha(\mathcal{S}, \mathcal{T}) \neq \emptyset$ can be considered to be maximally similar to $\mathcal{T}$. It is thus appropriate that Definition (153) designates precisely this class as relevant to success [as easily seen from Lemma (150)b]. It would thus be a satisfying state of affairs if the two alternatives discussed so far exhausted the possibilities. In this case, either $\mathcal{T}$ would possess a partially isomorphic, countable substructure, or there would be a maximal ordinal α satisfying (153)a. Definition (153) could then be interpreted as isolating just the countable substructures of $\mathcal{T}$ that are most similar to $\mathcal{T}$. However, the next lemma shows that matters are not so simple; neither alternative need obtain. Recall that ω_1 denotes the least, uncountable ordinal.

(159) LEMMA: Suppose that **Sym** is limited to a binary predicate. Then there is structure$_g$ $\mathcal{T}$ such that:

(a) for all countable substructures $\mathcal{S}$ of $\mathcal{T}$, $I_{\omega_1}(\mathcal{S}, \mathcal{T}) = \emptyset$;

(b) for all ordinals $\alpha < \omega_1$, there is a countable substructure $\mathcal{S}$ of $\mathcal{T}$ with $I_\alpha(\mathcal{S}, \mathcal{T}) \neq \emptyset$.

Proof: It is proven in [Fraissé, 1972] that:

(160) (a) for all ordinals α, if $0 < \alpha < \omega_1$ then $I_{\omega_1}(\alpha, \omega_1) = \emptyset$;

(b) for all ordinals $\alpha < \omega_1$, $I_\alpha(\omega^\alpha, \omega_1) \neq \emptyset$.

We take $\mathcal{T} = (\omega_1, \in)$ to witness the lemma.

Let countable substructure $\mathcal{S}$ of $\mathcal{T}$ be given. Since $\mathcal{S}$ is isomorphic to a countable ordinal, $I_{\omega_1}(\mathcal{S}, \mathcal{T}) = \emptyset$ by (160)a. This proves (159)(a). As for (159)(b), it follows immediately from (160)b and the fact that for all ordinals $\alpha < \omega_1$, ω^α is a countable substructure of ω_1. ∎

In summary, when $\mathcal{T}$ possesses "most similar," countable substructures, Definition (153) requires the scientist to behave appropriately on just this class. Otherwise, it is left to the scientist to choose the degree of similarity against which her performance will be measured.

3.7.4   The New Paradigm as a Generalization of the Original One

The first-order$_g$ paradigm is intended as a natural generalization of the original
one. At the least, such intention requires the two paradigms to coincide when
domains are limited to the countable case. The following proposition shows
this desideratum to be fulfilled. Recall that a problem **P** is the special kind of
problem$_g$ in which every $S \in \bigcup \mathbf{P}$ is countable.

(161)   PROPOSITION: Let problem **P** be given. Then **P** is solvable$_g$ if and only
if **P** is solvable.

Proof: Suppose that scientist$_g$ Ψ solves$_g$ **P**. Let scientist Ψ' be such that
for all $\sigma \in SEQ$, $\Psi'(\sigma) = \{S \in \Psi(\sigma) \mid S \text{ is countable}\}$. By Lemma (150)d,
$I_\alpha(S, S) \neq \emptyset$ for all ordinals α, so it follows directly from Definition (153)
that Ψ' solves **P**.

   Conversely, suppose that **P** is solvable, and let (countable) $\mathfrak{T} \in \bigcup \mathbf{P}$ be
given. By Lemma (155) choose ordinal α_0 such that for all countable sub-
structures S of $\mathfrak{T}$, $I_{\alpha_0}(S, \mathfrak{T}) \neq \emptyset$ if and only if S is partially isomorphic to $\mathfrak{T}$.
Since $\mathfrak{T}$ is countable, Lemma (150)d implies that for all countable substruc-
tures S of $\mathfrak{T}$, $I_{\alpha_0}(S, \mathfrak{T}) \neq \emptyset$ if and only if S is isomorphic to $\mathfrak{T}$. Obviously,
$I_{\alpha_0}(\mathfrak{T}, \mathfrak{T}) \neq \emptyset$, hence the first clause of Definition (153) is satisfied. The second
clause follows immediately from the choice of α_0 and the hypothesis that **P** is
solvable.   ∎

   The proof of the proposition reveals another way the first-order$_g$ paradigm
generalizes the original one. From the proof we see that if scientist$_g$ Ψ solves$_g$
problem **P**, then the scientist that results from restricting Ψ's outputs to (countable) structures solves **P**.

3.7.5   Problems$_g$ of Form $(T, \{\theta_0, \ldots, \theta_n\})$

Many of the results obtained in earlier sections carry over to the first-order$_g$
paradigm, at least in attenuated form. The theorem now to be discussed provides an illustration. To state it, we introduce a familiar notation, generalizing
in a straightforward way Definition (53). Note that for $T \subseteq \mathcal{L}_{sen}$, the class of
structures$_g$ that satisfy T is denoted by $MOD(T)_g$.

(162)   DEFINITION: Let problem$_g$ **P**, $T \subseteq \mathcal{L}_{sen}$, and $\theta_0 \ldots \theta_n \in \mathcal{L}_{sen}$ be given.
We say that **P** has the form $(T, \{\theta_0, \ldots, \theta_n\})$ just in case:

(a)   for every model$_g$ S of T there is exactly one $i \in \{0 \ldots n\}$ such that $S \models \theta_i$;

(b) $\mathbf{P} = \{MOD(T \cup \{\theta_i\})_g \mid 0 \leq i \leq n\}$.

The following theorem provides a finitary condition for the solvability$_g$ of problems$_g$ of the form $(T, \{\theta_0, \ldots, \theta_n\})$, provided that **Sym** is finite. It is the same condition seen earlier for the countable case (Section 3.3.2).

(163)   THEOREM: Suppose that **Sym** is finite. Let problem$_g$ of form $(T, \{\theta_0, \ldots, \theta_n\})$ be given. Then $(T, \{\theta_0, \ldots, \theta_n\})$ is solvable$_g$ if and only if for every $i \leq n$, θ_i is equivalent in T to an $\exists\forall$ sentence.

The theorem follows directly from Theorem (55) and the following lemma, of interest in its own right.

(164)   LEMMA: Suppose that **Sym** is finite. Let enumeration $\{X_i \mid i \in N\}$ of subsets of $pow(\mathcal{L}_{sen})$ be given. Then the problem $\{\bigcup_{E \in X_i} MOD(E) \mid i \in N\}$ is solvable if and only if the problem$_g$ $\{\bigcup_{E \in X_i} MOD(E)_g \mid i \in N\}$ is solvable$_g$.

Proof: For the right-to-left direction, suppose that $\{\bigcup_{E \in X_i} MOD(E)_g \mid i \in N\}$ is solvable$_g$. Then trivially, $\{\bigcup_{E \in X_i} MOD(E) \mid i \in N\}$ is solvable$_g$. So $\{\bigcup_{E \in X_i} MOD(E) \mid i \in N\}$ is solvable by Proposition (161).

For the other direction, suppose that scientist Ψ solves $\{\bigcup_{E \in X_i} MOD(E) \mid i \in N\}$. Let $i \in N$, $E \in X_i$ and $\mathcal{T} \in MOD(E)_g$ be given. Observe that Ψ is also a scientist$_g$. So by Definition (153) and the fact that $\bigcup_{E \in X_i} MOD(E) \subseteq \bigcup_{E \in X_i} MOD(E)_g$ it suffices to show:

(165)   (a)   there is countable substructure $\mathcal{S}$ of $\mathcal{T}$ with $I_\omega(\mathcal{S}, \mathcal{T}) \neq \emptyset$;

   (b)   for all countable substructures $\mathcal{S}$ of $\mathcal{T}$ with $I_\omega(\mathcal{S}, \mathcal{T}) \neq \emptyset$ and for all environments e for $\mathcal{S}$, $\emptyset \neq \Psi(e[k]) \subseteq \bigcup_{E \in X_i} MOD(E)$, for cofinitely many k.

By Lemma (150)c and the assumption on **Sym**, (165) follows directly from:

(166)   (a)   there is countable substructure $\mathcal{S}$ of $\mathcal{T}$ that is elementarily equivalent to $\mathcal{T}$;

   (b)   for all countable substructures $\mathcal{S}$ of $\mathcal{T}$ that are elementarily equivalent to $\mathcal{T}$, and for all environments e for $\mathcal{S}$, $\emptyset \neq \Psi(e[k]) \subseteq \bigcup_{E \in X_i} MOD(E)$, for cofinitely many k.

The downward Löwenheim-Skolem theorem implies (166)a. We obtain (166)b from the fact that Ψ solves $\bigcup_{E \in X_i} MOD(E)$, since the latter class is closed under elementarily equivalence [with respect to (countable) structures].   ∎

As an application of the lemma, we have the following.

(167) *Example:*   Suppose that **Sym** contains just two binary function symbols. For $k \in N$ either 0 or prime, let proposition$_g$ P_k be the collection of all fields$_g$ of characteristic k in this vocabulary. Then Lemma (164) and Exercise (68) imply that $\{P_k \mid k$ is either 0 or prime$\}$ is solvable$_g$. Exercise (69) can be transposed to the present setting in the same way.

Theorem (163) generalizes Theorem (55), but only in attenuated form because of the limitation to finite vocabulary. That this limitation is essential is shown by the following proposition and its corollary.

(168) PROPOSITION: Suppose that **Sym** consists of a denumerable set of unary predicates. Then there exists two structures$_g$ $\mathcal{S}$ and $\mathcal{T}$ and a universal sentence θ such that:

(a) $\mathcal{S} \models \theta$ and $\mathcal{T} \models \neg\theta$;

(b) $\{\{\mathcal{S}\}, \{\mathcal{T}\}\}$ is not solvable$_g$.

Proof: Let Q and P_i, $i \in N$ enumerate the unary predicates of **Sym**. Let structures$_g$ $\mathcal{S}$ and $\mathcal{T}$ be defined as follows.

(169) (a) $|\mathcal{S}| = |\mathcal{T}| = \text{pow}(N)$;

 (b) $Q^{\mathcal{S}} = \text{pow}(N)$;

 (c) $Q^{\mathcal{T}} = \{\emptyset\}$;

 (d) for all $i \in N$, $P_i^{\mathcal{S}} = P_i^{\mathcal{T}} = \{X \subseteq N \mid i \in N\}$.

From (169)a–c we have that $\mathcal{S} \models \forall x\, Qx$ and $\mathcal{T} \models \neg\forall x\, Qx$. From (169)d it follows that uncountably many elements are distinguished by the P_i's in both $\mathcal{S}$ and $\mathcal{T}$. It follows easily that for all (countable) structures $\mathcal{U}$, $I_1(\mathcal{S}, \mathcal{U}) = \emptyset$ and $I_1(\mathcal{T}, \mathcal{U}) = \emptyset$. This with Lemma (150)a,b implies that for all (countable) structures $\mathcal{U}$ and all ordinals α:

(170) $I_\alpha(\mathcal{S}, \mathcal{U}) \neq \emptyset$ iff $\alpha = 0$, and $I_\alpha(\mathcal{T}, \mathcal{U}) \neq \emptyset$ iff $\alpha = 0$.

Let (countable) structure $\mathcal{U}$ be defined as follows.

(a) $|\mathcal{U}| = \{\emptyset\}$;

(b) $Q^{\mathcal{U}} = \{\emptyset\}$;

(c) for all $i \in N$, $P_i^{\mathcal{U}} = \emptyset$.

Trivally, $\mathcal{U}$ is a substructure of $\mathcal{S}$ and $\mathcal{U}$ is a substructure of $\mathcal{T}$. Suppose for a contradiction that scientist$_g$ Ψ solves$_g$ $\{\{\mathcal{S}\}, \{\mathcal{T}\}\}$. Then Definition (153) together with (170) implies that for every environment e for $\mathcal{U}$, $\Psi(e[k])$ $= \{\mathcal{S}\}$ for cofinitely many k, and $\Psi(e[k]) = \{\mathcal{T}\}$ for cofinitely many k, contradiction.   ∎

As an easy consequence of Theorem (55) and Proposition (168), we obtain:

(171)   COROLLARY: Suppose that **Sym** consists of a denumerable set of unary predicates. Then there exists $\theta \in \mathcal{L}_{sen}$ such that:

(a) the problem $(\emptyset, \{\theta, \neg\theta\})$ is solvable;

(b) the problem$_g$ $(\emptyset, \{\theta, \neg\theta\})$ is not solvable$_g$.

3.7.6   Solvability$_g$ under Bounded Similarity

As discussed in Section 3.7.3, to solve$_g$ a problem$_g$ **P** requires proper behavior on environments for all countable substructures of $\mathcal{T} \in \bigcup \mathbf{P}$ that are sufficiently similar to $\mathcal{T}$. Choice of a high criterion of similarity can often lighten the scientist's load, by restraining the class of substructures for which she is responsible. It turns out, however, that there is a fixed level of similarity that is always high enough to solve$_g$ **P**, provided that all $\mathcal{T} \in \bigcup \mathbf{P}$ possess countable substructures with such similarity to $\mathcal{T}$. To state the matter precisely, we rely on the following definition.

(172)   DEFINITION: Let ordinal α and problem$_g$ **P** be given.

(a) We say that **P** is α-*saturated* just in case for every $\mathcal{T} \in \bigcup \mathbf{P}$ there is a countable substructure $\mathcal{S}$ of $\mathcal{T}$ such that $I_\alpha(\mathcal{S}, \mathcal{T}) \neq \emptyset$.

(b) We say that scientist$_g$ Ψ α-*solves*$_g$ **P** just in case:

(i) **P** is α-saturated, and

(ii) for all $P \in \mathbf{P}$, all $\mathcal{T} \in P$, all countable substructures $\mathcal{S}$ of $\mathcal{T}$ with $I_\alpha(\mathcal{S}, \mathcal{T}) \neq \emptyset$, and all environments e for $\mathcal{S}$, $\emptyset \neq \Psi(e[k]) \subseteq P$ for cofinitely many k.

In this case, we say that **P** is α-*solvable*$_g$.

So, α-solvability$_g$ is a special case of solvability$_g$. The next theorem shows, however, that 2ω-solvability$_g$ is equivalent to solvability$_g$ *tout court* for the class of 2ω-saturated problems$_g$.

(173)   THEOREM: Let problem$_g$ **P** be 2ω-saturated. Then **P** is solvable$_g$ if and only if **P** is 2ω-solvable$_g$.

Before proving the theorem let us apply it to the countable context.

(174)   COROLLARY: Let problem **P** be given. Then **P** is solvable if and only if **P** is 2ω-solvable$_g$.

Proof: Since **P** is a problem, every $S \in \bigcup P$ is countable. It follows immediately from Definition (172)a that **P** is α-saturated for all ordinals α. So the corollary follows directly from Proposition (161) and Theorem (173).   ∎

Proof of the theorem hinges on the following lemma.

(175)   LEMMA: Let structures$_g$ S and $\mathcal{T}$ be such that $I_{2\omega}(S, \mathcal{T}) \neq \emptyset$. Let X be any collection of formulas all of whose free variables are drawn from the same finite set. Then X is satisfiable in S iff X is satisfiable in $\mathcal{T}$.

Proof: Let $n \in N$ and variables $x_0 \dots x_n$ be such that for all $\varphi \in X$, $Var(\varphi) \subseteq \{x_0 \dots x_n\}$. Using standard techniques to eliminate constants and function symbols (see [Ebbinghaus *et al.*, 1994]), there exists symbol set **Sym*** which consists exclusively of predicates, **Sym***-structures $S^\star$ and $\mathcal{T}^\star$, and set $X^\star$ of **Sym***-formulas such that:

(176) (a) $|S| = |S^\star|$, $|\mathcal{T}| = |\mathcal{T}^\star|$, and $PI(S, \mathcal{T}) = PI(S^\star, \mathcal{T}^\star)$;

  (b) all free variables occuring in any member of $X^\star$ are drawn from $\{x_0 \dots x_n\}$;

  (c) X is satisfiable in S iff $X^\star$ is satisfiable in $S^\star$;

  (d) X is satisfiable in $\mathcal{T}$ iff $X^\star$ is satisfiable in $\mathcal{T}^\star$.

From (176)a and the hypothesis that $I_{2\omega}(S, \mathcal{T}) \neq \emptyset$, we infer that $I_{2\omega}(S^\star, \mathcal{T}^\star) \neq \emptyset$. From this and (176)b-d, it follows that we can, without loss of generality, suppose that **Sym** consists exclusively of predicates.

Now suppose that X is satisfiable in S. Choose $a_0 \dots a_n \in |S|$ such that for all $\varphi \in X$, $S \models \varphi[a_0/x_0 \dots a_n/x_n]$. Since $I_{2\omega}(S, \mathcal{T}) \neq \emptyset$, choose $p \in I_{\omega+n+1}(S, \mathcal{T})$. By Definition (148), p can be extended to $q \in I_\omega(S, \mathcal{T})$ with domain$(q) \supseteq \{a_0 \dots a_n\}$. Since the quantifier ranks of the formulas in X are all finite, it follows immediately from [Ebbinghaus *et al.*, 1994, Lemma XII.3.2] that $\mathcal{T} \models \varphi[q(a_0)/x_0 \dots q(a_n)/x_n]$, for all $\varphi \in X$.   ∎

Proof of Theorem (173):   We first note the following, immediate consequence of Definition (51) and Lemma (175).

(177)   Let structures$_g$ S and $\mathcal{T}$ be such that $I_{2\omega}(S, \mathcal{T}) \neq \emptyset$. Then S and $\mathcal{T}$ satisfy the same extended-$\exists\forall$ sentences of $\mathcal{L}_{\omega_1\omega}$.

Now we prove the theorem. Trivially, if $\mathbf{P}$ is 2ω-solvable$_g$ then $\mathbf{P}$ is solvable$_g$. For the other direction, suppose that scientist$_g$ Ψ solves$_g$ problem$_g$ $\mathbf{P}$. Since *SEQ* is countable, range(Ψ) is countable, so it follows immediately that $\mathbf{P}$ is countable. Because Ψ solves$_g$ $\mathbf{P}$, it follows from Definition (153) and Lemma (150)b that for each $P \in \mathbf{P}$ and $\mathcal{T} \in P$ we may choose a countable substructure $\mathcal{T}^\star$ of $\mathcal{T}$ such that: for some $\alpha \geq 2\omega$, $I_\alpha(\mathcal{T}^\star, \mathcal{T}) \neq \emptyset$ and for all environments e for $\mathcal{T}^\star$, $\emptyset \neq \Psi(e[k]) \subseteq P$ for cofinitely many k. For all $P \in \mathbf{P}$, set $P^\star = \{\mathcal{T}^\star \mid \mathcal{T} \in P\}$. Set $\mathbf{P}^\star = \{P^\star \mid P \in \mathbf{P}\}$. It follows immediately that $\mathbf{P}^\star$ is a problem (that is, the $P^\star \in \mathbf{P}^\star$ are disjoint sets of countable structures), and that $\mathbf{P}^\star$ is solvable. By Corollary (52), for all $P \in \mathbf{P}$ there is an extended-$\exists\forall$ sentence of $\mathcal{L}_{\omega_1\omega}$ that is true in all $\mathcal{T}^\star \in P^\star$, and false in all $\mathcal{T}^\star \in \bigcup \mathbf{P}^\star - P^\star$. So it follows from (177) and the choice of the $\mathcal{T}^\star$ that:

(178)   for all $P \in \mathbf{P}$ there is an extended-$\exists\forall$ sentence of $\mathcal{L}_{\omega_1\omega}$ that is true in all $\mathcal{T} \in P$, and false in all $\mathcal{T} \in \bigcup \mathbf{P} - P$.

For all $\mathcal{T} \in \bigcup \mathbf{P}$, define $P(\mathcal{T})$ to be the set of countable substructures $\mathcal{S}$ of $\mathcal{T}$ with $I_{2\omega}(\mathcal{S}, \mathcal{T}) \neq \emptyset$. For all $P \in \mathbf{P}$, set $P^\circ = \bigcup\{P(\mathcal{T}) \mid \mathcal{T} \in P\}$, and set $\mathbf{P}^\circ = \{P^\circ \mid P \in \mathbf{P}\}$. It follows from (177) and (178) that for all $P \in \mathbf{P}$ there is an extended-$\exists\forall$ sentence of $\mathcal{L}_{\omega_1\omega}$ that is true in all $\mathcal{T} \in P^\circ$, and false in all $\mathcal{T} \in \bigcup \mathbf{P}^\circ - P^\circ$. The $P^\circ \in \mathbf{P}^\circ$ are thus disjoint, so $\mathbf{P}^\circ$ is a problem. Moreover, since $\mathbf{P}$ is countable, so is $\mathbf{P}^\circ$. It follows from another application of Corollary (52) that $\mathbf{P}^\circ$ is solvable. With Definition (153), we conclude that $\mathbf{P}$ is 2ω-solvable$_g$. ∎

3.7.7   Exercises

(179)   EXERCISE: Structures$_g$ $\mathcal{S}$ and $\mathcal{T}$ are said to be *finitely isomorphic in the standard sense* just in case there is a sequence $\{J_n \mid n \in N\}$ of nonempty sets of partial isomorphisms from $\mathcal{S}$ to $\mathcal{T}$ such that for all $n \in N$ the following holds:

(a) For every $f \in J_{n+1}$ and $a \in |\mathcal{S}|$ there is $g \in J_n$ such that g extends f and $a \in \mathrm{domain}(g)$.

(b) For every $f \in J_{n+1}$ and $a \in |\mathcal{T}|$ there is $g \in J_n$ such that g extends f and $a \in \mathrm{domain}(g)$.

Show that $\mathcal{S}$ and $\mathcal{T}$ are finitely isomorphic in the standard sense iff they are finitely isomorphic in the sense of Definition (148).

(180)   EXERCISE: Structures$_g$ $\mathcal{S}$ and $\mathcal{T}$ are said to be *partially isomorphic in the standard sense* just in case there is a nonempty set J of partial isomorphisms from $\mathcal{S}$ to $\mathcal{T}$ such that the following holds.

(a) For every $f \in J$ and $a \in |\mathcal{S}|$ there is $g \in J$ such that g extends f and $a \in \text{domain}(g)$.

(b) For every $f \in J$ and $a \in |\mathcal{T}|$ there is $g \in J$ such that g extends f and $a \in \text{domain}(g)$.

Show that $\mathcal{S}$ and $\mathcal{T}$ are partially isomorphic in the standard sense iff they are partially isomorphic in the sense of Definition (148).

3.8   Notes

Let us follow up an idea due to [Kelly & Glymour, 1992], and increase the information offered in environments. For this purpose, choose some $n \geq 1$, and allow environments to include all the Σ_n formulas satisfied in a given structure and full assignment. In this new paradigm, the domain of scientists is modified accordingly, but the other definitions of the first-order paradigm may be preserved. We note that after an obvious change in the definition of tip-off, Theorem (40) remains true. That is, solvability using the richer environments is characterized essentially as before.

As already suggested by Corollary (52), the first-order paradigm can be generalized in various ways within the framework of abstract model theory [Barwise & Feferman, 1985]. A step in this direction is reported in [Osherson *et al.*, 1991b, Sec. 4].

An alternative approach to theoretical vocabulary is discussed in [Gaifman *et al.*, 1990].

In a sense somewhat different from that explored in Section 3.4.3, oracles have been the subject of investigation within Formal Learning Theory. See [Gasarch & Pleszkoch, 1989, Gasarch *et al.*, 1992] and [Stephan, 1996a, Stephan, 1996b].

As mentioned in Section 1.3.3, the PAC paradigm offers a perspective on inquiry that is substantially different from the one developed in this book. Nonetheless, there are already models of learning that combine PAC with elements of the first-order paradigm. See [Osherson *et al.*, 1989, Osherson *et al.*, 1991a, Maas & Turán, 1996].

A predecessor to Theorem (40) appears in [Osherson & Weinstein, 1989]. Earlier versions of Theorems (55) and (97), along with Exercise (104) appear in [Osherson *et al.*, 1991b]. We note that the "*r.e.*" hypothesis in Theorem (97) can be weakened to "Σ_2."

Much of Section 3.6 appears in preliminary form in [Osherson *et al.*, 1996]. Notice how our conception of probabilistic inquiry differs from that of [Gaifman & Snir, 1982]. Whereas we attach probabilities to the elements of each model, Gaifman and Snir attach them to collections of structures.

Theorem (122) says (roughly) that the solvability of a problem is equivalent to its probabilistic solvability with respect to "enough" distributions. An analogous result is proved for the numerical paradigm in [Angluin, 1988], and extended in [Montagna, 1996a].

See [Kelly, 1996, Ch. 13] for a penetrating and wide ranging discussion of Bayesianism and scientific inquiry. The latter work considers many issues not addressed here.

Section 3.6.9 skirted issues of computability, since it was shown that no Bayesian scientist (computable or not) solves $\mathbf{P}^*$. For discussion of the relation between Bayesianism and computable solvability, see [Osherson et al., 1988a, Juhl, 1993]. The matter is also taken up in [Howson & Urbach, 1993, pp. 434-6], but the important issue seems to be missed. What matters is not, as Howson and Urbach appear to believe, whether Bayesian behavior in some reasonable sense is always possible for computable scientists. The answer to this question is transparently negative inasmuch as the relevant probability distribution may be highly uncomputable. The poignant issue is whether Bayesian behavior *even when computationally possible* is always compatible with scientific discovery, when the latter, as well, is computationally possible. In at least one probabilistic set-up, the answer to this question appears to be negative. Indeed, it is shown in [Osherson et al., 1988a] that there are problems which (a) are computably solvable, (b) admit Bayesian behavior by a computable scientist, but (c) are not solved with probability 1 by any computable, Bayesian scientist. Clause (b) is evidently central to any evaluation of computable Bayesianism since there is no point criticizing a method where it cannot be applied. A related analysis of the use of probability calculations by computable scientists appears in [Martin & Osherson, 1995a].

4 Inquiry via Belief Revision

4.1 Belief Revision

4.1.1 The Search for Canonical Strategies

The role of fortune in science is apparent to everyone, and has led some to nihilism when questions arise about justifying a method of inquiry (e.g., [Feyerabend, 1975]). The chaotic face of successful science, however, is due in part to the hazards of data collection and to the mental limits of the scientists faced with the results. In contrast, the idealized paradigms of inductive logic are free of such factors, so they make it easier to compare the merits of different scientific strategies, with a view to justifying one or more of them. In this final chapter of the book we define a narrow class of strategies for choosing hypotheses, and investigate their utility for empirical inquiry.

To clarify the issues, let us take a scientific *strategy* to be no more or less than a class of scientists (this usage is familiar from Section 2.4.1). To justify reliance on a given strategy, it must be demonstrated that scientists within the strategy have sufficiently ample competence. To be more precise, we say that a strategy is *canonical* for a class C of problems (relative to a given choice of inductive paradigm) just in case every solvable problem in C is solved by some scientist belonging to the strategy. To illustrate, in Section 3.6.9 we saw that the class of Bayesian scientists is not canonical in the first-order paradigm for problems of the form $(T, \{\theta, \neg\theta\})$. We will not let this fact discourage us from seeking other canonical strategies, which indeed is the focus of the present chapter.

So that there is no ambiguity, we confirm that it is the nonprobabilistic version of the first-order paradigm that concerns us henceforth, that is, the paradigm defined in Section 3.1.3. In particular, we leave aside the developments in Section 3.6 (concerning probabilistic environments), as well as the developments in Section 3.7 (concerning uncountable structures$_g$).

4.1.2 Inquiry via the Revision of Belief States

Not every canonical strategy holds much interest. After all, the entire class of scientists is already canonical for the collection of all problems. So, our goal is not merely to define a canonical strategy, but to define one that has attractive properties. The properties of concern to us bear on the rationality of the scientist's choice of hypotheses. Specifically, we shall attempt to exhibit inquiry within the first-order paradigm as a process of rational belief revision in the light of data, starting from a background theory. In order to give substance to this idea, we conceive a scientist as initiating inquiry with a set X of formulas that represent her provisional beliefs prior to examining the environment. The arrival of data σ will then serve to modify X according to some fixed scheme of belief revision. The resulting set of beliefs is denoted $X \dotplus \sigma$, and signifies the impact of σ on X. We will be interested in revision schemes $\dotplus$ that are conducive to successful inquiry. To the extent that the scheme is justifiable on intuitive grounds, successful scientists based on it accredit the idea that inquiry can be carried out on a rational basis.

Crucial to the foregoing programme is the class of belief revision operators $\dotplus$. They will be introduced in the present section. Our conception of belief revision is indebted to the influential work of [Alchourrón et al., 1985, Gärdenfors, 1988], although we set things up somewhat differently. In particular, our approach may be considered a reformulation of [Hansson, 1994], which generalizes [Alchourrón & Makinson, 1985].

Throughout the chapter a set of formulas will be conceived as a potential *state of belief*, and such states will evolve with the accumulation of data. For now, deductive closure is not imposed on belief states; they are arbitrary subsets of $\mathcal{L}_{form}$. The effect on inquiry of deductive closure is the topic of Section 4.5, below.

Belief revision is represented as the application of a function of two arguments, belief state and data, yielding a new belief state. In conformity with the analysis in [Levi, 1980], revision is carried out in two steps. First, the original belief state is shrunk (if necessary) so as to be compatible with the data; then the data are added in. The first modification is called "belief contraction," and is the crucial step in the procedure. So we first consider contraction, followed by revision.

4.1.3   Contraction

If new data are consistent with one's current belief state, the natural reaction is to adopt them as new beliefs, leaving the rest untouched. This assumes that the data are credible, which will be our standing assumption.[1] Indeed, new data will be taken to be quasi-certain, meriting inclusion in one's stock of beliefs even at the expense of displacing other convictions. Such displacement calls for intelligent choice since there will in general be alternative subsets of the current belief state that can be sacrificed in favor of incorporating the data.

---

1. The assumption is of a piece with the conception of **Sym** as "observational" vocabulary (see the remarks in Section 3.3.4). A more realistic model would soften the impact of input from the environment.

Thus, to define the process of incorporating new information (namely, belief revision), we start with the process of abandoning old beliefs that contradict the new information (namely, belief contraction). Central to contraction is the following pair of definitions. They give substance to the idea of abandoning certain beliefs rather than others.

(1) DEFINITION: [Gärdenfors, 1988] Let $\phi \in \mathcal{L}_{form}$ and $B \subseteq \mathcal{L}_{form}$ be given. By a *maximal subset of B that fails to imply* ϕ is meant any subset B' of B with the following properties:

(a) $B' \not\models \phi$,

(b) there is no X with $B' \subset X \subseteq B$ and $X \not\models \phi$.

The class of all maximal subsets of B that fail to imply ϕ is denoted $B \perp \phi$. In particular, if $\models \phi$ then $B \perp \phi = \emptyset$, and $\bigcap(B \perp \phi) = B$.[2]

It is easy to establish the following lemma, which shows that $\bigcap(B \perp \phi)$ is the result of removing from B every formula that figures in a nonredundant proof of ϕ.

(2) LEMMA: $\bigcap(B \perp \phi) = \{\psi \in B \mid (\forall D \subseteq B)(\text{if } D \cup \{\psi\} \models \phi \text{ then } D \models \phi)\}$.

When contracting B to avoid implying ϕ, the suppression of formulas outside $\bigcap(B \perp \phi)$ seems capricious; so the contraction should leave behind at least $\bigcap(B \perp \phi)$. On the other hand, enough formulas must be removed from B to relieve the implication of ϕ (unless ϕ is valid). These two boundary conditions are the only ones we impose on contraction functions.

(3) DEFINITION: A mapping $\dot{-}$ from $\text{pow}(\mathcal{L}_{form}) \times \mathcal{L}_{form}$ to $\text{pow}(\mathcal{L}_{form})$ is a *contraction function* just in case for all $B \subseteq \mathcal{L}_{form}$ and $\phi \in \mathcal{L}_{form}$,

(a) $\bigcap(B \perp \phi) \subseteq B \dot{-} \phi \subseteq B$,

(b) if $\not\models \phi$ then $B \dot{-} \phi \not\models \phi$.

4.1.4 Special Kinds of Contraction

The class of contraction functions is broad enough to embrace some questionable policies, so we are led to define more respectable subsets. A familiar idea is to effect minimal change in a belief state, in the sense of abandoning no more

---

2. That is, following standard practice, the intersection over an empty field of sets is taken to be the universal set, in this case, B. In contrast, notice that if $B = \{\phi\}$ with $\not\models \phi$, then $B \perp \phi = \{\emptyset\} \neq \emptyset$, and $\bigcap(B \perp \phi) = \emptyset$.

than is necessary to avoid contradiction with new data. One way of capturing this idea is the following.

(4)   DEFINITION: A contraction function $\dot{-}$ is *maxichoice* just in case for every $B \subseteq \mathcal{L}_{form}$ and invalid $\phi \in \mathcal{L}_{form}$, $B \dot{-} \phi \in B \perp \phi$.

Thus, maxichoice contraction functions leave behind a $\subseteq$-maximal subset of the original beliefs that does not imply the unwanted formula. Although maxichoice contraction is pleasingly conservative, it is still compatible with disorderly selection of theories. For example, $\dot{-}$ can be maxichoice even if $B \dot{-} \phi \neq B \dot{-} \phi'$ for logically equivalent formulas ϕ, ϕ'. A more systematic approach to theory selection is embodied in the class of contraction functions defined next.

(5)   DEFINITION: A contraction function $\dot{-}$ is *stringent* just in case there is a strict total ordering $\prec$ of $\mathrm{pow}(\mathcal{L}_{form})$ such that for all $B \subseteq \mathcal{L}_{form}$ and invalid $\phi \in \mathcal{L}_{form}$, $B \dot{-} \phi$ is the $\prec$-least subset of B that does not imply ϕ.

Thus, stringent contraction proceeds by choosing the first subset of B in a fixed list that fails to imply the unwanted formula ϕ. It is clear that, among other things, $B \dot{-} \phi = B \dot{-} \phi'$ for logically equivalent formulas ϕ, ϕ'. It remains to show, of course, that stringent contraction functions exist [it is not evident that the needed ordering can be constructed in such a way as to satisfy Definition (3)]. Prior to the existence proof, let us demonstrate that stringent contraction (if there is any) implies maxichoice contraction.

(6)   PROPOSITION: Every stringent contraction function is maxichoice.

Proof: Let stringent contraction function $\dot{-}$ be given; let strict total ordering $\prec$ of $\mathrm{pow}(\mathcal{L}_{form})$ be such that for all $B \subseteq \mathcal{L}_{form}$ and invalid $\phi \in \mathcal{L}_{form}$, $B \dot{-} \phi$ is the $\prec$-least subset of B that does not imply ϕ. Suppose for a contradiction that $\dot{-}$ is not maxichoice. Then there is $B \subseteq \mathcal{L}_{form}$ and invalid $\phi \in \mathcal{L}_{form}$ such that $B \dot{-} \phi$ is the $\prec$-least subset of B that does not imply ϕ, and $B \dot{-} \phi \notin B \perp \phi$. Hence we may choose $B' \in B \perp \phi$ with $B \dot{-} \phi \subset B'$. By hypothesis:

(7) $B \dot{-} \phi \prec B'$.

From Definition (3) we infer immediately that:

(8) $B' \dot{-} \forall x (x \neq x) = B'$.

But since $B \dot{-} \phi$ is a strict subset of B' that does not imply $\forall x (x \neq x)$, it follows from (7) that B' is not the $\prec$-least subset of B' that does not imply $\forall x (x \neq x)$. This contradicts (8) and the definition of $\prec$.   ∎

We see that an agent that contracts in stringent fashion satisfies many formal criteria of rationality. Such an agent has established transitive and connected preferences among theories, and contracts his beliefs so as to simultaneously minimize belief change and maximize theory-acceptability.

In the remainder of the present subsection we establish the existence of stringent contraction functions. For this purpose we use a construction that will be pivotal in later proofs. Here and elsewhere we rely on elementary facts about ordinal numbers (see [Enderton, 1977, Levy, 1979] for background).

(9) DEFINITION: Let ordinal κ be given. Let enumeration $\mathbf{S} = \{S_\alpha \mid \alpha < \kappa\}$ of members of $\text{pow}(\mathcal{L}_{form})$ also be given. We specify a function $\dot{-}_{\mathbf{s}}$: $\text{pow}(\mathcal{L}_{form}) \times \mathcal{L}_{form} \to \text{pow}(\mathcal{L}_{form})$. Let $B \subseteq \mathcal{L}_{form}$ and $\phi \in \mathcal{L}_{form}$ be given. If $\models \phi$ then $B \dot{-}_{\mathbf{s}} \phi = B$. Suppose $\not\models \phi$. Then $B \dot{-}_{\mathbf{s}} \phi$ is defined by induction over $\alpha < \kappa$. Specifically, for each $\alpha < \kappa$ we define the set Y^α, and put $B \dot{-}_{\mathbf{s}} \phi = \bigcup_{\alpha < \kappa} Y^\alpha$. The inductive definition of Y^α is as follows.

Suppose that $\alpha < \kappa$ and that Y^β is defined for all $\beta < \alpha$. Then $Y^\alpha = (\bigcup_{\beta < \alpha} Y^\beta) \cup S_\alpha$ if $S_\alpha \subseteq B$ and $(\bigcup_{\beta < \alpha} Y^\beta) \cup S_\alpha \not\models \phi$; otherwise $Y^\alpha = \bigcup_{\beta < \alpha} Y^\beta$.

Intuitively, if $\kappa > \omega$ then $B \dot{-}_{\mathbf{s}} \phi$ is constructed as follows. At stage 1 you examine S_0. If $S_0 \subseteq B$ and $S_0 \not\models \phi$, then you dump S_0 into $B \dot{-}_{\mathbf{s}} \phi$; otherwise, you leave $B \dot{-}_{\mathbf{s}} \phi$ empty. At stage 2 you examine S_1. If $S_1 \subseteq B$ and $S_0 \cup S_1 \not\models \phi$, then you dump S_1 into $B \dot{-}_{\mathbf{s}} \phi$; otherwise, you leave $B \dot{-}_{\mathbf{s}} \phi$ as it was after the preceding stage. Keep going in this way. At stage ω, you just retain all the formulas already dumped into $B \dot{-}_{\mathbf{s}} \phi$ at earlier stages. Denote by S that portion of $B \dot{-}_{\mathbf{s}} \phi$ constructed so far. At stage $\omega + 1$, you examine S_ω. Suppose that $S_\omega \subseteq B$ and $S \cup S_\omega \not\models \phi$. Then you dump S_ω into $B \dot{-}_{\mathbf{s}} \phi$; otherwise, leave $B \dot{-}_{\mathbf{s}} \phi$ as S. Keep going . . .

The important fact about this construction is that it yields a stringent contraction function, provided only that the enumeration $\{S_\alpha \mid \alpha < \kappa\}$ contains all singleton sets $\{\varphi\}$, $\varphi \in \mathcal{L}_{form}$. Precisely:

(10) PROPOSITION: Let ordinal κ be given. Let enumeration $\mathbf{S} = \{S_\alpha \mid \alpha < \kappa\}$ of members of $\text{pow}(\mathcal{L}_{form})$ contain all singleton sets. Then $\dot{-}_{\mathbf{s}}$ is a stringent contraction function.

Proof: Let $B \subseteq \mathcal{L}_{form}$ and invalid $\phi \in \mathcal{L}_{form}$ be given. An easy application of compactness shows that $B \dot{-}_{\mathbf{s}} \phi \not\models \phi$. Moreover, because $\{\chi\} \in \mathbf{S}$ for all $\chi \in \mathcal{L}_{form}$, the construction in Definition (9) ensures that for all $\chi \in B - (B \dot{-}_{\mathbf{s}} \phi)$, $(B \dot{-}_{\mathbf{s}} \phi) \cup \{\chi\} \models \phi$. So $B \dot{-}_{\mathbf{s}} \phi \in B \perp \phi$, hence $\dot{-}_{\mathbf{s}}$ is a (maxichoice) contraction function.

Define the following strict total order $\prec^{\#}$ on $\text{pow}(\mathcal{L}_{form})$. Given $X, Y \subseteq \mathcal{L}_{form}$, $X \prec^{\#} Y$ iff: there is $\alpha < \kappa$ such that $S_\alpha \subseteq X$, $S_\alpha \not\subseteq Y$, and for all $\beta < \alpha$, $S_\beta \subseteq X$ iff $S_\beta \subseteq Y$. Indeed, $\prec^{\#}$ is a strict total ordering of $\text{pow}(\mathcal{L}_{form})$ because **S** contains all singleton sets. We need to show that $B \doteq_s \phi$ is the $\prec^{\#}$-least subset C of $\mathcal{L}_{form}$ with $C \subseteq B$ and $C \not\models \phi$. For a contradiction suppose that $C \subseteq \mathcal{L}_{form}$ is such that:

(11) (a) $C \prec^{\#} B \doteq_s \phi$,

     (b) $C \subseteq B$,

     (c) $C \not\models \phi$.

By (11)a, let $\alpha < \kappa$ be such that $S_\alpha \subseteq C$, $S_\alpha \not\subseteq (B \doteq_s \phi)$, and for all $\beta < \alpha$, $S_\beta \subseteq C$ iff $S_\beta \subseteq B \doteq_s \phi$. By (11)b, $S_\alpha \subseteq B$. Let $Y^\alpha = \bigcup\{S_\beta \mid S_\beta \subseteq C \text{ and } \beta < \alpha\}$ $= \bigcup\{S_\beta \mid S_\beta \subseteq B \doteq_s \phi \text{ and } \beta < \alpha\}$. By (11)c, $Y^\alpha \cup S_\alpha \not\models \phi$. These facts along with the construction in Definition (9) imply that $S_\alpha \subseteq B \doteq_s \phi$, contradiction. ∎

So, there are plenty of stringent contraction functions.

4.1.5 Revision Defined from Contraction

Contraction now in hand, we turn to revision. Revision is the integration of new information into a belief state in such a manner as to avoid contradiction. Contradiction is avoided by means of prior contraction, which makes logical room for the information to be added.

Whereas contraction is conceived as a function from $\text{pow}(\mathcal{L}_{form}) \times \mathcal{L}_{form}$ to $\text{pow}(\mathcal{L}_{form})$, we take revision to be a function from $\text{pow}(\mathcal{L}_{form}) \times SEQ$ to $\text{pow}(\mathcal{L}_{form})$. The appearance of SEQ is motivated by our desire to apply belief dynamics to scientific inquiry (since SEQ represents new information that arrives as data). In conformity with the usual practice, we define revision via the "Levi identity" [Levi, 1980], as follows.

(12) DEFINITION: A mapping $\dotplus$ from $\text{pow}(\mathcal{L}_{form}) \times SEQ$ to $\text{pow}(\mathcal{L}_{form})$ is a *revision function* just in case there is a contraction function $\doteq$ such that for all $B \subseteq \mathcal{L}_{form}$ and $\sigma \in SEQ$,

$$B \dotplus \sigma = \begin{cases} B & \text{if } \sigma = \emptyset \\ (B \doteq \neg \bigwedge \sigma) \cup \text{range}(\sigma) & \text{otherwise.} \end{cases}$$

In this case, we say that $\doteq$ *underlies* $\dotplus$.

As suggested earlier, revision of B by nonvoid σ proceeds in two steps. First, B is contracted so as to be consistent with $\bigwedge \sigma$ (that is, B must not imply $\neg \bigwedge \sigma$). Then, range(σ) is added in. Of course, if B is already consistent with $\bigwedge \sigma$ then the revision process is reduced to its second step, and the new belief state is $B \cup \text{range}(\sigma)$.

We call revision function $\dotplus$ "maxichoice" if it arises via the Levi identity from some maxichoice contraction function $\dotminus$, and similarly for "stringent." Since every $\sigma \in SEQ$ is consistent, we obtain the following lemma immediately from Definitions (3) and (12).

(13)   LEMMA: Let revision function $\dotplus$ be given. Then for all $B \subseteq \mathcal{L}_{form}$ and $\sigma \in SEQ$:

(a) $B \dotplus \sigma \models \bigwedge \sigma$.

(b) $B \dotplus \sigma \subseteq B \cup \text{range}(\sigma)$.

(c) If $B \not\models \neg \bigwedge \sigma$ then $B \dotplus \sigma = B \cup \text{range}(\sigma)$.

(d) If σ is nonvoid then $B \dotplus \sigma$ is consistent.

4.1.6   Exercises

(14)   EXERCISE: A mapping $\ominus$ from $\text{pow}(\mathcal{L}_{form}) \times \mathcal{L}_{form}$ to $\text{pow}(\mathcal{L}_{form})$ is a *partial meet contraction* just in case for all $B \subseteq \mathcal{L}_{form}$ and $\phi \in \mathcal{L}_{form}$, $B \ominus \phi$ is the intersection of some subset of $B \perp \phi$. (The partial meet idea is developed in [Gärdenfors, 1988].) Show that the class of contraction functions properly includes the class of partial meet contractions.

(15)   EXERCISE: We say that contraction function $\dotminus$ *respects implication* just in case for all $B_0, B_1 \subseteq \mathcal{L}_{form}$ and $\phi \in \mathcal{L}_{form}$, if $B_0 \models B_1$ then $B_0 \dotminus \phi \models B_1 \dotminus \phi$. Show that no contraction function respects implication.

(16)   EXERCISE: A contraction function $\dotminus$ is *definite* just in case there is a strict total ordering $\prec$ of $\text{pow}(\mathcal{L}_{form})$ such that for all $B \subseteq \mathcal{L}_{form}$ and invalid $\phi \in \mathcal{L}_{form}$, $B \dotminus \phi$ is $\prec$-least with $\bigcap(B \perp \phi) \subseteq B \dotminus \phi \subseteq B$ and $B \dotminus \phi \not\models \phi$.

(a) ♣   Let $\dotminus$ be the contraction function that maps (B, ϕ) to $\bigcap(B \perp \phi)$. Show that $\dotminus$ is definite but not maxichoice.

(b) ♠   Observe that any stringent contraction function is maxichoice, definite [c.f., Proposition (6)]. Show that there are maxichoice, definite contraction functions that are not stringent.

(17)   EXERCISE: ♥   A contraction function $\dotminus$ is *rigid* just in case there is a strict total ordering $\prec$ of $\text{pow}(\mathcal{L}_{form})$ such that for all $B \subseteq \mathcal{L}_{form}$ and

invalid $\phi \in \mathcal{L}_{form}$, $B \doteq \phi$ is the $\prec$-least member of $B \perp \phi$. Observe that any maxichoice, definite contraction function is rigid [see Exercice (16) for definite contraction functions]. Show that there are rigid contraction functions that are not definite (hence not stringent).

(18)   EXERCISE: ♣   Call a contraction function $\doteq$ *unified* just in case for all B, $B' \subseteq \mathcal{L}_{form}$, and all invalid $\phi, \phi' \in \mathcal{L}_{form}$, $B \perp \phi = B' \perp \phi'$ implies $B \doteq \phi = B' \doteq \phi'$. (This definition is adapted from [Hansson, 1993b].) Observe that all definite and all rigid contraction functions are unified [see Exercises (16) and (17)]. In contrast, show that there are maxichoice, unified contraction functions that are not rigid (hence not definite).

(19)   EXERCISE: Suppose that contraction functions $\doteq$ and $\doteq'$ underlie the same revision function. Show that the restrictions of $\doteq$ and $\doteq'$ to $\text{pow}(\mathcal{L}_{form}) \times \{\neg \bigwedge \sigma \mid \sigma \in SEQ\}$ are identical.

(20)   EXERCISE: Let $B \subseteq \mathcal{L}_{form}$, $\phi \in \mathcal{L}_{form}$, and contraction function $\doteq$ be given. Suppose there is $C \subseteq B$ such that: (a) for every $\chi \in C$, $\chi \models \phi$, and (b) $B - C \not\models \phi$. Prove that $B \doteq \phi = B - C$.

(.^{1})   EXERCISE: Let $B \subseteq \mathcal{L}_{form}$, invalid $\phi \in \mathcal{L}_{form}$ and contraction function $\doteq$ given. Suppose that for all $\psi, \chi \in B$, either $\psi \models \chi$, or $\chi \models \psi$. Show that $B \doteq \phi = B - Y$, where $Y = \{\psi \in B \mid \psi \models \phi\}$.

4.2   Hypothesis Selection via Belief Revision

4.2.1   From Revision to Science

In the present section we consider how belief revision can be harnassed for the purpose of inquiry. The matter is simple in essence. Given a set B of formulas and a revision operator $\dotplus$, the function $\lambda\sigma \,.\, B \dotplus \sigma$ maps SEQ into the power set of $\mathcal{L}_{form}$.[3] Intuitively, $\lambda\sigma \,.\, B \dotplus \sigma$ converts data into belief; data are represented by SEQ and beliefs by belief states, i.e., sets of formulas. With such a function we associate the scientist that maps $\sigma \in SEQ$ into the class of models of $B \dotplus \sigma$, and call such scientists *revision-based* (the official definition is given in Section 4.2.3, below). Scientists of this kind are special in two ways. First, their outputs are not arbitrary propositions, but must instead be definable by a

---

3. Following standard usage, λ in the expression "$\lambda\sigma \,.\, B \dotplus \sigma$" indicates that σ serves as argument to the function, whereas B and $\dotplus$ are adjustable parameters.

subset of $\mathcal{L}_{form}$. In other words, each conjecture must have the form $MOD(B)$ for some $B \subseteq \mathcal{L}_{form}$ instead of being just any collection of structures. Second, revision-based scientists are not arbitrary mappings from SEQ to the definable propositions. Instead, they are constrained by the definition of revision [notably, Definitions (3) and (12) of the previous section]. Additional constraint arises if we require the $\dot{+}$ in $\lambda\sigma$. $B \dot{+} \sigma$ to denote a restricted kind of revision function, such as maxichoice or stringent. We will see later that revision-based scientists employing stringent revision functions are canonical for a broad class of problems; any solvable problem in the class is solved by a scientist of this kind.

In the preceding discussion we distinguished two special properties of revision-based scientists. The first requires output propositions to be definable classes of structures. Isolating just this property yields the set of *linguistic* scientists. They are "linguistic" in the sense that the propositions they announce are expressible in our first-order language. Thus, the revision-based scientists are a subset of the linguistic scientists, and information about the competence of the latter also provides insight about the former. So we start our discussion with the linguistic scientists. Revision-based scientists are introduced subsequently, and their competence characterized in general terms.

As a preliminary, let us underline the fact that our model of inquiry is still the first-order paradigm introduced in Section 3.1. Our previous notations and definitions thus remain in vigor. The innovation of the present chapter consists just in defining and investigating special subsets of scientists.

4.2.2 Linguistic Scientists

Recall from Section 3.1 that scientists issue classes of structures in response to members of SEQ. When the classes are definable from a subset of $\mathcal{L}_{form}$, the scientists are called "linguistic." Officially:

(22) DEFINITION: Scientist Ψ is *linguistic* just in case there is $\psi : SEQ \to$ pow($\mathcal{L}_{form}$) such that for all $\sigma \in SEQ$, $\Psi(\sigma)$ is defined iff $\psi(\sigma)$ is defined, and when both are defined $\Psi(\sigma) = MOD(\psi(\sigma))$. In this case, we say that ψ *underlies* Ψ.

To lighten the notation we sometimes identify linguistic scientists with the functions $\psi : SEQ \to$ pow($\mathcal{L}_{form}$) that underlie them. Thus, if Ψ is linguistic, we allow ourselves to write $\Psi(\sigma) = B$, for $B \subseteq \mathcal{L}_{form}$, in place of the circumlocution: "$\psi(\sigma) = B$, where ψ underlies Ψ."

Are linguistic scientists a canonical form of inquiry? Of course not. Suppose that proposition P includes no isomorphically closed class of structures (for example, P might be a singleton). Then no problem that includes P is solvable by linguistic scientist since no nonempty subset of P can be picked out by a set of formulas. [This is because $MOD(B)$ is isomorphically closed, for any $B \subseteq \mathcal{L}_{form}$.] And there is no lack of solvable problems that hold propositions including no isomorphically closed class of structures; the trivial problem $\{\{S\}\}$, for any structure S, is an example.

On the other hand, linguistic scientists turn out to be adapted to a broad class of problems that will be of interest in this chapter. Recall Definition 3.(59), which is repeated here for convenience.

(23) DEFINITION: Let $T \subseteq \mathcal{L}_{sen}$ be given. We say that problem $\mathbf{P}$ has the form $(T, \{P_0, P_1, \ldots\})$ just in case $\mathbf{P} = \{P_0, P_1, \ldots\}$ and $\bigcup \mathbf{P} = MOD(T)$.

It is noteworthy that linguistic scientists can solve any solvable problem of form $(T, \{P_0, P_1, \ldots\})$. Indeed, for this purpose it suffices to consider just the linguistic scientists with range $\text{pow}(\mathcal{L}_{sen})$ [instead of their usual range $\text{pow}(\mathcal{L}_{form})$]. We record this fact in the next proposition, which follows immediately from Proposition 3.(61).

(24) PROPOSITION: Suppose that solvable problem $\mathbf{P}$ has the form $(T, \{P_0, P_1, \ldots\})$. Then some linguistic scientist Ψ with $\text{range}(\Psi) \subseteq \text{pow}(\mathcal{L}_{sen})$ solves $\mathbf{P}$.

For problems of the form $(T, \{P_0, P_1, \ldots\})$, Proposition (24) shows that no generality is lost by depriving linguistic scientists of the right to announce open formulas. However, the following fact shows this is not true for problems of other forms.

(25) LEMMA: Suppose that $\mathbf{Sym}$ is limited to a binary predicate. Then there is a proposition P such that $P = MOD(B_0)$ for some $B_0 \subseteq \mathcal{L}_{form}$, but for every consistent $B_1 \subseteq \mathcal{L}_{sen}$, $MOD(B_1) \not\subseteq P$.

Proof: Suppose that R is the binary predicate of $\mathbf{Sym}$. Let P be the class of structures that are elementarily equivalent to $\langle N, < \rangle$ and that satisfy $\{v_i \neq v_j \mid i \neq j\} \cup \{Rv_{i+1}v_i \mid i \in N\}$. The compactness and Löwenheim-Skolem theorems guarantee that $P \neq \emptyset$. Observe that all the models of P possess an infinite descending R-chain. Since there is no such R-chain in $\langle N, < \rangle$, for all consistent $B_1 \subseteq \mathcal{L}_{sen}$, $MOD(B_1) \not\subseteq P$.   ∎

For any proposition P of the kind described in the lemma, the degenerate problem $\{P\}$ is solvable by a linguistic scientist but not by any linguistic scientist with range in $\text{pow}(\mathcal{L}_{sen})$.

4.2.3 Scientists Based on Revision

We now pass from linguistic scientists to their revision-based subset.

(26) DEFINITION: Let revision function $\dotplus$ and $B \subseteq \mathcal{L}_{form}$ be given. Then $\lambda\sigma \cdot B \dotplus \sigma$ is a linguistic scientist, which we qualify as *revision-based*.

The theory of belief revision invites us to view empirical inquiry as the process of revising one's background theory by evidence that accumulates in the environment. We take the revision-based scientists to give formal substance to this idea.

Revision-based scientists have some claim to the title "rational" inasmuch as Lemma (13) guarantees them to be total and consistent in a sense close to that of Definition 2.(30). Furthermore, if we choose the revision function $\dotplus$ to be stringent, the rationality of $\lambda\sigma \cdot B \dotplus \sigma$ seems irreprochable, as observed just after Proposition (6). As a consequence, we shall be particularly interested in the competence of scientists of this form. We stress that our enterprise would be vitiated if B could be taken to be inconsistent, for there is nothing commendable in starting from a theory that can be shown false prior to examining data. So our positive results will only concern starting theories B that are consistent.

The next section is devoted to characterizing the competence of revision-based scientists with respect to problems of the form $(T, \{\theta_0, \ldots, \theta_n\})$. As a preliminary, the remainder of the present section brings out various facts that emerge at a more general level of analysis. Thus, we step back from problems of the special form $(T, \{\theta_0, \ldots, \theta_n\})$ or $(T, \{P_0, P_1, \ldots\})$, and consider the entire class of problems. In this more general setting, we hope to accomplish four things. First, we show that the stringent revision functions are canonical for the class of problems that can be solved by revision-based scientists, and that this canonicity also embraces efficient inquiry (Sections 4.2.4 and 4.2.5). Second, we show that stringent revision is no guarantee of successful inquiry, even for problems that are solvable in the revision-based way (Section 4.2.6). Third, we consider the interaction of computability and revision. Finally, we define a partial order on the competence of revision-based scientists.

These four topics have both mathematical and conceptual interest, but they are more technical and "theory internal" than the sequel. So the remainder of the present section can be omitted without loss of continuity, and the reader may resume at the beginning of Section 4.3, below.

4.2.4   The Inductive Power of Stringent Revision

In the present subsection we establish that a single, stringent revision function is sufficient to define successful, revision-based scientists. The matter may be stated as follows.

(27)   THEOREM: There is a stringent revision function $\dotplus$ with the following property. Let problem **P** be such that for some $Y \subseteq \mathcal{L}_{form}$ and revision function $\dot\oplus$, $\lambda\sigma \cdot Y \dot\oplus \sigma$ solves **P**. Then there is a consistent $X \subseteq \mathcal{L}_{form}$ such that $\lambda\sigma \cdot X \dotplus \sigma$ solves **P**.

To prove the theorem we rely on an evident fact, namely: for a revision-based scientist to solve a problem **P**, it must be possible to produce sets of formulas that imply the propositions of **P** while remaining consistent with the incoming data. The following definition states this condition precisely.

(28)   DEFINITION: Problem **P** is *feasible* just in case for all $P \in \mathbf{P}$, and all $\sigma \in SEQ$ for P, there is $Y \subseteq \mathcal{L}_{form}$ such that $\emptyset \neq MOD(Y \cup \text{range}(\sigma)) \subseteq P$.

We will now demonstrate two things. First, if a problem is not feasible then it is not solvable by a revision-based scientist. Second, there is a stringent revision function $\dotplus$ such that if a problem is both solvable and feasible then $\dotplus$ suffices to solve the problem in a revision-based way. First, revision-based solvability implies feasibility:

(29)   PROPOSITION: Suppose that problem **P** is not feasible. Then for all $B \subseteq \mathcal{L}_{form}$ and revision functions $\dotplus$, $\lambda\sigma \cdot B \dotplus \sigma$ does not solve **P**.

Proof: Let problem **P**, $P \in \mathbf{P}$, and $\sigma \in SEQ$ for P be such that for all $Y \subseteq \mathcal{L}_{form}$ either $MOD(Y \cup \text{range}(\sigma)) = \emptyset$ or $MOD(Y \cup \text{range}(\sigma)) \not\subseteq P$. Then for all $\tau \in SEQ$:

(30)   for all $Y' \subseteq \mathcal{L}_{form}$, either $MOD(Y' \cup \text{range}(\sigma * \tau)) = \emptyset$ or $MOD(Y' \cup \text{range}(\sigma * \tau)) \not\subseteq P$.

Let $X \subseteq \mathcal{L}_{form}$, revision function $\dotplus$, and environment e for P with $\sigma \subseteq e$ be given. Then (30) implies immediately that $\lambda\sigma \cdot X \dotplus \sigma$ does not solve P in e.   ∎

The second part of the proof of Theorem (27) is more demanding. It is formulated as follows.

(31)   PROPOSITION: There is a stringent revision function $\dotplus$ with the following property. For every solvable, feasible problem $\mathbf{P}$ there is consistent $X \subseteq \mathcal{L}_{form}$ such that $\lambda\sigma . X \dotplus \sigma$ solves $\mathbf{P}$.

Thus, the class of scientists of form $\lambda\sigma . X \dotplus \sigma$ with X consistent and $\dotplus$ stringent is a canonical strategy for the feasible problems.

Proof of Proposition (31):   Before we define a stringent revision function that satisfies the claim of the theorem, we need some notation. For all $i, m, n \in N$, let $\theta_{i,m,n}$ be the tautology $(v_i = v_i) \wedge (v_m = v_m) \wedge (v_n = v_n)$. Given $i, m \in N$ we say that $X \subseteq \mathcal{L}_{form}$ is *of type* (i, m) just in case there is an enumeration $\{\psi_n \mid n \in N\} \subseteq \mathcal{L}_{form}$ such that $X = \{\theta_{i,m,n} \wedge \psi_n \mid n \in N\}$. Note that there are 2^ω sets of type (i, m). Now we define an enumeration $\mathbf{S}$ of certain members of $\text{pow}(\mathcal{L}_{form})$, in view of defining a stringent revision function via Definition (9). Given $i, m \in N$, let $[i, m]$ denote some specific, well-ordering of all sets of type (i, m). Then, $\mathbf{S}$ begins this way:

$$[0, 0], [0, 1], [0, 2] \ldots [1, 0], [1, 1], [1, 2] \ldots \quad \ldots [i, 0], [i, 1], [i, 2] \ldots$$

$$[i, m] \ldots$$

That is, $\mathbf{S}$ begins with a well-ordering of all sets of type $(0, 0)$, continues with a well-ordering of all sets of type $(0, 1)$, etc. After $[i, m]$ has been enumerated for all $i, m \in N$, $\mathbf{S}$ ends with a well-ordering of all singletons.[4] Let $\dotminus_s$ be constructed from $\mathbf{S}$ as in Definition (9). We take $\dotplus$ to be defined from $\dotminus_s$ via Definition (12). So $\dotplus$ is stringent by Proposition (10).

   Let $0 < \kappa \le \omega$ and solvable, feasible problem $\mathbf{P} = \{P_j \mid j < \kappa\}$ be such that P_j and $P_{j'}$ are distinct for all $j < j' < \kappa$. To finish the proof we will define $X \subseteq \mathcal{L}_{form}$ such that X is consistent and $\lambda\sigma . X \dotplus \sigma$ solves $\mathbf{P}$. The definition of X proceeds in two stages. In the first stage we define $X_0 \subseteq \mathcal{L}_{form}$ such that:

(32) X_0 is of type $(0, 0)$.

In the second stage, given $i > 0$ and $m \in N$, we define $X_{i,m} \subseteq \mathcal{L}_{form}$ such that:

(33)   either $X_{i,m} = \emptyset$ or $X_{i,m}$ is of type (i, m).

---

4. The foregoing enumeration may be defined using ordinal notations, but no gain in clarity seems to be achieved thereby.

Then we set:

(34) $X = X_0 \cup \bigcup \{X_{i,m} \mid i > 0, m \in N\}$.

Here is the definition of X_0. By the feasibility assumption there is $Y \subseteq \mathcal{L}_{form}$ such that $\emptyset \neq MOD(Y) \subseteq P_0$. Let $Z \subseteq \mathcal{L}_{form}$ be the result of doubling the index of every variable appearing in Y. Then also:

(35) $\emptyset \neq MOD(Z) \subseteq P_0$.

So we may choose $\mathcal{S} \in P_0$ and full assignment h to $\mathcal{S}$ with $\mathcal{S} \models Z[h]$. (We may choose h to be onto $\mid \mathcal{S} \mid$ because of the spare variables created by the doubling of indexes.) By Proposition 3.(35), there exists a π-set π such that π belongs to a tip-off for P_0 in $\mathbf{P}$ and $\mathcal{S} \models \pi[h]$. Fix an enumeration $\{\xi_n \mid n \in N\}$ of $Z \cup \pi$. We set:

$X_0 = \{\theta_{0,0,n} \wedge \xi_n \mid n \in N\}$.

Note that $\mathcal{S} \models X_0[h]$ and $X_0 \models Z$. This with (35) implies that:

(36) $\emptyset \neq MOD(X_0) \subseteq P_0$.

Now we define the $X_{i,m}$, for $i > 0$, $m \in N$. By Proposition 3.(35), for all $j < \kappa$ let t_j be a tip-off for P_j in $\mathbf{P}$. By the countability of tip-offs, let the π-sets in $\bigcup_{j<\kappa} t_j$ be enumerated as $\{\pi_i \mid i > 0\}$. Given $i > 0$, fix an enumeration $\{\alpha_{i,m} \mid m \in N\}$ of all $\alpha \in \mathcal{L}_{basic}$ such that $\pi_i \models \alpha$. Fix an enumeration $\{\sigma_m \mid m \in N\}$ of SEQ. Let $i > 0$ and $m \in N$ be given. If $\pi_i \cup range(\sigma_m)$ is not satisfiable in $\bigcup \mathbf{P}$ then set $X_{i,m} = \emptyset$. Suppose otherwise. By Definition 3.(28) let unique $j < \kappa$ be such that $P_j \cap MOD(\pi_i) \neq \emptyset$. It is easy to see that there is $\sigma \in SEQ$ such that $range(\sigma) = \{\alpha_{i,0} \ldots \alpha_{i,m}\} \cup range(\sigma_m)$ and σ is for P_j. Hence, by the feasibility assumption there exists $Y_{i,m} \subseteq \mathcal{L}_{form}$ such that $\emptyset \neq MOD(Y_{i,m} \cup \{\alpha_{i,0} \ldots \alpha_{i,m}\} \cup range(\sigma_m)) \subseteq P_j$. Let $q \in N$ be the maximum index of a variable appearing in $\{\alpha_{i,0} \ldots \alpha_{i,m}\} \cup range(\sigma_m)$. Let $Z_{i,m} \subseteq \mathcal{L}_{form}$ be the result of doubling all indexes greater than q of any variable appearing in $Y_{i,m}$. Then also:

(37) $\emptyset \neq MOD(Z_{i,m} \cup \{\alpha_{i,0} \ldots \alpha_{i,m}\} \cup range(\sigma_m)) \subseteq P_j$.

So we may choose $\mathcal{S} \in P_j$ and full assignment h to $\mathcal{S}$ with $\mathcal{S} \models Z_{i,m} \cup \{\alpha_{i,0} \ldots \alpha_{i,m}\} \cup range(\sigma_m)[h]$. (Once again, we may choose h to be onto $\mid \mathcal{S} \mid$ because of the spare variables created by the doubling of indexes.) By Proposition 3.(35)

again, let $f(i, m) > 0$ be such that $S \models \pi_{f(i,m)}[h]$. Since $P_j \cap MOD(\pi_i) \neq \emptyset$, $S \in P_j$, and $S \models \pi_{f(i,m)}[h]$, we infer from Definition 3.(28) that:

(38)   for all $P \in \mathbf{P}$ with $P \cap MOD(\pi_i) = \emptyset$, $P \cap MOD(\pi_{f(i,m)}) = \emptyset$.

Fix an enumeration $\{\xi_{i,m,n} \mid n \in N\}$ of $Z_{i,m} \cup \{\alpha_{i,0} \ldots \alpha_{i,m}\} \cup \text{range}(\sigma_m) \cup \pi_{f(i,m)}$. Recall the enumeration $\{\xi_n \mid n \in N\}$ associated with the definition of X_0 above. We set:

$$X_{i,m} = \{\theta_{i,m,n} \wedge [(\xi_0 \wedge \ldots \wedge \xi_n) \vee (\xi_{i,m,0} \wedge \ldots \wedge \xi_{i,m,n})] \mid n \in N\}.$$

Since i, m were chosen arbitrarily, (38) and the definition of $X_{i,m}$ yields:

(39)   Let $i > 0$, $m \in N$, and $j < \kappa$ be given. If $X_{i,m} \neq \emptyset$ and $P_j \cap MOD(\pi_i) = \emptyset$, then $P_j \cap MOD(\pi_{f(i,m)}) = \emptyset$.

We also note the following consequence of (37) and the definition of $X_{i,m}$.

(40)   Let $\sigma \in SEQ$, $i > 0$ and $n, m \in N$ be such that $\text{range}(\sigma) \models \neg(\xi_0 \wedge \ldots \xi_n)$ and $X_{i,m} \cup \text{range}(\sigma)$ is consistent. Then $\emptyset \neq MOD(X_{i,m} \cup \text{range}(\sigma)) \subseteq P_j$, for the unique $j < \kappa$ with $P_j \cap MOD(\pi_i) \neq \emptyset$.

From (34), (36), the definition of X_0, and the definition of the $X_{i,m}$, we infer immediately that X is consistent. So it remains to prove that $\lambda\sigma . X \dotplus \sigma$ solves $\mathbf{P}$. Let $j < \kappa$, $S \in P_j$, full assignment h to S, and environment e for S and h be given. It suffices to show that $\lambda\sigma . X \dotplus \sigma$ solves P_j in e. We distinguish two cases.

Case 1: $\text{range}(e) \cup X_0$ is consistent. From (32), (33), (34), and the definition of $\dotplus$, we infer immediately that $X_0 \subseteq X \dotplus e[k]$ for all $k \in N$. This with (36) implies that $\emptyset \neq MOD(X \dotplus e[k]) \subseteq P_0$ for all $k \in N$. So to complete the proof in the current case it remains to show that $j = 0$. Recall from the definition of X_0 that there is a π-set π such that π belongs to a tip-off for P_0 in $\mathbf{P}$ and $X_0 \models \pi$. By Definition 3.(28) it suffices to show that $S \models \pi[h]$. Suppose otherwise. Let $\varphi \in \pi$ be such that $S \not\models \varphi[h]$. Since φ is a $\forall$ formula and e is an environment for S and h, there is $\alpha \in \mathcal{L}_{basic}$ such that $\alpha \in \text{range}(e)$ and $\varphi \models \neg\alpha$. Hence $\text{range}(e) \cup \{\varphi\}$ is inconsistent. Hence, since $X_0 \models \varphi$, $\text{range}(e) \cup X_0$ is inconsistent, contradiction.

Case 2: $\text{range}(e) \cup X_0$ is inconsistent. By compactness there exists $k_0, n \in N$ such that:

(41) $\text{range}(e[k_0]) \models \neg(\xi_0 \wedge \ldots \wedge \xi_n)$.

Let $i_0 > 0$ be such that $S \models \pi_{i_0}[h]$. We now show that the proof is complete if the following can be demonstrated.

(42)   (a)   for all $k, m \in N$, if $\sigma_m = e[k]$ then $X_{i_0,m} \neq \emptyset$ and $X_{i_0,m} \cup \text{range}(e[k])$
              is consistent;

        (b)   for all $i > 0$, if $MOD(\pi_i) \cap P_j = \emptyset$ then there exists $k \in N$ such that
              for all $m \in N$, either $X_{i,m} = \emptyset$ or $X_{i,m} \cup \text{range}(e[k])$ is inconsistent.

Demonstration that (42) suffices to complete the proof of the theorem: Suppose the truth of (42). Then there is $k_1 \in N$ such that for all $k \geq k_1$:

(43)   (a)   for all $m \in N$, if $\sigma_m = e[k]$ then $X_{i_0,m} \neq \emptyset$ and $X_{i_0,m} \cup \text{range}(e[k])$
              is consistent;

        (b)   for all $0 < i \leq i_0$, if $MOD(\pi_i) \cap P_j = \emptyset$ then for all $m \in N$, either
              $X_{i,m} = \emptyset$ or $X_{i,m} \cup \text{range}(e[k])$ is inconsistent.

Now let k be greater than both k_0 and k_1. It is enough to show that $\emptyset \neq MOD(X \dotplus e[k]) \subseteq P_j$. By (32), (33), and (34), X_0 is the only subset of X of type $(0, m)$ for some $m \in N$. So it follows from (41), $k > k_0$, and the definition of X_0 that for all $m \in N$, there is no subset of X of type $(0, m)$ which is consistent with range$(e[k])$. On the other hand, since $k > k_1$, from (32), (33), (34), and (43)b, we infer that for all $0 < i \leq i_0$ with $MOD(\pi_i) \cap P_j = \emptyset$ and for all $m \in N$, there is no subset of X of type (i, m) which is consistent with range$(e[k])$. By (33), (34), and (43)a, there is $m \in N$ and a subset of X of type (i_0, m) which is consistent with range$(e[k])$. From the definition of $\dotplus$, we conclude that there is $0 < i \leq i_0$ with $MOD(\pi_i) \cap P_j \neq \emptyset$, $m \in N$, and a subset Y of X of type (i, m) such that $(Y \cup \text{range}(e[k])) \subseteq X \dotplus e[k]$. This with (32), (33), and (34) implies the following.

(44)   There is $i > 0$ such that:

(a) for some $m \in N$, $X_{i,m} \neq \emptyset$ and $(X_{i,m} \cup \text{range}(e[k])) \subseteq X \dotplus e[k]$;

(b) $MOD(\pi_i) \cap P_j \neq \emptyset$.

From (40), (41), and (44), we infer that $\emptyset \neq MOD(X \dotplus e[k]) \subseteq P_j$.   ∎

So the proof is completed if we show that (42) is satisfied. Let $k, m \in N$ be such that $\sigma_m = e[k]$. Since $S \models \pi_{i_0} \cup \text{range}(\sigma_m)[h]$, it follows immediately from the definition of $X_{i_0,m}$ that $X_{i_0,m} \neq \emptyset$, and that $Z_{i_0,m} \cup \{\alpha_{i_0,0} \ldots \alpha_{i_0,m}\} \cup \text{range}(\sigma_m) \cup \pi_{f(i_0,m)}$ is consistent and implies $X_{i_0,m}$. This proves (42)a.

To prove (42)b, let $i > 0$ be such that $MOD(\pi_i) \cap P_j = \emptyset$. By (41) and the definition of $X_{i,m}$, $m \in N$, it suffices to show the existence of $k \geq k_0$ such that:

(45)   for all $m \in N$, either $X_{i,m} = \emptyset$ or $\{\alpha_{i,0} \ldots \alpha_{i,m}\} \cup \pi_{f(i,m)} \cup \mathrm{range}(e[k])$ is inconsistent.

Since $\mathcal{S} \in P_j$ and $MOD(\pi_i) \cap P_j = \emptyset$, Definition 3.(28) implies that π_i is not satisfiable in $\mathcal{S}$. Via (39), the unsatisfiability of π_i in $\mathcal{S}$ also implies that for all $m \in N$, either $X_{i,m} = \emptyset$ or $\pi_{f(i,m)}$ is not satisfiable in $\mathcal{S}$. Hence, since e is an environment for $\mathcal{S}$, since π_i is a π-set, and since $\pi_{f(i,m)}$ is a π-set for all $m \in N$ with $X_{i,m} \neq \emptyset$, we may choose $m_0 \in N$ and $k \geq k_0$ such that:

(46)   (a) $\mathrm{range}(e[k]) \models \neg\alpha_{i,m_0}$, and

      (b)   for all $m < m_0$, if $X_{i,m} \neq \emptyset$ then $\mathrm{range}(e[k]) \cup \pi_{f(i,m)}$ is inconsistent.

If $m \geq m_0$ and $X_{i,m} \neq \emptyset$, then $\{\alpha_{i,0} \ldots \alpha_{i,m}\} \cup \mathrm{range}(e[k])$ is inconsistent by (46)a. This with (46)b implies (45).   ∎

4.2.5   Efficient Revision

The present subsection is devoted to extending Theorem (27) to efficient inquiry. The conception of efficiency at issue is given in Definition 3.(73) and explored in Section 3.4.2. The proof of the following theorem tells us that the stringent revision function of Theorem (27) also behaves efficiently.

(47)   THEOREM: There is a stringent revision function $\dotplus$ with the following property. Let problem **P** be such that for some $Y \subseteq \mathcal{L}_{form}$ and revision function $\dotoplus$, $\lambda\sigma \cdot Y \dotoplus \sigma$ solves **P**. Then there is a consistent $X \subseteq \mathcal{L}_{form}$ such that $\lambda\sigma \cdot X \dotplus \sigma$ solves **P** efficiently.

Proof: Let $0 < \kappa \leq \omega$ and problem $\mathbf{P} = \{P_j \mid j < \kappa\}$ be such that:

(a) P_j and $P_{j'}$ are distinct for all $j < j' < \kappa$, and

(b) for some $Y \subseteq \mathcal{L}_{form}$ and revision function $\dotoplus$, $\lambda\sigma \cdot Y \dotoplus \sigma$ solves **P**.

Let revision function $\dotplus$, and $X \subseteq \mathcal{L}_{form}$ be the revision function and formula-set defined in the proof of Proposition (31). With Proposition (29), the proof of Proposition (31) shows that $\lambda\sigma \cdot X \dotplus \sigma$ solves **P**. Relying on the same notation, we extend the proof of Proposition (31) to show that $\lambda\sigma \cdot X \dotplus \sigma$ solves **P** efficiently.

Let $\sigma \in SEQ$ be for **P**. By Exercise 3.(87) it suffices to show the existence of $j < \kappa$, $\mathcal{S} \in P_j$, and full assignment h to $\mathcal{S}$ such that:

(48)   (a) $S \models \bigwedge \sigma[h]$;

(b) for all $\tau \in SEQ$, if τ extends σ and $S \models \bigwedge \tau[h]$, then $\emptyset \neq MOD(X \dotplus \tau) \subseteq P_j$.

We distinguish two cases.

Case 1: range(σ) $\cup X_0$ is consistent. With (35) and the fact that infinitely many spare variables occur in $\{\xi_n \mid n \in N\}$, it follows that there is $S \in P_0$ and full assignment h to S such that $S \models X_0 \cup$ range(σ)[h]. Let $\tau \in SEQ$ extend σ and be such that $S \models \bigwedge \tau[h]$. Hence $X_0 \cup$ range(τ) is consistent. From (32), (33), (34), and the definition of $\dotplus$, we infer that $X_0 \subseteq X \dotplus \tau$. (Informally, this is because X_0 belongs to the first class [0, 0] of subsets of $\mathcal{L}_{form}$ in the ordering that defines $\dotplus$, and no other subset of X belongs to [0, 0].) This with (36) implies that: $\emptyset \neq MOD(X \dotplus \tau) \subseteq P_0$. Hence (48) is satisfied.

Case 2: range(σ) $\cup X_0$ is inconsistent. By compactness and the definition of X_0, let $n \in N$ be such that:

(49)   range(σ) $\models \neg(\xi_0 \wedge \ldots \wedge \xi_n)$.

By (32), (33), and (34), X_0 is the only subset of X of type $(0, m)$ for some $m \in N$. So it follows that:

(50)   for all $m \in N$, there is no subset of X of type $(0, m)$ which is consistent with range(σ).

By Definition 3.(28), since σ is for **P**, there is $i > 0$ such that $\pi_i \cup$ range(σ) is consistent. Let $m \in N$ be such that $\sigma_m = \sigma$. It follows from the definition of $X_{i,m}$ that $X_{i,m} \neq \emptyset$. So let $i > 0$ be least such that for some $m \in N$, $X_{i,m} \neq \emptyset$ and $X_{i,m} \cup$ range(σ) is consistent. Let $m \in N$ be such that $X_{i,m} \neq \emptyset$ and $X_{i,m} \cup$ range(σ) is consistent. With (49) we infer immediately that $Z_{i,m} \cup \{\alpha_{i,0} \ldots \alpha_{i,m}\} \cup$ range(σ_m) $\cup \pi_{f(i,m)} \cup$ range(σ) is consistent. By Definition 3.(28) let unique $j < \kappa$ be such that $P_j \cap MOD(\pi_i) \neq \emptyset$. With (37) and the fact that infinitely many spare variables occur in $Z_{i,m}$, it follows that there is $S \in P_j$ and full assignment h to S such that $S \models Z_{i,m} \cup \{\alpha_{i,0} \ldots \alpha_{i,m}\} \cup$ range(σ_m) $\cup \pi_{f(i,m)} \cup$ range(σ)[h]. Let $\tau \in SEQ$ extend σ and be such that $S \models \bigwedge \tau[h]$. Hence $Z_{i,m} \cup \{\alpha_{i,0} \ldots \alpha_{i,m}\} \cup$ range(σ_m) $\cup \pi_{f(i,m)} \cup$ range(τ) is consistent. Hence $X_{i,m} \cup$ range(τ) is consistent. With (50) and the choice of i, this implies that i is least with the following property: there is $m \in N$ and a subset of X of type (i, m) which is consistent with range(τ). With (32), (33), (34), and the definition of $\dot{+}$, we infer that there is $m \in N$ such that $X_{i,m} \subseteq X \dot{+} \tau$. We deduce from (40), $P_j \cap MOD(\pi_i) \neq \emptyset$, (49), and the fact that τ extends σ that $\emptyset \neq MOD(X_{i,m} \cup \text{range}(\tau)) \subseteq P_j$. Hence (48) is satisfied.   ∎

4.2.6   Unsuccessful Inquiry Using Stringent Revision

Theorem (27) does not imply that every stringent revision function is adapted to feasible problems, only that some of them are. In fact, not every form of stringent revision can be made to work successfully, as revealed by the following result.

(51)   THEOREM: Suppose that **Sym** is limited to a binary predicate and countably many constants. Then there exists a stringent revision function $\dot{+}$, and propositions P_1, P_2 that meet the following conditions.

(a) $\{P_1, P_2\}$ is solvable and each of P_1, P_2 is closed under elementary equivalence (hence, $\{P_1, P_2\}$ is feasible).

(b) for all $B \subseteq \mathcal{L}_{form}$, $\lambda\sigma$. $B \dot{+} \sigma$ does not solve $\{P_1, P_2\}$.

Proof: Let R be the binary predicate of **Sym**. Let $\{\bar{n} \mid n \in N\}$ enumerate the constants of **Sym**. Let $\mathcal{S}$ be the structure with domain N that interprets $\bar{n}$ as n, for all $n \in N$, and such that $R^{\mathcal{S}}$ is the usual total order on N. Denote by θ a constant-free sentence such that $\mathcal{S} \models \theta$, and $\mathcal{U} \models \neg\theta$ for all finite structures $\mathcal{U}$. Define P_1 to be the set of all structures elementary equivalent to $\mathcal{S}$, P_2 the set of all finite structures. It is immediate that each of P_1, P_2 are closed under elementary equivalence. Moreover, it is straightforward to show that $\{P_1, P_2\}$ has tip-offs, so Proposition 3.(31) implies that $\{P_1, P_2\}$ is solvable.

Now we define $\dot{+}$. Fix an environment e for $\mathcal{S}$. For all $k \in N$, let $\{S_{(\omega \times k)+i} \mid i \in N\}$ be an enumeration of all finite $D \subseteq \mathcal{L}_{form}$ such that $D \cup \text{range}(e[k])$ is consistent and $D \cup \text{range}(e[k]) \models \theta$. Fix an enumeration $\{\varphi_i \mid i \in N\}$ of $\mathcal{L}_{form}$, and set $S_{\omega^2+i} = \{\varphi_i\}$ for all $i \in N$. Let $\mathbf{S} = \{S_\alpha \mid \alpha < \omega^2 + \omega\}$, and let $\dot{-}_{\mathbf{s}}$ be constructed from $\mathbf{S}$ as in Definition (9). We take $\dot{+}$ to be defined from $\dot{-}_{\mathbf{s}}$ via Definition (12). So $\dot{+}$ is stringent by Proposition (10). Let $B \subseteq \mathcal{L}_{form}$ be given, and suppose that $\lambda\sigma$. $B \dot{+} \sigma$ solves P_1 in e. To verify (51)b, it suffices to show that for some environment e' for P_2, $\lambda\sigma$. $B \dot{+} \sigma$ does not solve P_2 in e'. For this purpose, we need the following fact.

(52)   FACT: Let consistent and finite $D \subseteq \mathcal{L}_{form}$ be given. Then there is $\mathcal{U} \in P_2$ and full assignment h to $\mathcal{U}$ such that $\mathcal{U} \models \{\beta \in \mathcal{L}_{basic} \mid D \models \beta\}[h]$.

Proof of Fact (52):   Let consistent and finite $D \subseteq \mathcal{L}_{form}$ be given. Let $X_1 = D \cup \{\overline{0} = \overline{n} \mid \overline{n}$ does not appear in $D\}$. Then X_1 is consistent. Let $w_0, w_1 \ldots$ be an enumeration without repetition of the cofinitely many variables that do not appear in D. Then $X_2 = X_1 \cup \{w_n = \overline{n} \mid n \in N\}$ is consistent. Let X_3 be a maximally consistent extension of X_2 in $X_2 \cup \mathcal{L}_{basic}$. By a familiar construction due to Leon Henkin (used for proving the existence of models [Enderton, 1972]) it is clear that there is finite structure $\mathcal{U}$ and full assignment h to $\mathcal{U}$ such that $\mathcal{U} \models (X_3 \cap \mathcal{L}_{basic})[h]$. So, $\mathcal{U}$ and h satisfy Fact (52).   ∎

Since $\lambda\sigma \, . \, B \dotplus \sigma$ solves P_1 in e, $MOD(B \dotplus e[k]) \subseteq P_1$ for some $k \in N$. Hence there is smallest $k_0 \in N$ such that for some $Y \subseteq B$, $Y \cup range(e[k_0])$ is consistent and $Y \cup range(e[k_0]) \models \theta$. By compactness, let $i_0 \in N$ be smallest such that:

(53)   (a) $S_{(\omega \times k_0)+i_0} \subseteq B$ and $S_{(\omega \times k_0)+i_0} \cup range(e[k_0])$ is consistent;

       (b) $S_{(\omega \times k_0)+i_0} \cup range(e[k_0]) \models \theta$.

By (53)a and Fact (52), let structure $\mathcal{U} \in P_2$ and full assignment h to $\mathcal{U}$ be such that $\mathcal{U} \models \{\beta \in \mathcal{L}_{basic} \mid S_{(\omega \times k_0)+i_0} \cup range(e[k_0]) \models \beta\}[h]$. Let e' be an environment for $\mathcal{U}$ and h that extends $e[k_0]$. It follows from (53)a and the choice of e' that for all $k \geq k_0$, there is $Y \subseteq B$ such that $Y \cup range(e'[k])$ is consistent and $S_{(\omega \times k_0)+i_0} \subseteq Y$. This with the choice of k_0, i_0 implies for all $k \geq k_0$, $S_{(\omega \times k_0)+i_0} \subseteq B \dotplus e'[k]$. With (53)b we infer that for all $k \geq k_0$, $B \dotplus e'[k] \models \theta$. However, since $|\mathcal{U}|$ is finite, $\mathcal{U} \not\models \theta$ by hypothesis. Hence for all $k \geq k_0$, $B \dotplus e'[k] \not\subseteq P_2$. Hence $\lambda\sigma \, . \, B \dotplus \sigma$ does not solve P_2 in e', as required.   ∎

4.2.7   Scientists Based on Computable Revision

From Definition 3.(89) it follows that a revision-based scientist $\lambda\sigma \, . \, B \dotplus \sigma$ is computable if and only if there is computable $\psi : SEQ \to N$ such that for all $\sigma \in SEQ$, $B \dotplus \sigma = W_{\psi(\sigma)}^{form}$. In this subsection we consider the consequences of requiring revision-based scientists to be computer simulable. It will be seen that belief revision and computability do not mix well. The reason is the consistency requirement expressed in Lemma (13)d, which makes revision hard to calculate. Indeed, the proposition proved below shows that even some problems that are both feasible and computably solvable are outside the reach of computable revision.

Recall from Definition 3.(90) that a proposition is elementary just in case it has the form $MOD(\theta)$ for some $\theta \in \mathcal{L}_{sen}$. We have:

(54) PROPOSITION: Suppose that **Sym** is limited to the vocabulary of arithmetic (including $\overline{0}$ and a unary function symbol s) plus an additional constant.
Then there is a problem **P** with the following properties.

(a) Every member of **P** is elementary, hence **P** is feasible.

(b) **P** is computably solvable.

(c) For every $B \subseteq \mathcal{L}_{form}$ and revision function $\dotplus$, if $\lambda\sigma \,.\, B \dotplus \sigma$ is computable,
then it fails to solve **P**.

In view of Proposition (31), clauses (a) and (b) of the present proposition entail
that the promised **P** is solvable by a revision-based scientist.

Proof of Proposition (54):   Let a be the additional constant of **Sym**. For $n \in$
N, let $\overline{n}$ be the result of n applications of s to $\overline{0}$. Let Q be the seven axioms of
Robinson's arithmetic (see [Boolos & Jeffrey, 1989, Ch. 14]). Given $n \in N$, let
$\lceil n \rceil$ be the formula of arithmetic with gödel number n. By standard results, let
$\varphi(x) \in \mathcal{L}_{form}$ have one free variable x, exclude a, and be such that for all $n \in N$,
$Q \models \varphi(\overline{n})$ iff $\lceil n \rceil$ is derivable from Q. For $n \in N$, if $\lceil n \rceil$ is derivable from Q
then set proposition P_n equal to $MOD(\bigwedge Q \wedge a = \overline{n})$; otherwise, if $\lceil n \rceil$ is not
derivable from Q, set proposition P_n equal to $MOD(\bigwedge Q \wedge a = \overline{n} \wedge \neg\varphi(a))$.
We note that for all $n \in N$, $P_n \neq \emptyset$. In the case where $\lceil n \rceil$ is not derivable from
Q, $P_n \neq \emptyset$ follows from the following observation. If $P_n = \emptyset$ for such an n,
then $Q \models \varphi(\overline{n})$, contradicting the choice of n (given the choice of φ).

   Let $\mathbf{P} = \{P_n \mid n \in N\}$. (Since $P_n \neq \emptyset$ for all $n \in N$, **P** is well defined.) The
verification of (54)a is immediate. For (54)b, we describe a computable scien-
tist Ψ that solves **P**. Let $\sigma \in SEQ$ be given. If for all $n \in N$, $\bigwedge \sigma \not\models a = \overline{n}$, then
$\Psi(\sigma)$ is undefined. Otherwise, let $n \in N$ be least such that $\bigwedge \sigma \models a = \overline{n}$. Then
$\Psi(\sigma)$ is an index for $\{\bigwedge Q \wedge a = \overline{n} \wedge \neg\varphi(a)\}$ if $\lceil n \rceil$ does not appear in some
standard enumeration of $Cn(Q)$ within length(σ) steps; otherwise, $\Psi(\sigma)$ is an
index for $\{\bigwedge Q \wedge a = \overline{n}\}$. It is easy to see that Ψ solves every member of **P**.

   For (54)c, let S be the set of $\sigma \in SEQ$ such that $Q \cup range(\sigma)$ is consistent. It
is easy to verify that S is recursive. For a contradiction, suppose that $B \subseteq \mathcal{L}_{form}$
and revision function $\dotplus$ are such that $\lambda\sigma \,.\, B \dotplus \sigma$ is computable and solves **P**.
Let $n \in N$ be given with $\lceil n \rceil$ a consequence of Q. Suppose that the first member
of nonvoid $\sigma \in S$ is $a = \overline{n}$. Then $\bigwedge \sigma \models \bigwedge Q \rightarrow \varphi(a)$, so Lemma (13)d implies
that $B \dotplus \sigma \not\models \bigwedge Q \wedge \neg\varphi(a)$. This shows:

(55)   If $\lceil n \rceil$ is a consequence of Q, then for all $\sigma \in S$ that begin with $a = \overline{n}$,
$B \dotplus \sigma \not\models \bigwedge Q \wedge \neg\varphi(a)$.

Now suppose that $\lceil n \rceil$ is not a consequence of Q. Let $\mathcal{S} \in P_n$ be given, and let e be an environment for P_n that begins with $a = \bar{n}$. Then, since $\lambda\sigma \cdot B \dotplus \sigma$ solves P_n, there is $k \in N$ such that $B \dotplus e[k] \models \bigwedge Q \wedge \neg\varphi(a)$. This shows:

(56)   If $\lceil n \rceil$ is not a consequence of Q, then there is $\sigma \in S$ that begins with $a = \bar{n}$ and is such that $B \dotplus \sigma \models \bigwedge Q \wedge \neg\varphi(a)$.

However, in view of the recursivity of S and the computability of $\lambda\sigma \cdot B \dotplus \sigma$, the conjunction of (55) and (56) yields a positive test for the non-consequences of Q, which is known to be impossible (see [Boolos & Jeffrey, 1989, Ch. 15, Thm 1]).   ∎

The foregoing result prompts the search for interesting subclasses of problems that can be solved by computable, revision-based scientists. It also suggests the need to define weaker senses of "computer simulable" so as to attenuate the impact of the consistency requirement expressed in Lemma (13)d. We shall not pursue these projects here, however, preferring instead to work out the inductive logic of belief revision in the simpler context of potentially ineffective functions. Clarity in these matters can be expected to facilitate the subsequent analysis of computable, revision-based scientists.

4.2.8   Comparing the Competence of Revision-Based Scientists

A useful partial order on the competence of revision-based scientists may be defined as follows. We say that revision function $\dot\oplus$ "subsumes" revision function $\dotplus$ just in case for all $B \subseteq \mathcal{L}_{form}$ and $\sigma \in SEQ$, $B \dot\oplus \sigma \supseteq B \dotplus \sigma$. Definitions (4) and (12) imply immediately that the maxichoice revision functions are maximal elements of the subsumes relation. That is, every revision function is subsumed by some maxichoice revision function, and no maxichoice function is subsumed "properly" by any other function.

Subsumption between revision functions is related to scientific competence in the following way.

(57)   LEMMA: Suppose that revision functions $\dotplus$ and $\dot\oplus$ are such that $\dot\oplus$ subsumes $\dotplus$. Let problem **P** and $B \subseteq \mathcal{L}_{form}$ be given. If $\lambda\sigma \cdot B \dotplus \sigma$ solves **P** then $\lambda\sigma \cdot B \dot\oplus \sigma$ does also.

The lemma is easy to verify, and defines a clear sense in which maxichoice revision functions represent the most powerful scientific method based on hypothesis revision. If a problem is solvable by revision, then it is solvable by a revision-based scientist whose "$\dot{+}$" is maxichoice. Conversely, for a given starting point B, the weakest, revision-based scientist is defined from the contraction function that maps (B, ϕ) to $\bigcap(B \perp \phi)$. The latter function may be called "full meet." The revision function it underlies is subsumed by every other revision function. So the lemma implies that if the full meet scientist solves **P** then so does every scientist of form $\lambda\sigma \cdot B \dot{+} \sigma$. (The "full meet" terminology is adapted from [Gärdenfors, 1988].)

4.2.9   Exercises

(58)   EXERCISE: Suppose that **Sym** consists of a constant and a unary function symbol. Show that there is proposition P such that P is closed under isomorphism, $\{P\}$ is not feasible, and some linguistic scientist solves $\{P\}$.

(59)   EXERCISE: Provide a simple proof for the following weakening of Proposition (31).

There is a maxichoice revision function $\dot{+}$ with the following property. For every solvable, feasible problem **P** there is consistent $X \subseteq \mathcal{L}_{form}$ such that $\lambda\sigma \cdot X \dot{+} \sigma$ solves **P**.

4.3   Augmenting the Background Theory

A revision-based scientist $\lambda\sigma \cdot B \dot{+} \sigma$ has two working parts. On the one hand, there is the revision function $\dot{+}$ that determines how B will be revised under the impact of data σ. On the other hand, there is the belief state B itself. We may consider B as the scientist's *starting point* for inquiry, i.e., what she considers to be true prior to examining data.

Different choices of starting point and revision function yield scientists of varying competence. The present section is devoted to characterizing their ability with respect to problems of the form $(T, \{P_0, P_1, \ldots\})$, and especially its restricted variety $(T, \{\theta_0, \ldots, \theta_n\})$. It will be seen that the starting point of inquiry must be chosen carefully if success is to be achieved. To grasp the issue, consider a problem $(T, \{\theta_0, \ldots, \theta_n\})$, and let us view T as the "background theory" whose truth is known prior to inquiry. It is tempting to take T as the starting point of a revision-based scientist. We will shortly see cases, however, in which no scientist of form $\lambda\sigma \cdot T \dot{+} \sigma$ solves $(T, \{\theta_0, \ldots, \theta_n\})$. Thus, additional beliefs may need to be joined to T in order to prepare the way for successful revision.

The phenomenon just described (and documented below) leads us to frame the present section in terms of the needed extension of the background theory. We show that scientists of form $\lambda\sigma \, . \, B \dotplus \sigma$ have wide inductive powers for problems of form $(T, \{\theta_0, \ldots, \theta_n\})$ provided that B supplements T in the right way. Indeed, much of the work in our proofs consists in defining a successful supplement. Intuitively, a scientist whose starting point goes beyond the background theory is willing to engage in speculation "for the sake of the argument." The speculation will be tempered (in the typical case) by the arrival of data, but it is nonetheless essential to getting inquiry off the ground.

To begin the discussion, we consider a large class of theory extensions that lead revision-based scientists to failure. This will help motivate the choice of successful extension introduced subsequently.

4.3.1 Extensions of the Background Theory that Are Too Modest

To exploit belief revision for a problem $(T, \{\theta_0, \ldots, \theta_n\})$, it is natural to consider scientists of form $\lambda\sigma \, . \, T \dotplus \sigma$. Such scientists attempt to revise the background theory T in the face of data coming from the environment. Unfortunately, there are simple problems of this kind that lead all such scientists to failure. Indeed, the following proposition shows that failure also results if T is extended to any consistent subset of $\mathcal{L}_{sen}$. Recall that the latter set is limited to formulas without free variables.

(60) PROPOSITION: Suppose that **Sym** is limited to the binary predicate R. Let $T = \{\exists x \forall y \, Rxy \leftrightarrow \neg \exists y \forall x \, Rxy\}$, and $\theta = \exists x \forall y \, Rxy$. Then $(T, \{\theta, \neg\theta\})$ is solvable, but for all revision functions $\dotplus$ and all $B \subseteq \mathcal{L}_{sen}$ consistent with T, $\lambda\sigma \, . \, B \dotplus \sigma$ does not solve $(T, \{\theta, \neg\theta\})$.

Proof: Proposition 3.(54) shows that $(T, \{\theta, \neg\theta\})$ is solvable. Choose revision function $\dotplus$, and let $B \subseteq \mathcal{L}_{sen}$ be consistent with T. So we may choose $\mathcal{S} \in MOD(B \cup T)$. Suppose that $\mathcal{S} \models \neg\theta$ (the argument is parallel for the case $\mathcal{S} \models \theta$). We show that $\lambda\sigma \, . \, B \dotplus \sigma$ does not solve $MOD(T \cup \{\theta\})$. Let $\sigma \in SEQ$ be given, and suppose that $\mathcal{S}$ satisfies $\bigwedge \sigma$. Then there is a model of B and $\bigwedge \sigma$ that is not a model of θ. Along with Lemma (13)c, this proves:

(61) For all $\sigma \in SEQ$, if $\mathcal{S}$ satisfies $\bigwedge \sigma$, then $B \dotplus \sigma \not\models \theta$.

By the choice of T and the fact that $\mathcal{S} \models T \cup \{\neg\theta\}$, $\mathcal{S} \models \exists y \forall x \, Rxy$. Let $\mathcal{U}$ be such that $|\mathcal{U}| = |\mathcal{S}|$ and for all $x, y \in |\mathcal{U}|$, $(x, y) \in R^{\mathcal{U}}$ if and only if $(y, x) \in R^{\mathcal{S}}$. Then it can be seen that:

(62)   (a) $\mathcal{U} \models T \cup \{\theta\}$.

(b)   For all $\sigma \in SEQ$, $\mathcal{U}$ satisfies $\bigwedge \sigma$ if and only if $\mathcal{S}$ satisfies $\bigwedge \sigma$.

Let e be an environment for $\mathcal{U}$. By (61) and (62)b, for all $k \in N$, $B \dotplus e[k] \not\models \theta$. So by (62)a, $\lambda \sigma \, . \, B \dotplus \sigma$ does not solve $MOD(T \cup \{\theta\})$ in e.   ∎

Proposition (60) reveals that revision-based scientists cannot be made to work unless their starting points extend the background theory T to include open formulas. The formulas may be considered an "inductive leap" for the sake of inquiry, ready to be abandoned in whole or part according to the data encountered. Their free variables embody hypotheses about which of the objects shown to the scientist have special properties involved in the problem under investigation.

4.3.2   A Universal Method of Inquiry Based on Revision

If we are willing to extend our background theories with open formulas, then revision can be used successfully for inquiry. Indeed, it is possible to choose a single extension and a single revision function that works for any solvable problem of the form $(T, \{\theta_0, \ldots, \theta_n\})$. In this sense the next theorem exhibits a canonical method of inquiry. Its proof exhibits a theory-extender X and a stringent revision function $\dotplus$ with the following property. Every solvable problem $(T, \{\theta_0 \ldots \theta_n\})$ is solved by the scientist $\lambda \sigma \, . \, (T \cup X) \dotplus \sigma$. Moreover, this scientist has some claim to the title "rational." For, its starting belief state is a consistent extension of the background theory T, and revision occurs on the basis of a stringent function—that is, on the basis of transitive, connected preferences for theories that inflict minimal change on current opinion (as discussed in Section 4.1.4).

(63)   THEOREM: There exists $X \subseteq \mathcal{L}_{form}$ and stringent revision function $\dotplus$ such that for all consistent $T \subseteq \mathcal{L}_{sen}$, the following holds.

(a) $T \cup X$ is consistent.

(b) $\lambda \sigma \, . \, (T \cup X) \dotplus \sigma$ solves every solvable problem of the form $(T, \{\theta_0, \ldots, \theta_n\})$.

Proof: First we define X, then show it to satisfy (63)a. Next we define $\dotplus$, then show it to satisfy (63)b.

Definition of X.   Let $\{(\varphi_n^1(x), \varphi_n^2(x)) \mid n \in N\}$ enumerate all pairs of $\forall$ formulas having all free variables present in the sequence of variables x. For every

$n \in N$ we fix an enumeration $\{v_n^i \mid i \in N\}$ of all sequences of variables such that the following properties hold.

(64) (a) For all $i \in N$, v_n^i is of length equal to the length of x in $\varphi_n^1(x)$ (and hence of x in $\varphi_n^2(x)$).

   (b) No variable is repeated in v_n^0.

   (c) For all $m \in N$, if $n \neq m$ then v_n^0 and v_m^0 share no variables.

For all $n, i \in N$, we use the following abbreviations.

H_n is: $\exists x \varphi_n^1(x) \leftrightarrow \neg \exists x \varphi_n^2(x)$.

F_n^i is: $\bigvee_{r \leq i} [\, \varphi_n^1(v_n^r) \vee \varphi_n^2(v_n^r) \,]$.

Now we set $X = \{H_n \to F_n^i \mid n, i \in N\}$.

Proof of (63)a.   Let consistent $T \subseteq \mathcal{L}_{sen}$ be given. Let structure $\mathcal{S}$ be such that $\mathcal{S} \models T$. We exhibit an assignment h to $\mathcal{S}$ such that $\mathcal{S} \models X[h]$. (It is not assumed that h will end up being onto $|\mathcal{S}|$.) For all $n \in N$ such that $\mathcal{S} \models H_n$, if $p = \text{length}(v_n^0)$ and if variables $x_1 \ldots x_p$ are such that $v_n^0 = (x_1 \ldots x_p)$, then choose $s_1 \ldots s_p \in |\mathcal{S}|$ with $\mathcal{S} \models \varphi_n^1 \vee \varphi_n^2[s_1/x_1 \ldots s_p/x_p]$, and set $h(x_i) = s_i$ for all i, $0 < i \leq p$. This condition on h can be satisfied because of (64). For all variables x such that $h(x)$ is not defined by the foregoing, $h(x)$ may be selected arbitrarily. Hence for all $n \in N$, if $\mathcal{S} \models H_n$ then $\mathcal{S} \models F_n^0[h]$. Since for all $n, i \in N$, $F_n^0 \models F_n^i$, this proves that $\mathcal{S} \models X[h]$.

Definition of $\dotplus$.   Fix an enumeration $\{\theta_i \mid i \in N\}$ of $\mathcal{L}_{sen}$, an enumeration $\{\varphi_i \mid i \in N\}$ of X, and an enumeration $\{\psi_i \mid i \in N\}$ of $\mathcal{L}_{form}$. For all $i \in N$, set $S_i = \{\theta_i\}$, $S_{\omega+i} = \{\varphi_i\}$, and $S_{(\omega \times 2)+i} = \{\psi_i\}$. Let $\mathbf{S} = \{S_\alpha \mid \alpha < \omega \times 3\}$, and let $\dotdiv_S$ be constructed from $\mathbf{S}$ as in Definition (9). We take $\dotplus$ to be defined from $\dotdiv_S$ via Definition (12). So $\dotplus$ is stringent by Proposition (10). As an immediate consequence we have the following.

(65) Let $T \subseteq \mathcal{L}_{sen}$ and $\sigma \in SEQ$ be such that $T \cup \text{range}(\sigma)$ is consistent. Then $T \subseteq (T \cup X) \dotplus \sigma$.

Proof of (63)b.   Let solvable problem of form $(T, \{\theta_0, \ldots, \theta_n\})$ be given. Let $m \leq n$ also be given. By Theorem 3.(55), there is $n \in N$ such that $T \models \theta_m \leftrightarrow \exists x \varphi_n^1(x)$ and $T \models \neg \theta_m \leftrightarrow \exists x \varphi_n^2(x)$. Hence $T \models H_n$. Let $\mathcal{S} \in MOD(T \cup \{\theta_m\})$ and full assignment h to $\mathcal{S}$ be given. (The case $\mathcal{S} \in MOD(T \cup \{\neg \theta_m\})$

is parallel.) Then we may choose the least $i \in N$ such that $\mathcal{S} \models \varphi_n^1(v_n^i)[h]$. So $\mathcal{S} \models H_n \to F_n^i[h]$. Let e be an environment for $\mathcal{S}$ and h. To complete the proof it suffices to show:

(66) (a) for cofinitely many k, $(T \cup X) \dotplus e[k] \models H_n$;

     (b) for cofinitely many k, $(T \cup X) \dotplus e[k] \models H_n \to F_n^i$;

     (c) for cofinitely many k, $(T \cup X) \dotplus e[k] \models \neg F_n^{i-1} \wedge \neg \varphi_n^2(v_n^i)$.

Claim (66)c follows immediately from the choice of i and the choice of e. Claim (66)a follows immediately from (65), the choice of e, and the fact that $T \models H_n$. So it remains to demonstrate (66)b.

Let $j \in N$ be least such that $\varphi_j = H_n \to F_n^i$. Let X_0 be the set of all φ_p, $p < j$ such that $\mathcal{S} \models \varphi_p[h]$. By our choice of e, and the fact that $\mathcal{S} \models T$ and $\mathcal{S} \models H_n \to F_n^i[h]$, we have:

(67) For all $k > 0$, $T \cup X_0 \cup \{H_n \to F_n^i\} \not\models \neg \bigwedge e[k]$.

Because h is a full assignment to $\mathcal{S}$ we may choose $k_0 > 0$ such that:

(68) For all $p < j$, if $\varphi_p = H_{n'} \to F_{n'}^{i'}$ then the following holds: $\mathcal{S} \models F_{n'}^{i'}[h]$ iff $F_{n'}^{i'} \not\models \neg \bigwedge e[k_0]$.

Let $k \geq k_0$ be given. Consider $Y^{\omega+j}$ in the construction of $(T \cup X) \dotdiv_\mathcal{S} \neg \bigwedge e[k]$ via Definition (9). By (65), (68), and the choice of k, $Y^\beta \subseteq T \cup X_0$ for all $\beta < \omega + j$. So by (67), $Y^\beta \cup \{H_n \to F_n^i\} \not\models \neg \bigwedge e[k]$ for all $\beta < \omega + j$. Hence, $H_n \to F_n^i \in Y^{\omega+j} \subseteq (T \cup X) \dotdiv_\mathcal{S} \neg \bigwedge e[k]$. It follows from Definition (12) that $H_n \to F_n^i \in (T \cup X) \dotplus e[k]$, verifying (66)b. ∎

4.3.3 Supplementing as a Function of the Problem

Theorem (63) shows that some revision functions can be used for successful inquiry, but it does not show that all of them can. It thus leaves open the possibility that our conception of revision function is so lax as to embrace revision that is useless for inquiry. This would be true if some revision functions were led to failure on a solvable problem $(T, \{P_0, P_1, \ldots\})$ no matter how T is extended. But such is not the case. It will now be shown that every revision function can be made to succeed on every solvable problem of the form $(T, \{P_0, P_1, \ldots\})$.

(69) THEOREM: Suppose that solvable problem $\mathbf{P}$ is of form $(T, \{P_0, P_1, \ldots\})$. Then there is a consistent extension $X \subseteq \mathcal{L}_{form}$ of T such that for every revision function $\dotplus$, $\lambda \sigma . X \dotplus \sigma$ solves $\mathbf{P}$.

Proof: By Proposition 3.(35), for all $j \in N$ let t_j be a tip-off for P_j in **P**. By
the countability of tip-offs, let the π-sets in $\bigcup_{j \in N} t_j$ be enumerated as $\{\pi_i \mid i \in$
$N\}$. Without loss of generality we may assume that each π_i is consistent with
T. For all $i \in N$ fix an enumeration $\{\varphi_i^n \mid n \in N\}$ of the $\forall$ formulas in π_i. Then
set:

$$X = T \cup \{(\varphi_0^0 \wedge \ldots \wedge \varphi_0^{n_0}) \vee \ldots \vee (\varphi_i^0 \wedge \ldots \wedge \varphi_i^{n_i}) \mid i, n_0 \ldots n_i \in N\}.$$

We show that X satisfies the claim of the theorem.

Note that $T \cup \{\varphi_0^n \mid n \in N\}$ is consistent by hypothesis. This implies that X
is consistent.

Let revision function $\dot{+}$, $j \in N$, $\mathcal{S} \in P_j$, full assignment h to $\mathcal{S}$, and environment e for $\mathcal{S}$ and h be given. To finish the proof we must show that:

(70)   for cofinitely k, $\emptyset \neq MOD(X \dot{+} e[k]) \subseteq P_j$.

By Definition 3.(28), let i_0 be least with $\mathcal{S} \models \pi_{i_0}[h]$. For $i < i_0$, $\mathcal{S} \not\models \pi_i[h]$, so
for each $i < i_0$ we may choose $c(i) \in N$ such that:

(71)   (a) $\mathcal{S} \models (\varphi_i^0 \wedge \ldots \wedge \varphi_i^{c(i)-1})[h]$;

       (b) $\mathcal{S} \not\models \varphi_i^{c(i)}[h]$.

Since for all $i < i_0$, $\varphi_i^{c(i)}$ is a $\forall$ formula, and since h is onto $|\mathcal{S}|$, it follows from
(71)b that there is $k_0 > 0$ such that:

(72)   for all $k \geq k_0$ and for all $i < i_0$, range$(e[k]) \cup \{\varphi_m^{c(i)}\}$ is inconsistent.

Let $Y = \{(\varphi_0^0 \wedge \ldots \wedge \varphi_0^{n_0}) \vee \ldots \vee (\varphi_i^0 \wedge \ldots \wedge \varphi_i^{n_i}) \mid i < i_0, n_0 \geq c(0) \ldots n_i \geq$
$c(i)\}$. We deduce from (72) that:

(73)   for all $k \geq k_0$ and for all $\varphi \in Y$, range$(e[k]) \cup \{\varphi\}$ is inconsistent.

We deduce from (71)a and the fact that $\mathcal{S} \models \pi_{i_0}[h]$ that:

(74)   for all $k \in N$, range$(e[k]) \cup (X - Y)$ is consistent.

Suppose that contraction function $\dot{-}$ underlies $\dot{+}$. Then (73), (74), and Exercise (20) imply that for all $k \geq k_0$, $X \dot{-} \neg \bigwedge e[k] = X - Y$, hence:

(75)   for all $k \geq k_0$, $X \dot{+} e[k] = $ range$(e[k]) \cup (X - Y)$.

Now note that $Z = \{(\varphi_0^0 \wedge \ldots \wedge \varphi_0^{c(0)}) \vee \ldots \vee (\varphi_{i_0-1}^0 \wedge \ldots \wedge \varphi_{i_0-1}^{c(i_0-1)}) \vee (\varphi_{i_0}^0 \wedge$
$\ldots \wedge \varphi_{i_0}^n) \mid n \in N\} \subseteq X - Y$. Moreover, (72) implies that range$(e[k]) \cup Z \models$

$\{\varphi_{i_0}^n \mid n \in N\}$ for all $k \geq k_0$. This with (75) implies that $X \dotplus e[k] \models T \cup \{\varphi_{i_0}^n \mid n \in N\}$ for all $k \geq k_0$. Hence $X \dotplus e[k] \models T \cup \pi_{i_0}$ for all $k \geq k_0$. This with Definition 3.(28) and the hypothesis that $\mathcal{S} \models \pi_{i_0}[h]$ implies (70). ∎

Theorem (69) suggests that our definition of revision function is sufficiently strict, since it does not tolerate revision strategies that are inapt for science. The definition also appears not to be excessively strict, since Theorems (27) and (63) show that all solvable problems in a broad class can be solved by stringent revision. Let us therefore rejoice in the conviction that a sound and sufficient conception of revision has been achieved. Indeed, the moment for rejoicing is well chosen since our satisfaction will prove shortlived! In Section 4.4.2 we will encounter revision functions that cannot be used for the efficient solution of certain problems of form $(T, \{\theta, \neg\theta\})$. For now, however, all is well. Indeed, things get even better when we specialize our problems to those whose background theories are finitely axiomatizable.

4.3.4 Supplementing as a Function of Finitely Axiomatizable Theories

In this subsection and the next we look more closely at the way background theories can be successfully extended. The extension X of T foreseen in Theorem (69) depends not just on T but more generally on the problem $(T, \{P_0, P_1, \ldots\})$. If we limit attention to problems of the form $(T, \{\theta_0, \ldots, \theta_n\})$, and add the hypothesis that T is finitely axiomatizable, then T can be successfully extended without concern for the partition imposed by $\theta_0 \ldots \theta_n$. This is the burden of the next proposition. In the succeeding subsection we show that the hypothesis of finite axiomatizability cannot be lifted. In other words, in the absence of finite axiomatizability successful extension of a theory T depends on which problem $(T, \{\theta_0, \ldots, \theta_n\})$ is to be solved.

For the impatient reader we note that questions involving finite axiomatizability are not central to the message of this chapter. So the remainder of the present section (viz., 4.3.4 and 4.3.5) can be omitted on a first reading, and the discussion picked up in Section 4.4.

(76) PROPOSITION: Let $T \subseteq \mathcal{L}_{sen}$ be consistent and finitely axiomatizable. Then there is consistent $X \subseteq \mathcal{L}_{form}$ with $X \models T$ such that for every revision function $\dotplus$, $\lambda\sigma . X \dotplus \sigma$ solves every solvable problem of form $(T, \{\theta_0, \ldots, \theta_n\})$.

Proof: First we define X, using some notation. Fix an enumeration $\{\varphi_m \mid m \in N\}$ of all $\forall$ formulas. Let $(n_1 \ldots n_k) \in N^k$ be a strictly increasing sequence of k numbers. By induction, we associate a formula χ_{n_i} with each n_i appearing in

$(n_1 \ldots n_k)$. It will turn out that χ_{n_i} is a disjunction over $\{\varphi_m \mid m \in N\}$, and that φ_{n_i} is the disjunct of highest index in χ_{n_i}.

Basis step: χ_{n_1} is $\varphi_0 \vee \ldots \vee \varphi_{n_1}$. Observe that χ_{n_1} is a disjunction over $\{\varphi_m \mid m \in N\}$, and that φ_{n_1} is the disjunct of highest index in χ_{n_1}.

Induction step: Let $0 < i < k$ be given. Then $n_{i+1} > n_i$. Suppose that χ_{n_i} has been defined to be a disjunction over $\{\varphi_m \mid m \in N\}$ such that φ_{n_i} is the disjunct of highest index in χ_{n_i}. Let χ' be the result of suppressing φ_{n_i} in χ_{n_i}. Then $\chi_{n_{i+1}}$ is $\chi' \vee \varphi_{n_i+1} \vee \ldots \vee \varphi_{n_{i+1}}$. Observe that $\chi_{n_{i+1}}$ is a disjunction over $\{\varphi_m \mid m \in N\}$, and that $\varphi_{n_{i+1}}$ is the disjunct of highest index in $\chi_{n_{i+1}}$.

Finally, we associate the formula $\chi_{n_1} \wedge \ldots \wedge \chi_{n_k}$ with $(n_1 \ldots n_k)$. For example, the formula associated with $(2, 5, 7)$ is:

$$(\varphi_0 \vee \varphi_1 \vee \varphi_2) \wedge (\varphi_0 \vee \varphi_1 \vee \varphi_3 \vee \varphi_4 \vee \varphi_5) \wedge (\varphi_0 \vee \varphi_1 \vee \varphi_3 \vee \varphi_4 \vee \varphi_6 \vee \varphi_7).$$

Let $t \in \mathcal{L}_{sen}$ axiomatize T. Then we define X to be the set of all consistent formulas of form $t \wedge \chi$, where χ is any formula associated with some nonempty, increasing sequence of natural numbers.

The key to our construction is the following fact.

(77)   FACT: For every $\delta, \delta' \in X$, $\delta \models \delta'$ or $\delta' \models \delta$.

Proof of Fact (77):   Let γ and γ' be distinct, nonempty, finite, increasing sequences of numbers. Then there is $\lambda \in N^{<\omega}$ and $n \in N$ such that exactly one of the following holds:

(a) $\lambda * n$ is an initial segment of γ and for all $m \leq n$, $\lambda * m$ is not an initial segment of γ';

(b) $\lambda * n$ is an initial segment of γ' and for all $m \leq n$, $\lambda * m$ is not an initial segment of γ.

Suppose λ and n satisfy (a); the other case is exactly parallel. Let formula χ be associated with γ, and formula χ' be associated with γ'. Then it can be seen that the first length(λ) conjuncts in χ are the same as the first length(λ) conjuncts in χ', and that all of the disjuncts appearing in the length$(\lambda) + 1$st conjunct of χ appear as disjuncts in each of the conjuncts of χ' that come after the first length(λ) ones. So $\chi \models \chi'$. Hence $t \wedge \chi \models t \wedge \chi'$, and all $\delta \in X$ are of the form $t \wedge \chi$.   ∎

We now show that X satisfies the claim of the proposition. Since every member of X is consistent, it follows from Fact (77) that X is consistent. Let there be given solvable problem of form $(T, \{\theta_0, \ldots, \theta_n\})$, revision function $\dotplus$, and $i \leq n$. To finish the proof we show that $\lambda\sigma \ . \ X \dotplus \sigma$ solves $MOD(T \cup \{\theta_i\})$. Let $\mathcal{S} \in MOD(T \cup \{\theta_i\})$, full assignment h to $\mathcal{S}$, and environment e for $\mathcal{S}$ and h be given. By Theorem 3.(55), there is $\exists\forall$ sentence θ such that:

(78) $T \models \theta_i \leftrightarrow \theta$.

Since h is onto $|\mathcal{S}|$, this implies the existence of $m_0 \in N$ such that the existential closure of φ_{m_0} is θ, and $\mathcal{S} \models \varphi_{m_0}[h]$. From Exercise (21) and Fact (77) we deduce that:

(79)   for all $k \in N$, $X \dotplus e[k]$ includes the set of all formulas in X that are consistent with $\bigwedge e[k]$.

Denote by γ the increasing sequence of integers that ends with m_0 and such that for all $m \leq m_0$, m occurs in γ if and only if $\mathcal{S} \models \varphi_m[h]$. Denote by χ the formula associated with γ. It is immediate that $t \wedge \chi \in X$, and that for all $k \in N, t \wedge \chi$ is consistent with $\bigwedge e[k]$. We can thus deduce from (79) that:

(80)   for all $k \in N$, $t \wedge \chi$ belongs to $X \dotplus e[k]$.

Since h is onto $|\mathcal{S}|$, there is $k_0 \in N$ such that for all $k \geq k_0$ and for all $m \leq m_0$, if $m \notin \text{range}(\gamma)$ then $\bigwedge e[k] \models \neg\varphi_m$. From the definition of χ this can be seen to imply that for all $k \geq k_0$ and for all $m \in \text{range}(\gamma)$, $\text{range}(e[k]) \cup \{\chi\} \models \varphi_m$. In particular:

(81)   for all $k \geq k_0$, $\text{range}(e[k]) \cup \{\chi\} \models \varphi_{m_0}$.

From (78), (80), (81), and the definition of φ_{m_0} we deduce that for all $k \geq k_0$, $X \dotplus e[k] \models T \cup \{\theta_i\}$.   ∎

4.3.5   The Inability to Supplement Independently of T

It would be nice to find a method of extending a theory T that allows all revision functions to succeed on solvable problems of form $(T, \{\theta_0 \ldots \theta_n\})$. In this case the required extension would not depend on the sentences $\theta_0 \ldots \theta_n$, but only on the background theory T. Fulfilling this desire would amount to lifting the hypothesis of finite axiomatizability from Proposition (76). Unfortunately, the next proposition shows this hope to be unrealizable.

(82)   PROPOSITION: Suppose that **Sym** is limited to a binary predicate, a constant, and a unary function symbol. Then there exists $T \subseteq \mathcal{L}_{sen}$ and stringent revision function $\dotplus$ such that for all $B \subseteq \mathcal{L}_{form}, \lambda\sigma \ . \ B \dotplus \sigma$ fails to solve some solvable problem of form $(T, \{\theta, \neg\theta\})$.

Proof: Let R be the binary predicate, $\overline{0}$ the constant, s the unary function symbol of **Sym**. For $n \in N$, let $\overline{n}$ be the result of n applications of s to $\overline{0}$.

Let $\{\chi_i \mid i \in N\}$ enumerate $\mathcal{L}_{form}$, and for all $i \in N$ set $S_i = \{\chi_i\}$. Let $\mathbf{S} = \{S_i \mid i < \omega\}$, and let $\dot{-}_s$ be constructed from $\mathbf{S}$ as in Definition (9). We take $\dot{+}$ to be defined from $\dot{-}_s$ via Definition (12). So $\dot{+}$ is stringent by Proposition (10). Fix a nonrecursive subset E of N, and define T to be:

$$\{(\forall y \, R\overline{2n}y) \rightarrow \overline{2n} = \overline{2m+1} \mid n \in N \text{ and } m \in E\} \cup$$

$$\{(\forall y \, R\overline{2n}y) \rightarrow \overline{2n} \neq \overline{2m+1} \mid n \in N \text{ and } m \notin E\}.$$

Let $B \subseteq \mathcal{L}_{form}$ be given. By Theorem 3.(55), for every universal sentence θ, $(T, \{\theta, \neg\theta\})$ is solvable. So there would be nothing left to prove without assuming:

(83)   For every universal sentence θ, $\lambda\sigma$. $B \dot{+} \sigma$ solves $(T, \{\theta, \neg\theta\})$.

We use (83) to derive a contradiction. Specifically, an environment e for T will be exhibited such that for infinitely many k, there exists $n, m \in N$ with:

(84) $B \dot{+} e[k] \models (\forall y \, R\overline{2n}y) \wedge \overline{2n} \neq \overline{2m+1}$ if $m \in E$,
        $B \dot{+} e[k] \models (\forall y \, R\overline{2n}y) \wedge \overline{2n} = \overline{2m+1}$ if $m \notin E$.

This will imply that for infinitely many k, $B \dot{+} e[k] \not\models T$. So, trivially, $\lambda\sigma.B \dot{+} \sigma$ does not solve $(T, \{\forall x (x = x), \neg\forall x (x = x)\})$, contradicting (83).

Fix an enumeration $\{\alpha_i \mid i \in N\}$ of all atomic formulas. To exhibit the promised environment e we will build by induction on $i \in N$ a sequence $\{\tau_i \mid i \in N\}$ of members of SEQ and a sequence $\{n_i \mid i \in N\}$ of integers such that for all $i \in N$:

(85)   (a)   for all $j < i$, $\tau_j \subseteq \tau_i$;

        (b)   exactly one of α_i or $\neg\alpha_i$ occurs in τ_i;

        (c) $\bigwedge \tau_i \models \neg\forall y \, R\overline{2i}y$;

        (d) $B \dot{+} \tau_i \models \forall y \, R\overline{2n_i}y$;

        (e)   for some $m \in E$, $B \dot{+} \tau_i \models \overline{2n_i} \neq \overline{2m+1}$, or for some $m \notin E$, $B \dot{+} \tau_i \models \overline{2n_i} = \overline{2m+1}$.

Then we will set $e = \bigcup_{i \in N} \tau_i$. It follows from (85)a,b that e is an environment, which by (85)c is for T. It follows from (85)d,e that (84) is satisfied. Let $i \in N$ be given. Suppose that τ_j and n_j have been defined for all $j < i$ and satisfy

(85) for $i = j$. We define τ_i and n_i that satisfy (85). Choose $n_i \neq i$ such that $\overline{2n_i}$ does not appear in τ_j for any $j < i$. We may choose an environment d for $T \cup \{\neg \forall y R \overline{2i} y, \forall y R \overline{2n_i} y\}$ that extends τ_j for all $j < i$. By (83) we deduce the existence of $k_0 \in N$ with:

(86)   (a)   for all $j < i$, $d[k_0]$ extends τ_j;

     (b)   exactly one of α_i, $\neg \alpha_i$ occurs in $d[k_0]$;

     (c) $\bigwedge d[k_0] \models \neg \forall y R \overline{2i} y$;

     (d) $B \dotplus d[k_0] \models \forall y R \overline{2n_i} y$.

By (86)d and compactness, let finite $D \subseteq (B \mathbin{\dot{-}_s} \neg \bigwedge d[k_0])$ be such that $D \cup \{\bigwedge d[k_0]\} \models \forall y R \overline{2n_i} y$. Denote by p the greatest integer such that $\chi_p \in D$, and set $D' = \{\chi_0 \dots \chi_p\} \cap (B \mathbin{\dot{-}_s} \neg \bigwedge d[k_0])$. So:

(87) $D \subseteq D' \subseteq B \dotplus d[k_0]$.

Since E is non recursive there is $m \in E$ with $D' \cup \{\bigwedge d[k_0]\} \not\models \overline{2n_i} = \overline{2m + 1}$, or there is $m \notin E$ with $D' \cup \{\bigwedge d[k_0]\} \not\models \overline{2n_i} \neq \overline{2m + 1}$. Choose such an m, and set $\tau_i = d[k_0] * (\overline{2n_i} \neq \overline{2m + 1})$ if $m \in E$, $\tau_i = d[k_0] * (\overline{2n_i} = \overline{2m + 1})$ if $m \notin E$. From (86)a-c we infer (85)a-c immediately. By (87), the definition of $\mathbin{\dot{-}_s}$ and the choice of m we obtain: $D' \subseteq B \mathbin{\dot{-}_s} \neg \bigwedge \tau_i$, hence $B \dotplus \tau_i \models \forall y R \overline{2n_i} y$, verifying (85)d. By Lemma (13)a, $B \dotplus \tau_i \models (\overline{2n_i} \neq \overline{2m + 1})$ if $m \in E$, and $B \dotplus \tau_i \models (\overline{2n_i} = \overline{2m + 1})$ if $m \notin E$, verifying (85)e.   ∎

The proposition shows that in the general case, theory extensions must be chosen as a function of the problem to be solved. Otherwise, some revision functions will be led to needless failure.

4.3.6   Exercises

(88)   EXERCISE: Let $T \subseteq \mathcal{L}_{sen}$ and $\theta \in \mathcal{L}_{sen}$ be such that θ is equivalent in T to a $\forall$ sentence φ. Show that for every revision function $\dotplus$, $\lambda \sigma \,.\, (T \cup \{\varphi\}) \dotplus \sigma$ solves $(T, \{\theta, \neg\theta\})$.

(89)   EXERCISE: ♦   Let decidable, recursively axiomatisable $T \subseteq \mathcal{L}_{sen}$ and solvable problem of form $(T, \{\theta_0, \dots, \theta_n\})$ be given. Show that there is consistent $X \subseteq \mathcal{L}_{form}$ with $X \models T$ such that the following holds.

(a)   There is stringent revision function $\dotplus$ such that $\lambda \sigma \,.\, X \dotplus \sigma$ is computable.

(b)   For every revision function $\dotplus$ (computable or not), $\lambda \sigma \,.\, X \dotplus \sigma$ solves $(T, \{\theta_0, \dots, \theta_n\})$.

(90)   EXERCISE: A mapping $\pm : \mathrm{pow}(\mathcal{L}_{form}) \times SEQ \rightarrow \mathrm{pow}(\mathcal{L}_{form})$ is an *external revision function* just in case there is contraction function $\dot{-}$ such that for all $B \subseteq \mathcal{L}_{form}$ and $\sigma \in SEQ$,

$$B \pm \sigma = \begin{cases} B & \text{if } \sigma = \emptyset \\ (B \cup \mathrm{range}(\sigma)) \dot{-} \neg \bigwedge \sigma & \text{otherwise.} \end{cases}$$

(This definition is based on [Hansson, 1993b].) Prove the following variant of Theorem (69).

Let solvable problem **P** be of form $(T, \{P_0, P_1, \ldots\})$. Then there is a consistent extension $X \subseteq \mathcal{L}_{form}$ of T such that for every external revision function $\pm$, $\lambda \sigma . X \pm \sigma$ solves **P**.

(91)   EXERCISE: ♥   Suppose that **Sym** consists of a binary predicate, a constant, and a unary function symbol. Prove the following.

There exists $\theta \in \mathcal{L}_{sen}$ and stringent revision function $\dotplus$ with the following property. For all $B \subseteq \mathcal{L}_{form}$ there is $T \subseteq \mathcal{L}_{sen}$ such that:

(a) $(T, \{\theta, \neg\theta\})$ is solvable;

(b) $\lambda \sigma . (T \cup B) \dotplus \sigma$ does not solve $(T, \{\theta, \neg\theta\})$.

4.4   Efficient Inquiry via Belief Revision

Recall that the goal of the present chapter is to represent inquiry as a process of rational belief revision. The theorems of the preceding section show that for a wide class of problems, revision-based scientists can be made to succeed whenever success is possible in principle. We now consider efficient inquiry in the same terms. Our question is whether and to what extent efficient inquiry can be carried out using belief revision.

First it will be shown that belief revision is compatible with efficient solvability for problems of the form $(T, \{P_0, P_1, \ldots\})$; that is, for every solvable problem of this form, some belief revision function can be made to work efficiently. In contrast, it will then be demonstrated that not all belief revision functions are suitable for efficient discovery. Indeed, even some stringent ones cannot be used to efficiently solve certain "easy" problems. The consequences of this finding for our theory of inquiry are then discussed.

Throughout the section we rely on the conception of efficient inquiry introduced in Definitions 3.(73) and 3.(74). In rough summary, scientist Ψ is *efficient* on problem **P** if Ψ succeeds faster than any rival in many environments for **P**, where a "rival" is a scientist that succeeds faster than Ψ on even one environment for **P**. In contrast, Ψ is *dominated* on **P** if a rival scientist succeeds at least as quickly as Ψ on all environments for **P**, and more quickly on some.

4.4.1   Efficiency by Revision Is Possible

Theorem (47) of Section 4.2.5 provides a strong sense in which belief revision is compatible with efficient solvability. Indeed, it was there shown that a sole, stringent revision function suffices to efficiently solve every problem, provided only that the problem be solvable in the revision-based way. The theorem can be used in conjunction with Propositions (31) and 3.(61) to demonstrate the fact that occupies the present subsection, namely: efficient inquiry can be carried out on a broad class of problems by revision-based scientists whose starting points preserve the background theory. To prove this result, however, we prefer not to rely on the difficult arguments of Sections 4.2.5 and 4.2.4. Instead, we offer a simpler demonstration, which builds on the proof of Theorem (69).

(92)   THEOREM: For every solvable problem **P** of the form $(T, \{P_0, P_1, \ldots\})$ there is a consistent extension $X \subseteq \mathcal{L}_{form}$ of T and a stringent revision function $\dotplus$ such that $\lambda\sigma . X \dotplus \sigma$ solves **P** efficiently.

Proof: Let solvable problem $\mathbf{P} = (T, \{P_0, P_1, \ldots\})$ be given. Recall from the proof of Theorem (69) the enumeration $\{\pi_i \mid i \in N\}$ of the π-sets consistent with T, and the enumeration $\{\varphi_i^n \mid n \in N\}$ of π_i. Recall also the definition of $X \subseteq \mathcal{L}_{form}$. For all $i \in N$, set $S_i = T \cup \{(\varphi_0^0 \wedge \ldots \wedge \varphi_0^{n_0}) \vee \ldots \vee (\varphi_i^0 \wedge \ldots \wedge \varphi_i^{n_i}) \mid n_0 \ldots n_i \in N\}$. Fix an enumeration $\{\psi_i \mid i \in N\}$ of $\mathcal{L}_{form}$, and set $S_{\omega+i} = \{\psi_i\}$ for all $i \in N$. Let $\mathbf{S} = \{S_\alpha \mid \alpha < \omega \times 2\}$, and let $\dotminus_S$ be constructed from $\mathbf{S}$ as in Definition (9). We take $\dotplus$ to be defined from $\dotminus_S$ via Definition (12). So, $\dotplus$ is stringent by Proposition (10). The proof of Theorem (69) shows that $\lambda\sigma . X \dotplus \sigma$ solves **P**. To complete the proof of the theorem we show that $\lambda\sigma . X \dotplus \sigma$ solves **P** efficiently.

Let $\sigma \in SEQ$ be for **P**. By Definition 3.(28) there is least $i_0 \in N$ such that $T \cup \pi_{i_0} \cup \text{range}(\sigma)$ is consistent. Let $P \in \mathbf{P}$, $\mathcal{S} \in P$, and full assignment h to $\mathcal{S}$ be such that $\mathcal{S} \models \pi_{i_0} \cup \text{range}(\sigma)[h]$. Let $\tau \in SEQ$ extend σ and be such that $\mathcal{S} \models \bigwedge \tau[h]$. As an immediate consequence of our choice of $\dotplus$, we have: $S_{i_0} \subseteq X \dotplus \tau$. Moreover by compactness, for all $i < i_0$ we may choose $c(i) \in N$

such that $T \cup \text{range}(\sigma) \cup \{\varphi_i^0 \wedge \ldots \wedge \varphi_i^{c(i)}\}$ is inconsistent. Since $\{(\varphi_0^0 \wedge \ldots \wedge \varphi_0^{c(0)}) \vee \ldots \vee (\varphi_{i_0-1}^0 \wedge \ldots \wedge \varphi_{i_0-1}^{c(i_0-1)}) \vee (\varphi_{i_0}^0 \wedge \ldots \wedge \varphi_{i_0}^n) \mid n \in N\} \subseteq S_{i_0}$, this implies that $S_{i_0} \cup \text{range}(\sigma) \models \{\varphi_{i_0}^0 \wedge \ldots \wedge \varphi_{i_0}^n \mid n \in N\}$. With $S_{i_0} \subseteq X \dot{+} \tau$ we deduce that $X \dot{+} \tau \models T \cup \pi_{i_0}$ which implies via Definition 3.(28) (and for the case $\sigma = 0$, the fact that X is consistent) that $\emptyset \neq MOD(X \dot{+} \tau) \subseteq P$. It follows immediately from Exercise 3.(87) that $\lambda\sigma . X \dot{+} \sigma$ solves $\mathbf{P}$ efficiently. ∎

4.4.2   Efficiency by Revision Is Not Inevitable

Theorem (92) shows that there are enough revision functions to carry out efficient inquiry on problems of form $(T, \{P_0, P_1, \ldots\})$. Moreover, just the stringent subclass is sufficient. It would be pleasing to report the converse fact as well, namely, that every revision function can be used efficiently. Theorem (69) would be reinforced thereby, since the latter states that every revision function can be used to solve any solvable problem of form $(T, \{P_0, P_1, \ldots\})$. Unfortunately, there is no such counterpart to Theorem (69). Some revision functions cannot be brought to efficiently solve certain solvable problems. Worse, the guilty revision functions can be taken to be stringent, and the recalcitrant problem has a particularly simple form. Here is a precise statement of the situation.

(93)   THEOREM: Suppose that $\mathbf{Sym}$ is limited to countably many constants. Then there is a problem of form $(T, \{\theta, \neg\theta\})$ and a stringent revision function $\dot{+}$ with the following properties.

(a) T is recursive.

(b) $(T, \{\theta, \neg\theta\})$ is solvable.

(c) For all $B \subseteq \mathcal{L}_{form}$, if $\lambda\sigma . B \dot{+} \sigma$ solves $(T, \{\theta, \neg\theta\})$ then $\lambda\sigma . B \dot{+} \sigma$ is dominated on $(T, \{\theta, \neg\theta\})$.

Proof: Let a, b, and $\bar{n}$, $n \in N$, be distinct constants. (In view of the assumptions on $\mathbf{Sym}$ made in Section 3.1.2, we assume that this can be done via a total computable isomorphism between N and the set of constants.) Choose a recursive subset E of N that is not primitive recursive. We take θ to be $a = b$, and T to be the following set of sentences:

$$\begin{cases} \theta \leftrightarrow a = \bar{n}, & \text{for all } n \in E, \\ \theta \leftrightarrow a \neq \bar{n}, & \text{for all } n \notin E. \end{cases}$$

It is immediate that (93)a,b are satisfied. For (93)c, we say that $B \subseteq \mathcal{L}_{form}$ *disagrees* with T if there is $n \in N$ with either ($n \in E$ and $B \models a \neq \bar{n}$) or ($n \in$

$\overline{E}$ and $B \models a = \overline{n}$); otherwise, we say that B *agrees* with T. Fix an enumera-
tion $\{\varphi_i \mid i \in N\}$ of $\mathcal{L}_{form}$, and set $S_i = \{\varphi_i\}$ for all $i \in N$. Let $\mathbf{S} = \{S_i \mid i < \omega\}$,
and let $\dot{-}_\mathbf{S}$ be constructed from $\mathbf{S}$ as in Definition (9). We take $\dotplus$ to be de-
fined from $\dot{-}_\mathbf{S}$ via Definition (12). So, $\dotplus$ is stringent by Proposition (10). To
verify (93)c, let $B \subseteq \mathcal{L}_{form}$ be given, and define $\Psi = \lambda\sigma \,.\, B \dotplus \sigma$. We suppose
that Ψ solves $(T, \{\theta, \neg\theta\})$ since otherwise there is nothing left to prove. To
complete the proof it must be shown that Ψ is dominated on $(T, \{\theta, \neg\theta\})$. Let
$P_1 = MOD(T \cup \{\theta\})$, $P_2 = MOD(T \cup \{\neg\theta\})$, and let scientist Θ be defined as
follows.

(94)   For all $\sigma \in SEQ$,

$$\Theta(\sigma) = \begin{cases} P_1 & \text{if } T \cup \text{range}(\sigma) \models \theta. \\ P_2 & \text{if } T \cup \text{range}(\sigma) \models \neg\theta. \\ P_1 & \text{if } T \cup \text{range}(\sigma) \not\models \theta, T \cup \text{range}(\sigma) \not\models \neg\theta, \text{ and neither} \\ & \emptyset \neq \Psi(\sigma) \subseteq P_1, \text{ nor } \emptyset \neq \Psi(\sigma) \subseteq P_2. \\ \Psi(\sigma) & \text{otherwise.} \end{cases}$$

It follows that Θ solves $(T, \{\theta, \neg\theta\})$. Hence, for all environments e for $P \in$
$\{P_1, P_2\}$, $SP(\Theta, e, P)$ and $SP(\Psi, e, P)$ are well defined, and $SP(\Theta, e, P) \leq$
$SP(\Psi, e, P)$, as easily verified. Thus, by Definition 3.(74) it suffices to show
that there is an environment e for some $P \in \{P_1, P_2\}$ such that $SP(\Theta, e, P) <$
$SP(\Psi, e, P)$. This is demonstrated via the following, exhaustive cases.

(95)   (a) $B \models \theta$ iff $B \models \neg\theta$.

    (b) B is consistent, $B \models \theta$, and B disagrees with T.

    (c) B is consistent, $B \models \theta$, and B agrees with T.

    (d)   Same as (b) except that θ is replaced by $\neg\theta$.

    (e)   Same as (c) except that θ is replaced by $\neg\theta$.

If (95)a, let e be any environment for P_1 such that for some $n \in E$, $e(0) =$
$(a = \overline{n})$. By Definition (12), $B \dotplus e[0] = B \dotplus \emptyset = B$, so by (95)a, neither $\emptyset \neq$
$\Psi(e[0]) \subseteq P_1$ nor $\emptyset \neq \Psi(e[0]) \subseteq P_2$. Since $T = T \cup \text{range}(e[0]) \not\models \theta$ and $T \cup$
$\text{range}(e[0]) \not\models \neg\theta$, the third clause of (94) implies that $\Theta(e[0]) = \Theta(\emptyset) = P_1$.
Moreover, for all $k > 0$, $T \cup \text{range}(e[k])$ is consistent and implies θ. Hence,
the first clause of (94) implies $SP(\Theta, e, P_1) = 0$. On the other hand, since it is
not the case that $\emptyset \neq \Psi(e[0]) \subseteq P_1$, $SP(\Psi, e, P_1) > 0$.

If (95)b, then either there is $n \in E$ with $B \models a \neq \overline{n}$ or $n \in \overline{E}$ with $B \models$
$a = \overline{n}$. We consider these subcases in turn. Suppose n is such that $n \in E$ and

$B \models a \neq \bar{n}$. Let e be an environment for P_2 such that $e(0) = (a \neq \bar{n})$. Then by Lemma (13)c and the fact that B is consistent, $\Psi(e[1]) = B \dotplus e[1] \models B$, so $\Psi(e[1]) \models \theta$. It follows that $\mathrm{SP}(\Psi, e, P_2) > 1$. In contrast, the definition (94) of Θ implies that $\mathrm{SP}(\Theta, e, P_2) = 1$. Now suppose n is such that $n \in \bar{E}$ and $B \models a = \bar{n}$. Let e be an environment for P_2 such that $e(0) = (a = \bar{n})$. Then again by Lemma (13)c and the fact that B is consistent, $\Psi(e[1]) = B \dotplus e[1] \models B$, so $\Psi(e[1]) \models \theta$. It follows that $\mathrm{SP}(\Psi, e, P_2) > 1$, whereas (94) implies $\mathrm{SP}(\Theta, e, P_2) = 1$.

Suppose that (95)c holds. We rely on the following fact, an immediate corollary to the proof that the monadic predicate calculus is decidable (see [Boolos & Jeffrey, 1989, Ch. 25]).

(96)   FACT: Suppose that **Sym** is limited to constants. Then the function that associates to every $\phi \in \mathcal{L}_{form}$ the value 1 if $\models \phi$, and the value 0 if $\not\models \phi$, is primitive recursive.

Let $i_0 \in N$ be least such that $D = B \cap \{\varphi_i \mid i < i_0\} \models \theta$. Then there exists $n \in E$ such that $D \not\models a = \bar{n}$, or there exists $n \in \bar{E}$ such that $D \not\models a \neq \bar{n}$. Indeed, if this were not the case then (96) would imply that the characteristic function of E is primitive recursive, contradicting the choice of E. We consider the two sub-cases in turn. Assume n is such that $n \in E$ and $D \not\models a = \bar{n}$. Let e be an environment for P_2 such that $e(0) = (a \neq \bar{n})$. Then by the choice of i_0 and the definition of $\dotplus$, $D \subseteq B \dotplus e[1] = \Psi(e[1])$, hence $\Psi(e[1]) \models \theta$. So $\mathrm{SP}(\Psi, e, P_2) > 1$. In contrast, $\mathrm{SP}(\Theta, e, P_2) = 1$ since $T \cup \mathrm{range}(e[1]) \models \neg\theta$. Now assume n is such that $n \in \bar{E}$ and $D \not\models a \neq \bar{n}$. Let e be an environment for P_2 such that $e(0) = (a = \bar{n})$. Then, once again, by the choice of i_0 and the definition of $\dotplus$, $D \subseteq B \dotplus e[1] = \Psi(e[1])$, hence $\Psi(e[1]) \models \theta$. So $\mathrm{SP}(\Psi, e, P_2) > 1$, whereas $\mathrm{SP}(\Theta, e, P_2) = 1$.

The argument for (95)d is parallel to that for (95)b, and the argument for (95)e is parallel to that for (95)c.   ∎

Although the stringent revision functions possess strong credentials of rationality, their properties do not guarantee that they can be used for efficient inquiry. This is what the foregoing theorem shows. To state the matter more generally, let us call a revision function $\dotplus$ "efficient" just in case for every solvable problem **P** of the form $(T, \{\theta, \neg\theta\})$ there is consistent $X \subseteq \mathcal{L}_{form}$ such that $\lambda\sigma \,.\, X \dotplus \sigma$ solves **P** efficiently. Theorems (47) and (93) show that some but not all stringent revision functions are efficient.

The inefficient revision functions are in some way defective, since it is not rational to conduct inquiry in needlessly dilatory fashion. In particular,

Theorem (92) shows that inefficient revision can always be accelerated by a rival method of belief change, hence there would seem to be no justification for using the slower function. The question thus arises as to additional conditions that can be imposed on revision functions in view of guaranteeing efficiency. Such conditions should be intuitively justifiable in terms of modifying belief in a rational way, and also pick out the efficient subset of stringent functions. At present we have no proposal to offer, and are reduced to noting that the conditions need to be strict, since the proof of Theorem (93) reveals a broad class of inefficient, stringent functions.

4.4.3 Exercises

(97) EXERCISE: ◆ Let $T \subseteq \mathcal{L}_{sen}$ be consistent and finitely axiomatizable. Let solvable problem of form $(T, \{\theta_0, \ldots, \theta_n\})$ be given. Show that there is consistent $X \subseteq \mathcal{L}_{form}$ with $X \models T$ such that for every revision function $\dotplus$, $\lambda\sigma \, . \, X \dotplus \sigma$ efficiently solves $(T, \{\theta_0, \ldots, \theta_n\})$.

4.5 Closure

Recall that belief states in our theory are interpreted as arbitrary subsets of formulas, not necessarily closed under deduction. In other words, belief states need not include all of their logical consequences. The present section explores the impact on revision-based inquiry of requiring the starting point B of scientist $\lambda\sigma \, . \, B \dotplus \sigma$ to be closed under some fragment of logic. We shall see that closed starting points are consistent with revision-based success. However, even a very weak form of closure perturbs Theorem (69). That is, not every revision function can be used to solve every solvable problem of the form $(T, \{P_0, P_1, \ldots\})$ if the starting point is required to include its own logical consequences.

We begin our discussion with some remarks about the epistemological issues surrounding deductive closure of belief states. The following notation will be used. Given $B \subseteq \mathcal{L}_{form}$, we use $\mathrm{Cn}(B)$ to denote the set of logical consequences of B. Thus, $B \subseteq \mathrm{Cn}(B)$, and B is closed under Cn if and only if $B = \mathrm{Cn}(B)$.

4.5.1 Closure Operators

Within the literature on belief revision, two conceptions of belief states have emerged. According to the "coherence" view, belief states are deductively closed, and the agent is committed in the same way to statements whose conviction is rooted directly in experience as she is to their logical consequences.

The coherence approach is elaborated and defended in [Gärdenfors, 1988]. In contrast, belief states within the "foundationalist" approach contain only those formulas which the agent has extra-logical reason to believe, and thus are not in general deductively closed. The foundationalist approach is developed in [Fuhrmann, 1991, Hansson, 1992, Hansson, 1993b, Hansson, 1993a, Nayak, 1994a] and in work cited there. Both approaches advocate logical integrity in the following sense. A rational agent confronted with information that her belief state implies a falsehood is required to abandon some beliefs in order to remove the implication. However, foundationalism provides more guidance than coherence regarding what to abandon and what to conserve. To see this, let ϕ, ψ be logically independent formulas, and consider $A_1 = \{\phi, \psi\} \dotplus \neg\phi$ versus $A_2 = \text{Cn}(\{\phi, \psi\}) \dotplus \neg\phi$. For every revision function $\dotplus$, $A_1 = \{\neg\phi, \psi\}$. In contrast, A_2 can vary widely. Thus, for one choice of $\dotplus$, A_2 includes ψ but not $\psi \rightarrow \phi$, whereas the situation is reversed for another choice. Giving such liberty to revision functions leads some of them to mischief, as will be seen shortly.

Closure under Cn versus no closure at all can be viewed as extremes along a gradient. In order to consider intermediate points we rely on the following definition.

(98) DEFINITION: By a *closure operator* is meant any function cl to and from subsets of $\mathcal{L}_{form}$ such that for all $B \subseteq \mathcal{L}_{form}$: $B \subseteq \text{cl}(B) \subseteq \text{Cn}(B)$.

At one end, we have the identity closure operator cl defined by $\text{cl}(B) = B$ for every $B \subseteq \mathcal{L}_{form}$. At the other end is deductive consequence. When belief states are closed under a closure operator cl, then revision-based scientists have the form $\lambda\sigma . \text{cl}(B) \dotplus \sigma$, for some $B \subseteq \mathcal{L}_{form}$ and revision function $\dotplus$. If B and $\dotplus$ are chosen astutely, revision-based scientists can still be made to work under these conditions. Indeed, examination of the proof of Theorem (63) shows that it can be strengthened to the following.

(99) THEOREM: There exists $X \subseteq \mathcal{L}_{form}$ and stringent revision function $\dotplus$ such that for all closure operators cl and all consistent $T \subseteq \mathcal{L}_{sen}$, the following holds.

(a) $T \cup X$ is consistent.

(b) $\lambda\sigma . \text{cl}(T \cup X) \dotplus \sigma$ solves every solvable problem of the form $(T, \{\theta_0, \ldots, \theta_n\})$.

The foregoing theorem provides a sense in which closure does no harm to revision-based inquiry. Closure nonetheless complicates matters inasmuch as closed starting points require the revision function to be chosen with care if inquiry is to succeed. This is explained next.

4.5.2 Discovery under Coherence-like Starting Points

Theorem (69) showed that every revision function can be used to solve any solvable problem of the form $(T, \{P_0, P_1, \ldots\})$. Requiring belief states to be deductively closed, however, perturbs this result. In fact, the following theorem shows that it is enough to close belief states under the single rule: $\psi, \phi \,/\, \psi \to \phi$. Belief state $B \subseteq \mathcal{L}_{form}$ is closed under the latter rule just in case $\psi, \phi \in B$ implies $\psi \to \phi \in B$. Notice how weak and innocent this condition appears to be!

(100) THEOREM: Let **Sym** consist of a binary predicate. Then there exists a problem of form $(T, \{\theta, \neg\theta\})$ (with T finite), and maxichoice revision function $\dotplus$ such that:

(a) $(T, \{\theta, \neg\theta\})$ is solvable;

(b) for all $B \subseteq \mathcal{L}_{form}$, if B is closed under $\psi, \phi \,/\, \psi \to \phi$, then $\lambda\sigma \,.\, B \dotplus \sigma$ does not solve $(T, \{\theta, \neg\theta\})$.

Proof of the theorem relies on the following lemma.

(101) LEMMA: Suppose that $Y \subseteq \mathcal{L}_{form}$ is closed under $\psi, \phi \,/\, \psi \to \phi$. Let $\sigma \in SEQ$, and $\theta \in \mathcal{L}_{sen}$ be given with $Y \cup \text{range}(\sigma)$ inconsistent and $\text{range}(\sigma) \not\models \theta$. Then there is $Z \in Y \perp \neg \bigwedge \sigma$ such that $Z \cup \text{range}(\sigma) \not\models \theta$.

Proof: Suppose that $Y \subseteq \mathcal{L}_{form}$ is closed under $\psi, \phi \,/\, \psi \to \phi$, and let σ, θ be such that:

(102) (a) $Y \cup \text{range}(\sigma)$ is inconsistent.

(b) $\text{range}(\sigma) \not\models \theta$.

We shall demonstrate the existence of a sequence $\{\chi_i \mid i \in N\}$ of formulas with the following properties.

(103) (a) $\{\chi_i \mid i \in N\} \subseteq Y$.

(b) $\{\neg\chi_i \mid i \in N\} \cup \text{range}(\sigma)$ is consistent.

(c)   For every finite subset D of Y, if $D \cup \text{range}(\sigma) \models \theta$, then $D \cup \{\neg\chi_i \mid i \in N\}$ is inconsistent.

This will suffice to prove the lemma, as shown by the following argument.

Proof that the existence of $\{\chi_i \mid i \in N\}$ *satisfying (103) implies Lemma (101):*
Let $A = \{\chi_i \to \delta \mid i \in N,\ \delta \in Y\}$. By (103)a and the hypothesis that Y is
closed under $\psi, \phi \,/\, \psi \to \phi$, we obtain $A \subseteq Y$. By (103)b, A is consistent
with range(σ). So, relying on compactness, A can be built up into $Z \subseteq Y$ such
that:

$(*) A \subseteq Z.$

$(**) Z \in Y \perp \neg \bigwedge \sigma$ [hence, $Z \not\models \neg \bigwedge \sigma.]$

By $(**)$, to finish the proof that (103) implies (101) it suffices to show that
$Z \cup \text{range}(\sigma) \not\models \theta$. For a contradiction, suppose that $Z \cup \text{range}(\sigma) \models \theta$. Then
(103)c and compactness imply that $Z \cup \{\neg\chi_i \mid i \in N\}$ is inconsistent. Hence
by compactness again, there is $i_0 \in N$ such that $Z \models \chi_0 \vee \ldots \vee \chi_{i_0}$. By $(*)$,
$\{\chi_i \to \delta \mid i \le i_0,\ \delta \in Y\} \subseteq A \subseteq Z$, hence $Z \models Y$. In conjunction with (102)a,
however, this contradicts $(**)$.   ∎

To finish the proof of the lemma, we now demonstrate the existence of $\{\chi_i \mid$
$i \in N\}$ satisfying (103). Let $\{D_n \mid n \in N\}$ enumerate the finite subsets of Y,
with $D_0 = \emptyset$. We construct in stages a sequence $\{S_n \mid n \in N\}$ of finite sets of
formulas such that for all $n \in N$:

(104) (a) $S_n \subseteq Y.$

     (b) $\{\neg\xi \mid \xi \in S_n\} \cup \text{range}(\sigma) \not\models \theta.$

     (c)   If $D_n \cup \text{range}(\sigma) \models \theta$ then $D_n \cup \{\neg\xi \mid \xi \in S_n\}$ is inconsistent.

     (d) $S_n \supseteq S_{n-1}$ for $n > 0.$

By enumerating $\bigcup_{n \in N} S_n$, it is clear that we obtain $\{\chi_i \mid i \in N\}$ satisfying
(103).

Basis step: Set $S_0 = \emptyset$. Then (104)a is trivial and (104)b,c follow from $D_0 = \emptyset$
and (102)b.

Induction step: Suppose that S_m has been defined, satisfying (104) at stage
$n = m$. Choice of S_{m+1} depends on the truth of:

(105) $D_{m+1} \cup \text{range}(\sigma) \models \theta.$

If (105) is false then we set $S_{m+1} = S_m$, and (104) is obviously verified at stage
$n = m + 1$. So suppose (105). We show that for some $\delta \in D_{m+1}$:

(106) $\{\neg\xi \mid \xi \in S_m\} \cup \{\neg\delta\} \cup \text{range}(\sigma) \not\models \theta.$

For, suppose that $\{\neg\xi \mid \xi \in S_m\} \cup \{\neg\delta\} \cup \text{range}(\sigma) \models \theta$ for all $\delta \in D_{m+1}$. Then $\{\neg\xi \mid \xi \in S_m\} \cup \{\neg\theta\} \cup \text{range}(\sigma) \models D_{m+1}$, hence by (105), $\{\neg\xi \mid \xi \in S_m\} \cup \{\neg\theta\} \cup \text{range}(\sigma)$ is inconsistent, contradicting (104)b at stage $n = m$. So, choose $\delta \in D_{m+1}$ satisfying (106), and set $S_{m+1} = S_m \cup \{\delta\}$. It is immediate that S_{m+1} satisfies (104) at stage $n = m + 1$. ∎

We now exploit Lemma (101) to prove the theorem.

Proof of Theorem (100): Let R be the binary predicate of **Sym**. Let $T = \{\exists x \forall y\, Rxy \leftrightarrow \forall xy(x = y)\}$ and $\theta = \exists x \forall y\, Rxy$. We claim that $(T, \{\theta, \neg\theta\})$ witnesses the proposition. Clause (100)a is evident. For Clause (100)b, observe:

(107)   for every $\sigma \in SEQ$, $\text{range}(\sigma) \not\models \neg\theta$.

To finish the proof we define a maxichoice revision function $\dotplus$ such that for all $B \subseteq \mathcal{L}_{form}$, if B is closed under $\psi, \phi \mathbin{/} \psi \to \phi$, then $\lambda\sigma . B \dotplus \sigma$ does not solve $(T, \{\theta, \neg\theta\})$. It follows directly from (107) and Lemma (101) that there exists a maxichoice contraction function $\dotminus$ with the following property.

(108)   Suppose that $B \subseteq \mathcal{L}_{form}$ is closed under $\psi, \phi \mathbin{/} \psi \to \phi$. Then for every $\sigma \in SEQ$, if $B \cup \text{range}(\sigma)$ is inconsistent, $(B \dotminus \neg \bigwedge \sigma) \cup \text{range}(\sigma) \not\models \neg\theta$.

Let $\dotplus$ be the maxichoice revision function that $\dotminus$ underlies, via Definition (12). Then immediately from (108):

(109)   Suppose that $B \subseteq \mathcal{L}_{form}$ is closed under $\psi, \phi \mathbin{/} \psi \to \phi$. Then for every nonvoid $\sigma \in SEQ$, if $B \cup \text{range}(\sigma)$ is inconsistent, $B \dotplus \sigma \not\models \neg\theta$.

Let environment e be for $MOD(T \cup \{\theta\})$. By compactness:

(110) $T \cup \text{range}(e) \not\models \theta$.

Let $B \subseteq \mathcal{L}_{form}$ be closed under $\psi, \phi \mathbin{/} \psi \to \phi$. If $B \dotplus e[k] \not\models \theta$ for all $k \in N$, then $\lambda\sigma . B \dotplus \sigma$ does not solve $MOD(T \cup \{\theta\})$, and we are done. So let $\sigma \subset e$ be such that

(111) $B \dotplus \sigma \models \theta$.

By (110) there is a structure $\mathcal{U} \in MOD(T \cup \{\neg\theta\})$ and a full assignment h to $\mathcal{U}$ such that $\mathcal{U} \models \text{range}(\sigma)[h]$. So we can choose environment e' for $\mathcal{U}$ via h such that $\sigma \subset e'$. Let $k > \text{length}(\sigma)$ be given. It suffices to show that $B \dotplus e'[k] \not\models \neg\theta$. For this purpose we distinguish two cases.

Case 1: $B \cup \text{range}(e'[k])$ is consistent. Then by (111), Lemma (13)c, and the fact that $\sigma \subset e'$, $B \dotplus e'[k] \models \theta$. Hence, by Lemma (13)d, $B \dotplus e'[k] \not\models \neg\theta$.

Case 2: $B \cup \text{range}(e'[k])$ is inconsistent. Then (109) implies $B \dotplus e'[k] \not\models \neg\theta$.

■

4.5.3   Remarks on Closure and Successful Revision

Let us consider what conclusions may be drawn from Theorem (100). The theorem does not show that deductive closure of belief is incompatible with scientific success via belief revision. After all, Theorem (99) assures us that some revision function succeeds no matter what closure operator is in force. The lesson of Theorem (100) seems to bear more on the issue of commitment than belief. That the beliefs of an ideally rational agent should be deductively closed seems difficult to challenge. However, an agent with closed beliefs may nonetheless be more attached to her background theory T and to its extension X than to their nontrivial, logical consequences. Indeed, this kind of situation arises whenever we discover an implausible consequence of our strong beliefs, and suffer doubt about which to abandon. Often our desire is to conserve as much as possible of the original belief, with no attempt to measure how large a fragment of their consequences will be lost. With this in mind, Theorem (100) suggests that when an agent faces contradictory data, she should seek to salvage a subset of her starting beliefs X and then let the logical consequences fall where they may. For, as the theorem shows, a revision function that does not adopt this policy may be doomed to fail on some solvable problems, starting from any belief state that is deductively closed.

Alternatively, if closed starting points have high epistemological priority, then further conditions should be placed on revision functions. The conditions should ensure that all conforming revision functions can be used successfully on solvable problems, starting from closed starting points. Otherwise, one is left in the uncomfortable position of insisting that starting beliefs be deductively closed, while admitting that some acceptable means of revising those beliefs lead to needless scientific failure.

4.5.4   Exercises

(112)   EXERCISE: ♠   Closure operator cl is *foundation-like* just in case for all $B \subseteq \mathcal{L}_{form}$ and $\phi \in \mathcal{L}_{form}$, $\phi \in cl(B)$ only if ϕ can be obtained from B by finitely many uses of the inference rules:

(a) $\cdot \, / \, \psi$ if $\models \psi$

(b) $\psi \, / \, \phi$ if $\models \psi \leftrightarrow \phi$

(c) $\psi, \phi \, / \, \psi \wedge \phi$

(d) $\psi, \phi \ / \ \psi \lor \phi$

(e) $\psi \ / \ \exists x_1 \ldots x_k \psi$ if $\exists x_1 \ldots x_k \psi \in \mathcal{L}_{sen}$

Prove the following.

(a) Let solvable problem of form $(T, \{\theta_0, \ldots, \theta_n\})$ be given. Then there is a consistent extension $X \subseteq \mathcal{L}_{form}$ of T such that for all foundation-like closure operators cl and revision functions $\dotplus$, $\lambda\sigma \ . \ \mathrm{cl}(X) \dotplus \sigma$ solves $(T, \{\theta_0, \ldots, \theta_n\})$.

(b) Suppose that consistent $T \subseteq \mathcal{L}_{sen}$ is finitely axiomatizable. Then there is consistent $X \subseteq \mathcal{L}_{form}$ with $X \models T$ such that for all foundation-like closure operators cl, and revision functions $\dotplus$, $\lambda\sigma \ . \ \mathrm{cl}(X) \dotplus \sigma$ solves every solvable problem of form $(T, \{\theta_0, \ldots, \theta_n\})$.

(113)   EXERCISE: ♥   Given closure operator cl, a mapping $\pm : \mathrm{pow}(\mathcal{L}_{form})$ $\times SEQ \rightarrow \mathrm{pow}(\mathcal{L}_{form})$ is a cl-*external revision function* just in case there is contraction function $\dotdiv$ such that for all $B \subseteq \mathcal{L}_{form}$ and $\sigma \in SEQ$,

$$B \pm \sigma = \begin{cases} B & \text{if } \sigma = \emptyset \\ \mathrm{cl}(B \cup \mathrm{range}(\sigma)) \dotdiv \neg \bigwedge \sigma & \text{otherwise.} \end{cases}$$

If $\dotdiv$ is stringent, we say that $\pm$ is stringent.

Suppose that **Sym** is limited to a binary predicate. Let closure operator cl be closed under the inference rules:

(a) $\psi, \phi \ / \ \psi \land \phi$

(b) $\psi \ / \ \exists x_1 \ldots x_k \psi$ if $\exists x_1 \ldots x_k \psi \in \mathcal{L}_{sen}$

Exhibit solvable problem of form $(T, \{\theta, \neg\theta\})$ with the following property. There is stringent cl-external revision function $\pm$ such that for all $B \subseteq \mathcal{L}_{form}$, the scientist $\lambda\sigma \ . \ B \pm \sigma$ does not solve $(T, \{\theta, \neg\theta\})$. [Compare Exercise (112).]

4.6   Scientists Defined via Course-of-Values

4.6.1   Iterated Revision

The scientists discussed so far in this chapter revise their starting point by confronting it anew with each initial segment of their environment. An alternative conception allows starting points to evolve with time, confronting only the latest datum at any given stage of inquiry. To explain, let $B \subseteq \mathcal{L}_{form}$ be a belief state. Faced with data $\langle \delta_1, \delta_2 \rangle \in SEQ$, the scientist $\lambda\sigma \ . \ B \dotplus \sigma$ conjectures $B \dotplus \delta_1$ at the first step of inquiry and $B \dotplus \langle \delta_1, \delta_2 \rangle$ at the second. In contrast, an iterated model of inquiry conceives the scientist as first conjecturing $B \dotplus \delta_1$, as before, but then conjecturing $(B \dotplus \delta_1) \dotplus \delta_2$. In general, at step $n + 1$ of iterated inquiry, the belief state facing the $(n + 1)$th datum has already been modified n times. To formalize this idea we define the "iterated extension" of a revision function.

(114)   DEFINITION: Let revision function $\dotplus$ be given. The *iterated extension of* $\dotplus$ is defined to be the function $\dotplus_{ie}$: $\text{pow}(\mathcal{L}_{form}) \times SEQ \rightarrow \text{pow}(\mathcal{L}_{form})$ such that for all $B \subseteq \mathcal{L}_{form}$, $\sigma \in SEQ$ and $k \in N$,

$$B \dotplus_{ie} \sigma = \begin{cases} B & \text{if } \sigma = \emptyset, \\ [B \dotplus_{ie} (\beta_0 \ldots \beta_{k-1})] \dotplus \beta_k & \text{if } \sigma = (\beta_0 \ldots \beta_k). \end{cases}$$

Iterative revision resembles memory-limitation within the numerical paradigm (Section 2.4). In both cases the latest datum plays a special role in hypothesis selection, whereas earlier data are preserved only through their impact on the hypothesis of the preceding stage.

Let us observe that iterated revision naturally represents the influence of earlier hypotheses on the choice of later ones. In particular, the fragment of the starting point that survives long enough to confront late data depends on decisions about what to preserve in the face of early data. In contrast, non-iterative revision allows the following kind of cut-and-paste operation. Let Ψ_1 and Ψ_2 be revision-based scientists with the same starting point, B. Then the following scientist Ψ_3 is also revision-based. For all $\sigma \in SEQ$:

$$\Psi_3(\sigma) = \begin{cases} \Psi_1(\sigma) & \text{if length}(\sigma) < 1000; \\ \Psi_2(\sigma) & \text{otherwise.} \end{cases}$$

Scientist Ψ_3 is revision-based because revision functions $\dotplus$ impose no constraint on the relation between belief states of the forms $B \dotplus \sigma$ and $B \dotplus (\sigma * \tau)$. [See Definitions (3) and (12).]

We now consider what can and cannot be achieved using iterative revision. Our first two propositions show that iterative revision is compatible with successful inquiry, and even with efficient inquiry. However, it will then be demonstrated that some revision functions cannot be used in iterative fashion to solve certain simple problems, no matter what starting point is chosen.

4.6.2   Achievements of Iterative Revision

The following propositions show that iterative revision can be used successfully for inquiry. In particular (and similarly to before), a sole revision function and theory-extender suffice to solve every solvable problem of form $(T, \{\theta_0, \ldots, \theta_n\})$.

(115)   PROPOSITION: There exists $X \subseteq \mathcal{L}_{form}$ and stringent revision function $\dotplus$ such that for all consistent $T \subseteq \mathcal{L}_{sen}$, the following holds.

(a) $T \cup X$ is consistent.

(b) $\lambda\sigma \, . \, (T \cup X) \dotplus_{ie} \sigma$ solves every solvable problem of the form $(T, \{\theta_0, \ldots, \theta_n\})$.

Proof: Let X be defined as in the proof of Theorem (63). Fix an enumeration $\{\chi_i \mid i \in N\}$ of $\mathcal{L}_{basic} \cup \mathcal{L}_{sen}$, an enumeration $\{\varphi_i \mid i \in N\}$ of X, and an enumeration $\{\psi_i \mid i \in N\}$ of $\mathcal{L}_{form}$. For all $i \in N$, set $S_i = \{\chi_i\}$, $S_{\omega+i} = \{\varphi_i\}$, and $S_{(\omega\times2)+i} = \{\psi_i\}$. Let $\mathbf{S} = \{S_\alpha \mid \alpha < \omega \times 3\}$, and let $\dotminus_\mathbf{S}$ be constructed from $\mathbf{S}$ as in Definition (9). We take $\dotplus$ to be defined from $\dotminus_\mathbf{S}$ via Definition (12). So $\dotplus$ is stringent by Proposition (10). Then it is straightforward to adapt the proof of Theorem (63) for the present proposition.   ∎

Proposition (115) is thus the iterative counterpart to Theorem (63). The analogy extends to efficient inquiry insofar as the next proposition shows that iterative revision can be used to efficiently solve every solvable problem of form $(T, \{P_0, P_1, \ldots\})$. [Compare Theorem (92).]

(116)   PROPOSITION: For every solvable problem $\mathbf{P}$ of the form $(T, \{P_0, P_1, \ldots\})$ there is a consistent extension $X \subseteq \mathcal{L}_{form}$ of T and a stringent revision function $\dotplus$ such that $\lambda\sigma \, . \, X \dotplus_{ie} \sigma$ solves $\mathbf{P}$ efficiently.

Proof: Let solvable problem $\mathbf{P} = (T, \{P_0, P_1, \ldots\})$ be given. Recall from the proof of Theorem (69) the enumeration $\{\pi_i \mid i \in N\}$ of the π-sets consistent with T, and the enumeration $\{\varphi_i^n \mid n \in N\}$ of π_i. Without loss of generality, we may suppose that $\varphi_0^0 \notin \mathcal{L}_{basic}$. Recall from the proof of Theorem (92) the enumeration $\{S_i \mid i \in N\}$ of subsets of $\mathcal{L}_{form}$. Fix an enumeration $\{\beta_i \mid i \in N\}$ of $\mathcal{L}_{basic}$, an enumeration $\{\psi_i \mid i \in N\}$ of $\mathcal{L}_{form}$, and for all $i \in N$ set $S_i' = \{\beta_i\}$, $S_{\omega+i}' = S_i$, and $S_{(\omega\times2)+i}' = \{\psi_i\}$. Let $\mathbf{S} = \{S_\alpha' \mid \alpha < \omega \times 3\}$, and let $\dotminus_\mathbf{S}$ be constructed from $\mathbf{S}$ as in Definition (9). We take $\dotplus$ to be defined from $\dotminus_\mathbf{S}$ via Definition (12). So, $\dotplus$ is stringent by Proposition (10). Then it is straightforward to adapt the proof of Theorem (92) for the present proposition.   ∎

Despite the achievements recorded in the foregoing propositions we shall now see that hypothesis revision is less successful in the iterative context than before.

4.6.3   Iterative Revision Is Not Robust

Given $B \subseteq \mathcal{L}_{form}$ and nonvoid $\sigma \in SEQ$, Lemma (13)a shows that $B \dotplus \sigma \models \bigwedge \sigma$. In contrast, it is easy to construct examples to show that $B \dotplus_{ie} \sigma$ may contradict $\bigwedge \sigma$, although it must imply σ's last member. Such license to ignore past data gives iterative revision so much flexibility that some revision functions can no longer be used to solve simple problems. The infirmity afflicts even the cherished class of stringent functions. This is shown by the following theorem, which may be contrasted with Theorem (69). To state our result, let **two** be the sentence $\exists xy[x \neq y \wedge \forall z(z = x \vee z = y)]$, asserting that there are exactly two individuals. It is easy to verify that $(\emptyset, \{\textbf{two}, \neg\textbf{two}\})$ is solvable. However:

(117)   THEOREM: Suppose that $\textbf{Sym} = \emptyset$. Then there exists a stringent revision function $\dotplus$ such that for all $B \subseteq \mathcal{L}_{form}$, $\lambda\sigma$. $B \dotplus_{ie} \sigma$ does not solve $(\emptyset, \{\textbf{two}, \neg\textbf{two}\})$.

Proof of the theorem relies on two lemmas. The first is a general fact about iterative revision, straightforward to verify.

(118)   LEMMA: Let $B \subseteq \mathcal{L}_{form}$ and $\sigma \in SEQ$ be given. Then $B \dotplus_{ie} \sigma \subseteq B \cup \text{range}(\sigma)$.

(119)   LEMMA: Let finite, consistent $D \subseteq \mathcal{L}_{form}$ be such that $D \models \textbf{two}$. Then $\{\beta \in \mathcal{L}_{basic} \mid D \models \beta \text{ and } \not\models \beta\}$ is a finite set of equalities and inequalities among variables.

Proof of Lemma (119):   Suppose for a contradiction that Lemma (119) is false. Then since D is finite there is $\beta \in \{\beta \in \mathcal{L}_{basic} \mid D \models \beta \text{ and } \not\models \beta\}$ that contains a variable x that does not appear in D. Up to logical equivalence, β has either the form $x = y$ or $x \neq y$, with y distinct from x. In the first case $D \models \forall x(x = y)$, which is a contradiction since D is consistent and $D \models \textbf{two}$. In the second case $D \models \forall x(x \neq y)$, which is also a contradiction since D is consistent. ∎

Proof of Theorem (117):   Fix an enumeration $\{D_i \mid i \in N\}$ of all finite, consistent $D \subseteq \mathcal{L}_{form}$ such that $D \models \textbf{two}$, an enumeration $\{\beta_i \mid i \in N\}$ of $\mathcal{L}_{basic}$, and an enumeration $\{\varphi_i \mid i \in N\}$ of $\mathcal{L}_{form}$. For all $i, j \in N$, set $S_{(\omega \times i)+j} = D_i - \{\beta_j, \beta_{j+1} \ldots\}$. Set $S_{\omega^2+i} = \{\varphi_i\}$ for all $i \in N$. Let $\textbf{S} = \{S_\alpha \mid \alpha < \omega^2 + \omega\}$, and let $\dotminus_{\textbf{s}}$ be constructed from $\textbf{S}$ as in Definition (9). We take $\dotplus$ to be defined from $\dotminus_{\textbf{s}}$ via Definition (12). So $\dotplus$ is stringent by Proposition (10). Let $B \subseteq \mathcal{L}_{form}$ be given. To prove that $\lambda\sigma \cdot B \dotplus_{ie} \sigma$ does not solve $(\emptyset, \{\mathbf{two}, \neg\mathbf{two}\})$, we distinguish two cases.

Case 1: For all $i \in N$, $D_i - \mathcal{L}_{basic} \nsubseteq B$. Then by Lemma (118) and compactness, $\lambda\sigma \cdot B \dotplus_{ie} \sigma$ solves no environment for $MOD(\mathbf{two})$.

Case 2: For some $i \in N$, $D_i - \mathcal{L}_{basic} \subseteq B$. Let $i_0 \in N$ be least such that $D_{i_0} - \mathcal{L}_{basic} \subseteq B$. Let $n \in N$ and integers $j_0 < \ldots < j_{n-1}$ be such that $D_{i_0} = (D_{i_0} - \mathcal{L}_{basic}) \cup \{\beta_{j_0} \ldots \beta_{j_{n-1}}\}$. With the definition of $\dotplus$ and the choice of i_0, it is easy to verify by induction on $m \leq n$ that $(D_{i_0} - \mathcal{L}_{basic}) \cup \{\beta_{j_0} \ldots \beta_{j_{m-1}}\} \subseteq B \dotplus_{ie} (\beta_{j_0} \ldots \beta_{j_{m-1}})$. In particular:

$$(120) \quad D_{i_0} \subseteq B \dotplus_{ie} (\beta_{j_0} \ldots \beta_{j_{n-1}}).$$

By Lemma (119) there exists structure $\mathcal{S}$ and full assignment h to $\mathcal{S}$ such that $\mathcal{S} \nvDash \mathbf{two}$ and $\mathcal{S} \vDash \{\beta \in \mathcal{L}_{basic} \mid D_{i_0} \vDash \beta\}[h]$. We may choose environment e for $\mathcal{S}$ and h that begins with $(\beta_{j_0} \ldots \beta_{j_{n-1}})$. Since $D_{i_0} \cup \text{range}(e)$ is consistent, it follows from the definition of $\dotplus$, the choice of i_0, and (120) that for all $k \geq n$, $D_{i_0} \subseteq B \dotplus_{ie} e[k]$. Hence $B \dotplus_{ie} e[k] \vDash \mathbf{two}$ for all $k \geq n$. We conclude that $\lambda\sigma \cdot B \dotplus_{ie} \sigma$ does not solve $MOD(\neg\mathbf{two})$ in e.   ■

4.6.4   Iterative Scientists under Deductive Closure

Having examined deductive closure of belief states in Section 4.5, and iterated belief revision in the present section, it is natural to wonder about the consequences of combining the two features of inquiry.[5] To give substance to this idea requires closing each belief state that emerges at successive steps of iterated revision. It is not enough to close only the starting belief state since its successors may be left open by the revision process. We therefore formalize the matter as follows.

(121)   DEFINITION: Let revision function $\dotplus$ be given. The *closed iterated extension of* $\dotplus$ is defined to be the function $\dotplus_{ie}^{Cn}: \text{pow}(\mathcal{L}_{form}) \times SEQ \to \text{pow}(\mathcal{L}_{form})$ such that for all $B \subseteq \mathcal{L}_{form}$, $\sigma \in SEQ$ and $k \in N$,

$$B \dotplus_{ie}^{Cn} \sigma = \begin{cases} B & \text{if } \sigma = \emptyset, \\ \text{Cn}(B \dotplus_{ie}^{Cn} (\beta_0 \ldots \beta_{k-1})) \dotplus \beta_k & \text{if } \sigma = (\beta_0 \ldots \beta_k). \end{cases}$$

It will now be seen that iterative revision under closure, like its unclosed counterpart, suffers from the existence of stringent revision functions unable to

---

5. The present subsection is not central to our theory, and may be omitted on a first reading.

solve simple problems. As before, let **two** be the sentence $\exists xy[x \neq y \wedge \forall z(z = x \vee z = y)]$. Then we have the following parallel to Theorem (117).

(122) PROPOSITION: Suppose that $\mathbf{Sym} = \emptyset$. Then there exists a stringent revision function $\dotplus$ such that for all $B \subseteq \mathcal{L}_{form}$, $\lambda\sigma \cdot B \dotplus_{ie}^{Cn} \sigma$ does not solve $(\emptyset, \{\mathbf{two}, \neg\mathbf{two}\})$.

Proof: Fix an enumeration $\{\varphi_i \mid i \in N\}$ of $\mathcal{L}_{form}$ with $\varphi_0 = \mathbf{two}$, and set $S_i = \{\varphi_i\}$ for all $i \in N$. Let $\mathbf{S} = \{S_i \mid i < \omega\}$, let $\dot{-}_S$ be constructed from $\mathbf{S}$ as in Definition (9), and take $\dotplus$ to be defined from $\dot{-}_S$ via Definition (12). So Proposition (10) implies that $\dotplus$ is stringent. Let $B \subseteq \mathcal{L}_{form}$ be given, and let e_0 be an environment for $MOD(\mathbf{two})$. Suppose that $\lambda\sigma \cdot B \dotplus_{ie}^{Cn} \sigma$ solves $MOD(\mathbf{two})$ in e_0 (otherwise we are done). Hence we may choose $k_0 \in N$ such that $B \dotplus_{ie}^{Cn} e_0[k_0] \models \mathbf{two}$. This implies:

(123) $\mathbf{two} \in Cn(B \dotplus_{ie}^{Cn} e_0[k_0])$.

It easy to verify the existence of an environment e such that:

(124) (a) $e_0[k_0] \subset e$.

     (b) e is for the structure $\{0, 1, 2\}$.

Now for all $\beta \in \mathcal{L}_{basic}$, $\{\mathbf{two}, \beta\}$ is consistent (recall that $\mathbf{Sym} = \emptyset$). So Definition (121), (123), (124)a, and the definition of $\dotplus$ imply that for all $k > k_0$, $\mathbf{two} \in B \dotplus_{ie}^{Cn} e[k]$. Thus, (124)b implies that $\lambda\sigma \cdot B \dotplus_{ie}^{Cn} \sigma$ does not solve $MOD(\neg\mathbf{two})$ in e.   ∎

4.6.5   Exercises

(125) EXERCISE: Let $\pm$ be an external revision function, in the sense of Exercise (90). We say that $\pm$ is stringent if it arises from a stringent contraction function. Parallel to Definition (114), the iterated extension of $\pm$ is the function $\pm_{ie} : \text{pow}(\mathcal{L}_{form}) \times SEQ \to \text{pow}(\mathcal{L}_{form})$ such that for all $B \subseteq \mathcal{L}_{form}$, $\sigma \in SEQ$ and $k \in N$,

$$B \pm_{ie} \sigma = \begin{cases} B & \text{if } \sigma = \emptyset, \\ [B \pm_{ie} (\beta_0 \ldots \beta_{k-1})] \pm \beta_k & \text{if } \sigma = (\beta_0 \ldots \beta_k). \end{cases}$$

Suppose that $\mathbf{Sym} = \emptyset$.

(a) Show that for every solvable problem $\mathbf{P}$ of the form $(T, \{P_0, P_1, \ldots\})$, there is a consistent extension $X \subseteq \mathcal{L}_{form}$ of T and a stringent, external revision function $\pm$ such that $\lambda\sigma \cdot X \pm_{ie} \sigma$ solves $\mathbf{P}$ efficiently.

(b) Show that there exists a stringent, external revision function $\pm$ such that for all $B \subseteq \mathcal{L}_{form}$, $\lambda\sigma \,.\, B \pm_{ie} \sigma$ does not solve $(\emptyset, \{\mathbf{two}, \neg\mathbf{two}\})$.

4.7   Summary and Prospects

Let us summarize some of the results in this chapter using a suggestive terminology. We say that revision function $\dotplus$ "succeeds" on problem $\mathbf{P}$ just in case there is consistent $X \subseteq \mathcal{L}_{form}$ such that $\lambda\sigma \,.\, X \dotplus \sigma$ solves $\mathbf{P}$. Otherwise, $\dotplus$ "fails" on $\mathbf{P}$. Thus, $\dotplus$ succeeds on $\mathbf{P}$ if and only if some revision-based scientist that relies on $\dotplus$ solves $\mathbf{P}$. In the contrary case, $\dotplus$ is unsuited for inquiry on $\mathbf{P}$ since no starting point induces it to perform correctly on $\mathbf{P}$. This terminology will be adapted to the contexts of efficiency, closure, and iteration in the obvious way. We offer glosses for some, but not all, of the results formulated earlier as propositions or theorems; additional material may be gleaned from the exercises. As a preliminary, let us recall from Proposition (6) that every stringent revision function is maxichoice.

Concerning problems that can be solved by revision-based scientists (Section 4.2)

• Theorem (47): Some stringent revision function succeeds efficiently on every problem that can be solved by a revision-based scientist.

• Theorem (51): Some stringent revision function fails on some problem that can be solved by a revision-based scientist.

• Theorem (54): There is a computably solvable problem that can be solved by a revision-based scientist on which every computable revision function fails.

Concerning success on problems of the form $(T, \{P_0, P_1, \ldots\})$ (Section 4.3)

• Theorem (69): Every revision function succeeds on every solvable problem of the form $(T, \{P_0, P_1, \ldots\})$.

Concerning efficiency on problems of the form $(T, \{P_0, P_1, \ldots\})$ (Section 4.4)

• Theorem (92): For every solvable problem of the form $(T, \{P_0, P_1, \ldots\})$ there is a stringent revision function that succeeds efficiently.

• Theorem (93): Some stringent revision function does not succeed efficiently on some solvable problem of form $(T, \{\theta, \neg\theta\})$.

Concerning deductive closure (Section 4.5)

• Theorem (99): Some stringent revision function succeeds under any closure operator on every solvable problem of the form $(T, \{\theta_0, \ldots, \theta_n\})$.

• Theorem (100): Some maxichoice revision function fails under all coherence-like closure operators on some solvable problem of the form $(T, \{\theta, \neg\theta\})$.

Concerning iterative inquiry (Section 4.6)

• Proposition (116): For every solvable problem of the form $(T, \{P_0, P_1, \ldots\})$ there is a stringent revision function that succeeds iteratively and efficiently.

• Theorem (117): Some stringent revision function fails iteratively on some solvable problem of the form $(T, \{\theta, \neg\theta\})$.

• Proposition (122): Some stringent revision function fails iteratively under deductive closure on some solvable problem of the form $(T, \{\theta, \neg\theta\})$.

We have seen that revision-based scientists offer a nonvacuous canonical form for several kinds of inquiry. The importance of this finding, of course, is no greater than the intuitive appeal of the revision functions introduced in Section 4.1. For simplicity, only a few classes of functions were there defined. It remains to characterize new species of revision, to justify them intuitively, and then to explore their utility for inquiry.

Now let us make a substantial and non-evident assumption, namely, that the first-order paradigm offers genuine (albeit exiguous) insight into empirical inquiry. Then it is reasonable to expect that further progress in characterizing the abilities of revision-based scientists will clarify the nature of successful theory choice in science, and assist in the evaluation of policies for modifying beliefs.

4.8   Notes

Much of Chapter 4 is based on [Martin & Osherson, 1995b, Martin & Osherson, in press]. A different perspective on belief revision and inquiry is reported in [Kelly *et al.*, 1995].

There are several attempts to unite the alternative perspectives on belief kinematics offered by probability theory versus belief revision, for example, [Lindström & Rabinowicz, 1989, Darwiche & Pearl, 1994, Gärdenfors, 1988]. Whether such a unified approach can throw fresh light on the power of revision-based scientists remains an unexplored issue.

We do not know whether Theorem (100) remains true if the revision function there garanteed is upgraded from maxichoice to stringent. Thus, we have the following question.

Is there a solvable problem of form $(T, \{\theta, \neg\theta\})$ and a stringent revision function $\dot{+}$ such that for all $B \subseteq \mathcal{L}_{form}$, if B is closed under $\psi, \phi \mathbin{/} \psi \rightarrow \phi$, then $\lambda\sigma \,.\, B \dot{+} \sigma$ does not solve $(T, \{\theta, \neg\theta\})$?

We do not know whether Theorem (51) can be demonstrated via an example with finite vocabulary. Similarly, it would be nice to find a version of Theorem (93) that involves a finite vocabulary. In that case, it would no longer be necessary to take **Sym** decidable.

Consider again the search for conditions to impose on stringent revision functions, in view of guaranteeing efficiency (see Section 4.4.2). Some candidates arise in the penetrating analysis [Darwiche & Pearl, 1994] of iterated revision. However, when the Darwiche-Pearl conditions are forced into the present framework, many turn out to be true of no revision function at all. For example, the following principle seems innocuous enough.

If $\bigwedge \sigma \models \bigwedge \tau$ then $(X \dot{+} \tau) \dot{+} \sigma$ is logically equivalent to $X \dot{+} \sigma$. (That is: more specific data cancel the effect of less specific, earlier data.)

However, it can be shown that no revision function $\dot{+}$ satisfies it.

Call a closure operator cl "healthy" just in case for all solvable problems of the form $(T, \{\theta, \neg\theta\})$, there is a consistent extension $X \subseteq \mathcal{L}_{form}$ of T such that for all revision functions $\dot{+}$, $\lambda\sigma \,.\, cl(X) \dot{+} \sigma$ solves $(T, \{\theta, \neg\theta\})$. Theorem (100) and Exercise (112) provide some information about healthy and unhealthy closure operators, but an informative characterization of the boundary between them eludes us. Particularly satisfying would be the discovery that the healthy closure operators are just those representing "relevant implication" in some motivated sense. A well motivated definition of the latter concept is presented in [Schurz & Weingartner, 1987, Schurz, 1996], and it is noteworthy that the unhealthy rule $\psi, \phi \mathbin{/} \psi \rightarrow \phi$ is irrelevant according to the definition. Unfortunately, the healthy foundation-like closure operators of Exercise (112) harbor irrelevant implications in the same sense, for example: $\psi, \phi \mathbin{/} \psi \vee \phi$. The latter inference rule is acceptable according to the relevance criterion advanced in [Tzouvaras, 1996], but the first two rules mentioned in Exercise (112) are not. Yet a different theory of relevant implication is advanced in [Gemes, 1994], but once again it does not square with the healthy/unhealthy distinction between closure operators. For an earlier attempt to connect relevant implication to scientific inquiry, see [Osherson & Weinstein, 1993].

For more discussion of iterated belief revision, see [Nayak, 1994b, Boutilier, 1996], and references cited there.

A Solutions to Exercises for Chapter 1

Section 1.2

Solution to Exercise 1.(13):

Let $0 < \kappa \le \omega$ be given. Let $\{\prec_n \mid n < \kappa\}$ be an enumeration, without repetition, of strict total orders over N. Define scientist Ψ as follows. For every $\sigma \in SEQ$, let $n < \kappa$ be smallest such that for all $i, j \in N$, if $R(i, j)$ appears in σ then $i \prec_n j$. In case n is defined and $\prec_n$ has a least element, set $\Psi(\sigma) = \mathbf{y}$; otherwise set $\Psi(\sigma) = \mathbf{n}$.

Let $n_0 < \kappa$ and environment e for $\prec_{n_0}$ be given. If follows from Fact 1.(4) that for every $n < n_0$, there exists $i, j, k \in N$ such that $R(i, j)$ appears in $e[k]$ whereas $i \prec_n j$ is false. Hence there is $k_0 \in N$ such that for every $k \ge k_0$, n_0 is smallest such that for all $i, j \in N$, if $R(i, j)$ appears in $e[k]$ then $i \prec_{n_0} j$. This implies that Ψ solves $\{\prec_n \mid n < \kappa\}$. ∎

Solution to Exercise 1.(14):

Let Ψ be any scientist that satisfies the following condition. Suppose that $\sigma, \tau \in SEQ$ are such that τ extends σ and $\text{length}(\tau) = \text{length}(\sigma) + 1$. If the L-score of τ equals the L-score of σ then $\Psi(\tau) = \mathbf{y}$; otherwise, $\Psi(\tau) = \mathbf{n}$. [The definition of "L-score" is given in the proof of Proposition 1.(7).] Let environment e be for order $\prec$. It easy to verify the following.

(a) If $\prec$ has a least element then for cofinitely many k, the L-score of $e[k]$ is equal to the L-score of $e[k + 1]$.

(b) If $\prec$ has no least element then for infinitely many k, the L-score of $e[k]$ is smaller than the L-score of $e[k + 1]$.

This implies that $\Psi(e[k]) = \mathbf{y}$ for cofinitely many k if and only if $\prec$ has a least element. Hence Ψ half solves the entire collection of orders. ∎

Solution to Exercise 1.(15):

Suppose that scientist Ψ solves problem $\mathbf{P}$. Then for every $\prec \in \mathbf{P}$ and environment e for $\prec$,

$$\lim_{k \to \infty} \frac{\text{card}(\{k' \le k \mid \Psi(e[k']) = \mathbf{y}\})}{k}$$

exists and equals 1 if $\prec$ has a smallest element, and

$$\lim_{k \to \infty} \frac{\text{card}(\{k' \le k \mid \Psi(e[k']) = \mathbf{n}\})}{k}$$

exists and equals 1 otherwise. Hence Ψ gradually solves $\mathbf{P}$.

Now suppose that scientist Ψ gradually solves every environment for every $\prec \in \mathbf{P}$. Let scientist Ψ' be such that for every $k > 0$ and every environment e, $\Psi'(e[k]) = \mathbf{y}$ if

$$\frac{\operatorname{card}(\{k' \le k \mid \Psi(e[k']) = \mathbf{y}\})}{k} > .5,$$

and $\Psi'(e[k]) = \mathbf{n}$ otherwise. Let $\prec \in \mathbf{P}$ be given. By the assumption on Ψ it follows that for every environment e for $\prec$, there is $k_0 \in N$ such that $k \ge k_0$ implies

$$\frac{\operatorname{card}(\{k' \le k \mid \Psi(e[k']) = \mathbf{y}\})}{k} > .5$$

if $\prec$ has a least element, and

$$\frac{\operatorname{card}(\{k' \le k \mid \Psi(e[k']) = \mathbf{y}\})}{k} = 1 - \frac{\operatorname{card}(\{k' \le k \mid \Psi(e[k']) = \mathbf{n}\})}{k} < .5$$

otherwise. Hence Ψ' solves $\mathbf{P}$. ∎

Solution to Exercise 1.(16):
It is easy (but tedious) to explicitly define a function $f : SEQ \to SEQ$ with the following properties:

(a) for all $\sigma \in SEQ$, $f(\sigma)$ is the initial segment of some canonical environment;

(b) for all $\sigma, \tau \in SEQ$, $\sigma \subseteq \tau$ implies $f(\sigma) \subseteq f(\tau)$;

(c) for all environments e for order $\prec$, $\bigcup_{k \in N} f(e[k])$ is the canonical environment for $\prec$.

Now suppose that scientist Ψ solves problem $\mathbf{P}$ on canonical environments. We define scientist $\Psi' = \Psi \circ f$. That is, for all $\sigma \in SEQ$, $\Psi'(\sigma) = \Psi(f(\sigma))$. It is easy to see that for all environments e for an order $\prec$, if $\Psi(e[k]) = \mathbf{y}$ for cofinitely many k, then $\Psi'(e[k]) = \mathbf{y}$ for cofinitely many k, and if $\Psi(e[k]) = \mathbf{n}$ for cofinitely many k, then $\Psi'(e[k]) = \mathbf{n}$ for cofinitely many k. This suffices to show that Ψ' solves every order that Ψ solves. ∎

Solution to Exercise 1.(17):
In the proof of Proposition 1.(7), replace "environment" by "imperfect environment". This yields a proof of 1.(17)a.

For 1.(17)b, let $N' = N - \{0, 1\}$. Let problem $\mathbf{P}_0$ consist of every order of one of the two forms:

(1)   (a) an ordering of N' that is isomorphic to ω, followed by the numbers 0 then 1;

      (b) an ordering of N' that is isomorphic to $\omega^*\omega$, followed by the numbers 1 then 0.

Define scientist Ψ such that for all $\sigma \in SEQ$, $\Psi(\sigma) = \mathbf{y}$ if $R(0, 1)$ appears in σ; $= \mathbf{n}$ otherwise. It is easy to see that Ψ solves $\mathbf{P}_0$.

Now suppose for a contradiction that scientist Ψ' solves $\mathbf{P}_0$ on imperfect environments. It is easy to specify a function $f : SEQ \to SEQ$ with the following properties:

(a) for all $\sigma, \tau \in SEQ$, $\sigma \subseteq \tau$ implies $f(\sigma) \subseteq f(\tau)$;

(b) if e is an environment for an order isomorphic to ω then $\bigcup_{k \in N} f(e[k])$ is an imperfect environment for an order of type (1)a;

(c) if e is an environment for an order isomorphic to $\omega^*\omega$ then $\bigcup_{k \in N} f(e[k])$ is an imperfect environment for an order of type (1)b.

Hence, $\Psi' \circ f$ solves $\mathbf{P}$ of Proposition 1.(9), contradiction.　∎

Section 1.3

Solution to Exercise 1.(19):
Let m be given. Let $\mathbf{P}_y$ be the class of orders isomorphic to ω. Let $\mathbf{P}_n$ be the class of orders isomorphic to ω^*. We must show that $\mathbf{P} = \mathbf{P}_y \cup \mathbf{P}_n$ is not m-solvable. It is easy to verify the following fact.

(2)   Let $k \in N$ and environment e for some order in $\mathbf{P}$ be given. Then:

(a) there is $\prec \in \mathbf{P}_y$ and environment e' for $\prec$ such that $e'[k] = e[k]$;

(b) there is $\prec \in \mathbf{P}_n$ and environment e' for $\prec$ such that $e'[k] = e[k]$.

Suppose for a contradiction that scientist Ψ satisfies:

(3) Ψ m-solves $\mathbf{P}$.

Then in particular:

(4) Ψ solves $\mathbf{P}$.

Using (4) we show how to construct $k_0 \in N$, and environment e_0 for some order in $\mathbf{P}$ such that:

(5) $\text{card}(\{k \leq k_0 \mid \Psi(e_0[k]) \neq \Psi(e_0[k+1])\}) = m+1,$

which contradicts (3).

Construction of k_0 and e_0 to satisfy (5)

Let e_1 be an environment for some order in $\mathbf{P}_y$. By (4), there is $k_1 \in N$ such that $\Psi(e_1[k_1]) = \mathbf{y}$. By (2)a, let e_2 be an environment for some order in $\mathbf{P}_n$ that extends $e_1[k_1]$. By (4), there is $k_2 > k_1$ such that $\Psi(e_2[k_2]) = \mathbf{n}$. By (2)b, let e_3 be an environment for some order in $\mathbf{P}_y$ that extends $e_2[k_2]$. By (4), there is $k_3 > k_2$ such that $\Psi(e_3[k_3]) = \mathbf{y}$. Continue the construction until (5) is satisfied.

■

Solution to Exercise 1.(20):
We rely on the following fact, easy to verify.

(6)   Let $\sigma \in SEQ$ be given. Then there is an environment that extends σ and is for an order with a least point, and there is another environment that extends σ and is for an order without a least point.

Let scientist Ψ be given. We construct an environment e_Ψ for an order $\prec_\Psi$ such that:

(7)   either $\prec_\Psi$ has a least point and $\Psi(e_\Psi[k]) = \mathbf{n}$ for infinitely many k, or $\prec_\Psi$ has no least point and $\Psi(e_\Psi[k]) = \mathbf{y}$ for infinitely many k.

We construct e_Ψ in stages. At each stage n we will either have finished constructing e_Ψ, or else we will have constructed an initial segment σ^n of e_Ψ. Denote the length of $\sigma \in SEQ$ by $\text{length}(\sigma)$.

Construction of e_Ψ:

Stage 0: Let e be an environment for N under its natural order. If $\Psi(e[k]) = \mathbf{n}$ for infinitely many k, then $e_\Psi = e$ and the construction is over. Otherwise, let k_0 be such that $\Psi(e[k_0]) = \mathbf{n}$, and set $\sigma^0 = e[k_0]$.

Stage $2n+1$: Suppose that σ^{2n} has been constructed. Let $\tau \in SEQ$ extend σ^{2n} and be such that for all $0 \leq i < j \leq n$, either $R(i,j)$ or $R(j,i)$ appears in τ. By (6), let environment e extend τ and be for an order without least point. If $\Psi(e[k]) = \mathbf{y}$ for infinitely many k, then $e_\Psi = e$ and the construction is over. Otherwise, let $k_{2n+1} > \text{length}(\tau)$ be such that $\Psi(e[k_{2n+1}]) = \mathbf{n}$, and set $\sigma^{2n+1} = e[k_{2n+1}]$.

Stage $2n+2$: Suppose that σ^{2n+1} has been constructed. By (6), let environment e extend σ^{2n+1} and be for an order with least point. If $\Psi(e[k]) = \mathbf{n}$

for infinitely many k, then $e_\Psi = e$ and the construction is over. Otherwise, let $k_{2n+2} > \text{length}(\sigma^{2n+1})$ be such that $\Psi(e[k_{2n+2}]) = \mathbf{y}$, and set $\sigma^{2n+2} = e[k_{2n+2}]$.

If the construction enters infinitely many stages, then take $e_\Psi = \bigcup_{n \in N} \sigma^n$. It is clear that whether or not the construction enters infinitely many stages, e_Ψ is an environment for some order $\prec_\Psi$, and (7) is satisfied. It follows immediately that:

(8)   Let scientist Ψ be given. Then Ψ does not solve any problem that contains $\prec_\Psi$.

Let $\mathbf{P} = \{\prec_\Psi | \Psi$ is a computable scientist$\}$. Then $\mathbf{P}$ is countable, so solvable by Exercise 1.(13). On the other hand, (8) implies that no computable scientist solves $\mathbf{P}$.   ∎

Solution to Exercise 1.(21):

Define scientist Ψ as follows. For every $\sigma \in SEQ$, $\Psi(\sigma) = \mathbf{y}$ if $\text{length}(\sigma) \geq 1,000,000$, the G-score of σ is greater than the G-score of $e[1,000,000]$ and the L-score of σ is lower than the G-score of σ; $\Psi(\sigma) = \mathbf{n}$ otherwise. [The definition of "L-score" and "G-score" is given in the proof of Proposition 1.(7).] It is immediate that Ψ is dotty. Let environment e for $\prec \in \mathbf{P}$ be given. If $\prec$ has no least element and the G-score of $e[1,000,000]$ is equal to the greatest element of $\prec$, then $\Psi(e[k]) = \mathbf{n}$ for all $k \in N$, and Ψ solves e. If $\prec$ has a least element, or if $\prec$ has no least element but the G-score of $e[1,000,000]$ is lower than the greatest element of $\prec$, then for some $k_0 > 1,000,000$, the G-score of $e[k]$ is greater than the G-score of $e[1,000,000]$ for all $k \geq k_0$, and it follows from the proof of Proposition 1.(7) that Ψ solves e. So we have proved that Ψ solves $\mathbf{P}$.

Let scientist Ψ' that solves $\mathbf{P}$ and environment e for an order in $\mathbf{P}$ be such that $CP(\Psi, e) > CP(\Psi', e)$ (if there is no such Ψ' and e then there is nothing left to prove). Hence there is $k_0 \in N$ such that:

(9) $\Psi'(e[k_0]) \neq \Psi(e[k_0])$.

To finish the proof of the proposition it suffices to show that there exists $\prec \in \mathbf{P}$ such that:

(10)   (a) some environment for $\prec$ extends $e[k_0]$, and

      (b) for all environments e' for $\prec$ that extend $e[k_0]$, for all $k \geq k_0$, $\Psi(e'[k]) = \Psi(e[k_0])$.

Indeed, (9), (10), and the fact that Ψ solves $\mathbf{P}$ imply that $\mathrm{CP}(\Psi, e') \leq k_0 < \mathrm{CP}(\Psi', e')$ for every environment e' for $\prec$ that extends $e[k_0]$. We may obviously choose $\prec \in \mathbf{P}$ such that:

(a) $\prec$ has the G-score of $e[k_0]$ as greatest element if $\Psi(e[k_0]) = \mathbf{n}$;

(b) $\prec$ has the L-score of $e[k_0]$ as least element if $\Psi(e[k_0]) = \mathbf{y}$;

(c) $\prec$ agrees with $e[k_0]$.

With the definition of Ψ, it is easy to verify that (10) is satisfied.   ■

B Solutions to Exercises for Chapter 2

Section 2.1

Solution to Exercise 2.(5):

For part (a), define scientist Ψ as follows. For every $\sigma \in SEQ$, $\Psi(\sigma) = \text{range}(\sigma)$. Let environment e for nonempty, finite $D \subseteq N$ be given. Since there is $k \in N$ such that $\text{range}(e[k']) = D$ for all $k' \geq k$, it follows that Ψ converges on e to $\text{range}(e)$, hence solves e.

For part (b), define scientist Ψ as follows. For every $\sigma \in SEQ$, $\Psi(\sigma)$ contains all members of N except the first m ones that do not occur in $\text{range}(\sigma)$. Let environment e for $N - D$ be given, where $D \subseteq N$ has cardinality m. Choose $n \in N$ greater than every member of D. It is easy to see that there exists $k \in N$ such that for every $k' \geq k$, $\text{range}(e[k']) \cap [0 \ldots n] = [0 \ldots n] - D$. This implies that Ψ converges on e to $\text{range}(e)$, hence solves e. ∎

Solution to Exercise 2.(6):

We first consider the problem described in Exercise 1.(5)a, then the collection of problems described in 1.(5)b.

For 1.(5)a, let scientist Ψ solve the collection of all nonempty, finite subsets of N. Suppose there were a sequence $\{\sigma_n \mid n \in N\}$ of members of SEQ that satisfy the following, for every $n \in N$:

(1) (a) $\sigma_n \subset \sigma_{n+1}$;

  (b) $\Psi(\sigma_n) = \{0 \ldots n\}$.

By (1)a, $e = \bigcup_{n \in N} \sigma_n$ is an environment, and by (1)b, Ψ does not converge on e. Hence Ψ is not confident. So to prove that no confident scientist solves the collection of all nonempty, finite subsets of N, it suffices to build by induction on $n \in N$ a sequence $\{\sigma_n \mid n \in N\}$ of members of SEQ that satisfy (1). Since Ψ solves $\{0\}$, we can choose $k_0 \in N$ such that $\Psi(e_0[k_0]) = \{0\}$, where e_0 is the unique environment for $\{0\}$. Set $\sigma_0 = e_0[k_0]$. Given $n \in N$, suppose that $\sigma_0 \ldots \sigma_n$ have been defined. Choose any environment e_{n+1} for $\{0 \ldots n+1\}$ that extends σ_n. Since Ψ solves $\{0 \ldots n+1\}$, we can choose $k_{n+1} > \text{length}(\sigma_n)$ such that $\Psi(e_{n+1}[k_{n+1}]) = \{0 \ldots n+1\}$. Set $\sigma_{n+1} = e_{n+1}[k_{n+1}]$. This completes the construction.

For 1.(5)b, fix $m \in N$. Let scientist Ψ solve the collection of all sets of the form $N - D$, where $\text{card}(D) = m$. Suppose there were a sequence $\{\sigma_n \mid n \in N\}$ of members of SEQ and a sequence $\{p_n \mid n \in N\}$ of distinct integers that satisfy the following, for all $n \in N$:

(2)   (a) $\sigma_n \subset \sigma_{n+1}$;

        (b) $\Psi(\sigma_n) = N - \{p_n \ldots p_n + m - 1\}$.

By (2)a, $e = \bigcup_{n \in N} \sigma_n$ is an environment, and Ψ does not converge on e by (2)b. Hence Ψ is not confident. So to prove that no confident scientist solves the collection of all sets of form $N - D$, where card$(D) = m$, it suffices to build by induction on $n \in N$ a sequence $\{\sigma_n \mid n \in N\}$ of members of SEQ and a sequence $\{p_n \mid n \in N\}$ of distinct integers that satisfy (2). Let $p_0 = 0$, and choose any environment e_0 for $L = \{m, m + 1 \ldots\}$. Since Ψ solves L, we can choose $k_0 \in N$ such that $\Psi(e[k_0]) = \{m, m + 1 \ldots\}$. Set $\sigma_0 = e_0[k_0]$. Given $n \in N$, suppose that $\sigma_0 \ldots \sigma_n, p_0 \ldots p_n$ have been defined. Let $p_{n+1} > p_n$ be least such that for all $p \geq p_{n+1}$, $p \notin \text{range}(\sigma_n)$, and choose any environment e_{n+1} for $L = N - \{p_{n+1} \ldots p_{n+1} + m - 1\}$ that extends σ_n. Since Ψ solves L, we can choose $k_{n+1} > \text{length}(\sigma_n)$ such that $\Psi(e_{n+1}[k_{n+1}]) = N - \{p_{n+1} \ldots p_{n+1} + m - 1\}$. Set $\sigma_{n+1} = e_{n+1}[k_{n+1}]$. This completes the construction.   ■

Section 2.2

Solution to Exercise 2.(18):
For 2.(18)a, in the proof of Lemma 2.(8), modify 2.(9) as follows:

for every $\sigma, \sigma' \in SEQ$ with range$(\sigma * \sigma') \subseteq L$, there is $\tau \in SEQ$ with range$(\tau) \subseteq L$ such that $\Psi(\sigma * \sigma' * \tau) \neq \Psi(\sigma * \sigma')$,

and define γ_0 to be $\sigma * e_1(0)$ instead of $\langle e_1(0) \rangle$.

For 2.(18)b, let scientist Ψ be defined as follows, for all $\sigma \in SEQ$. If σ has the form $\langle 1, 0, 0, \ldots, 0, 0, 1 \rangle$, then $\Psi(\sigma) = N$; otherwise $\Psi(\sigma) = \{0, 1\}$. It is easy to see that Ψ solves $\{0, 1\}$ but the environment $1, 0, 0, 0, \ldots$ is not a locking environment for Ψ and $\{0, 1\}$.   ■

Solution to Exercise 2.(19):
Suppose that reliable scientist Ψ solves $\emptyset \neq L \subseteq N$. Let nonempty $\sigma \in SEQ$ be a locking-sequence for Ψ and L. Choose $n \in \text{range}(\sigma)$. Hence Ψ converges on $\sigma * n * n * \ldots$ to L. Since Ψ is reliable, $L = \text{range}(\sigma * n) = \text{range}(\sigma)$, and L is finite.   ■

Solution to Exercise 2.(20):
By Corollary 2.(17)b, $\{N\} \cup \{N - \{n\} \mid n \in N\}$ is not solvable. We show that it is solvable on ascending environments. Let scientist Ψ be such that for all

$\sigma \in SEQ$, the following holds. If $\sigma = \langle 0 \ldots n \rangle$ for some $n \in N$, then $\Psi(\sigma) = N$; if $\langle 0 \ldots n - 1, n + 1 \rangle \subseteq \sigma$ for some $n \in N$, then $\Psi(\sigma) = N - \{n\}$. It is easy to verify that Ψ solves $\{N\} \cup \{N - \{n\} \mid n \in N\}$ on ascending environments.

We show that $\{N\} \cup \{D \subseteq N \mid D$ finite and nonempty $\}$ is not solvable on ascending environments. It is easy to adapt the proof of Lemma 2.(8) to show the following:

(3)   Suppose that scientist Ψ solves language L. Then there is $\sigma \in SEQ$ such that σ begins some ascending environment for L and for all $\tau \in SEQ$, if $\sigma * \tau$ begins some ascending environment for L then $\Psi(\sigma * \tau) = \Psi(\sigma)$.

Now suppose that scientist Ψ solves N on ascending environments. Choose $\sigma \in SEQ$ that satisfies (3) for $L = N$, and denote by n the least element not occurring in σ. Since for all $m \in N$,

$$\sigma * \overbrace{\langle n \ldots n \rangle}^{m}$$

begins some ascending environment for N, it follows that

$$\Psi(\sigma * \overbrace{\langle n \ldots n \rangle}^{m}) = N$$

for all $m \in N$. Hence Ψ does not solve range$(\sigma) \cup \{n\}$ on $\sigma * n * n * \ldots$. Hence Ψ does not solve $\{N\} \cup \{D \subseteq N \mid D$ finite and nonempty $\}$.   ■

Solution to Exercise 2.(21):

Let $n \in N$ and enumeration $\{L_0 \ldots L_n\}$ of languages be given. For every $m \leq n$, define D_m to be any finite subset of L_m such that for all $i \leq n$, if $L_i \subset L_m$ then $D_m \cap (L_m - L_i) \neq \emptyset$. It is easy to verify that for all $m \leq n$, D_m is a tip-off for L_m in $\{L_0 \ldots L_n\}$. Theorem 2.(16) then implies that $\{L_0 \ldots L_n\}$ is solvable.   ■

Solution to Exercise 2.(22):

By Lemma 2.(3) it suffices to show that some uncountable problem has tip-offs. So it suffices to exhibit an uncountable problem $\mathbf{P}$ such that for every distinct $L, L' \in \mathbf{P}$, neither $L \subseteq L'$ nor $L' \subseteq L$, since then $\emptyset$ is trivially a tip-off for every language in $\mathbf{P}$. Given $f \in 2^\omega$ denote by L_f the language such that for all $n \in N$, $2n \in L_f$ iff $f(n) = 0$, and $2n + 1 \in L_f$ iff $f(n) = 1$. We show that $\mathbf{P} = \{L_f \mid f \in 2^\omega\}$ is as required. It is immediate that $\mathbf{P}$ is uncountable. Let distinct $f, f' \in 2^\omega$ be given. Choose $n \in N$ such that $f(n) \neq f'(n)$. Then $2n$

belongs to one of L_f and $L_{f'}$ but not both, and $2n + 1$ belongs to one of L_f and $L_{f'}$ but not both. Hence neither $L_f \subseteq L_{f'}$ nor $L_{f'} \subseteq L_f$.   ∎

Solution to Exercise 2.(23):

By Exercise 2.(5)a the collection **P** of all finite, nonempty subsets of N is solvable. **P** is saturated since for every infinite $L \subseteq N$, $\mathbf{P} \cup \{L\}$ is not solvable by Corollary 2.(17)a. Let **P** be any solvable problem that contains at least one infinite $L \subseteq N$. By Corollary 2.(17)a again, there exists finite, nonempty $D \subseteq L$ such that $D \notin \mathbf{P}$. Let scientist Ψ solve **P**. Define scientist Ψ' as follows. For every $\sigma \in SEQ$, $\Psi'(\sigma) = D$ if range$(\sigma) = D$, and $\Psi'(\sigma) = \Psi(\sigma)$ otherwise. It is easy to verify that Ψ' solves $\mathbf{P} \cup \{D\}$. Hence **P** is not saturated. So the class of all finite, nonempty subsets of N is the only saturated problem.   ∎

Section 2.3

Solution to Exercise 2.(39):

Suppose that Ψ is not dominated on **P**. Let σ be for **P**, and suppose for a contradiction that:

(4)   for all environments e for **P** that extend σ, there is $k \geq$ length(σ) such that either $\Psi(e[k])$ is not defined or $\Psi(e[k]) \neq$ range(e).

Since some environment for **P** extends σ and Ψ solves **P**, there is least $k_0 \in N$ with the following property:

(5)   For some environment e for **P**, e extends σ and $\Psi(e[k]) = \Psi(e[k_0])$ for all $k \geq k_0$.

It follows from (4), (5), and the fact that Ψ solves **P** that:

(6) $k_0 >$ length(σ).

Let scientist Ψ' satisfy the following conditions, for all $\tau \in SEQ$. If $\sigma \subseteq \tau \subseteq e[k_0]$ then $\Psi'(\tau) =$ range(e). If $\Psi(\tau)$ is defined but either $\sigma \not\subseteq \tau$ or $\tau \not\subseteq e[k_0]$, then $\Psi'(\tau) = \Psi(\tau)$.

  It is immediate that Ψ' solves **P**. Let environment e' for **P** be given. If e' does not extend σ then CP$(\Psi', e') =$ CP(Ψ, e') by the definition of Ψ'. If e' extends σ and $e' \neq e$, then CP$(\Psi, e') \geq k_0$ by the definition of k_0, and CP$(\Psi', e') \leq$ CP(Ψ, e') by the definition of Ψ'. Finally from (5), (6), and the definition of Ψ' it is easy to see that CP$(\Psi', e) \leq$ length$(\sigma) < k_0 =$ CP(Ψ, e).

This contradicts the hypothesis. So we have shown that if Ψ is not dominated on $\mathbf{P}$, then conditions (a) and (b) are satisfied.

For the opposite suppose that conditions (a) and (b) are satisfied. Let scientist Ψ' solve $\mathbf{P}$, and let environment e for $\mathbf{P}$ be such that $CP(\Psi', e) < CP(\Psi, e)$ (if there is no such Ψ' and e, then there is nothing left to prove). Since $CP(\Psi', e) < CP(\Psi, e)$ there is $k_0 \in N$ such that:

(7) $\Psi(e[k_0]) \neq \Psi'(e[k_0])$.

By hypothesis there exists an environment e' for $\mathbf{P}$ such that e' extends $e[k_0]$ and $\Psi(e'[k]) = \Psi(e'[k_0])$ for all $k \geq k_0$. This, (7), and the fact that Ψ solves $\mathbf{P}$ imply that $CP(\Psi, e') \leq k_0 < CP(\Psi', e')$. With Definition 2.(28) we conclude that Ψ is not dominated on $\mathbf{P}$.   ∎

Solution to Exercise 2.(40):
Suppose that scientist Ψ solves problem $\mathbf{P}$. Exercise 2.(39) implies that if Ψ is not dominated on $\mathbf{P}$ then Ψ is $\mathbf{P}$-total, $\mathbf{P}$-bound, and $\mathbf{P}$-consistent. So if Ψ is either not $\mathbf{P}$-total or not $\mathbf{P}$-bound or not $\mathbf{P}$-consistent, then Ψ is dominated on $\mathbf{P}$.

For the converse, let $\mathbf{P} = \{\{0\}, \{0, 1\}\}$, and define scientist Ψ as follows.

(a) $\Psi(\emptyset) = \{0\}$,

(b) $\Psi(\langle 0 \rangle) = \{0, 1\}$, and

(c)   for all $\sigma \in SEQ - \{\emptyset, \langle 0 \rangle\}$, $\Psi(\sigma) = \{0\}$ if $1 \notin range(\sigma)$, and $\Psi(\sigma) = \{0, 1\}$ otherwise.

It is immediate that Ψ is a $\mathbf{P}$-total, $\mathbf{P}$-bound, and $\mathbf{P}$-consistent scientist that solves $\mathbf{P}$. We show that Ψ is dominated on $\mathbf{P}$. Define scientist Ψ' as follows. $\Psi'(\emptyset) = \{0, 1\}$ and for all $\sigma \in SEQ - \{\emptyset\}$, $\Psi'(\sigma) = \Psi(\sigma)$. It is immediate that Ψ' solves $\mathbf{P}$. If e is the unique environment for $\{0\}$ then $CP(\Psi, e) = CP(\Psi', e) = 2$. If e is an environment for $\{0, 1\}$, let $k \in N$ be least such that $e(k) = 1$. If $k \geq 2$ then $CP(\Psi, e) = CP(\Psi', e) = k + 1$. If $k \leq 1$ then $CP(\Psi, e) = 1$ and $CP(\Psi', e) = 0$. Hence Ψ is dominated on $\mathbf{P}$.   ∎

Solution to Exercise 2.(41):
Suppose that Ψ solves $\mathbf{P}$ efficiently. Let $\sigma \in SEQ$ be for $\mathbf{P}$. We show that conditions (a)-(c) are satisfied. By Exercise 2.(40) there is $L \in \mathbf{P}$ such that $range(\sigma) \subseteq L$ and $\Psi(\sigma) = L$. Suppose that for all $L' \in \mathbf{P}$, if $L' \neq L$ then $range(\sigma) \not\subseteq L'$. Let scientist Ψ' have the following properties, for all $\tau \in SEQ$. If $\sigma \subseteq \tau$ then $\Psi'(\tau) = L$. If $\sigma \not\subseteq \tau$ and $\Psi(\tau)$ is defined, then $\Psi'(\tau) = \Psi(\tau)$. It is easy to see that Ψ' solves $\mathbf{P}$, and that for every environment e for $\mathbf{P}$:

(a) $CP(\Psi', e) \leq CP(\Psi, e)$, and

(b) if e extends σ and $\Psi(e[k]) \neq L$ for some $k > \text{length}(\sigma)$, then $CP(\Psi', e) < CP(\Psi, e)$.

It follows from Definition 2.(28) and Lemma 2.(29) that $\Psi(e[k]) = L$ for every environment e for $\mathbf{P}$ that extends σ and for every $k \geq \text{length}(\sigma)$. So, conditions (a)–(c) of the exercise are satisfied for $\tau = \sigma$. Thus, to conclude this direction of the proof, suppose that there is $L' \in \mathbf{P}$ with $L' \neq L$ and $\text{range}(\sigma) \subseteq L'$. By Exercise 2.(40) there exists $L' \in \mathbf{P}$ with $L' \neq L$, $\tau \in SEQ$, and $n \in N$ such that:

(a) $\sigma \subseteq \tau$ and $\tau * n$ begins some environment for L';

(b) for every $\gamma \in SEQ$, if $\sigma \subseteq \gamma \subseteq \tau$ then $\Psi(\gamma) = L$;

(c) $\Psi(\tau * n) = L'$.

By Exercise 2.(39) and Lemma 2.(29) again, there exists some environment e' for L' that extends $\tau * n$ and such that for all $k > \text{length}(\tau)$, $\Psi(e'[k]) = L'$. Now let scientist Ψ' satisfy the following. $\Psi'(\tau) = L'$. For every $\gamma \in SEQ$, if $\gamma \neq \tau$ and $\Psi(\gamma)$ is defined, then $\Psi'(\gamma) = \Psi(\gamma)$. Then:

(a) $CP(\Psi', e') < CP(\Psi, e') = \text{length}(\tau) + 1$, and

(b) for every environment e for $\mathbf{P}$, $CP(\Psi, e) < CP(\Psi', e)$ only if e extends τ, e is for L, and for all $k \geq \text{length}(\tau)$, $\Psi(e[k]) = L$.

From this and Definition 2.(24), we deduce that there is $\gamma \in SEQ$ with $\tau \subseteq \gamma$, $\text{range}(\gamma) \subseteq L$, and $\Psi(e[k]) = L$ for every environment e for L that extends γ and for every $k \geq \text{length}(\tau)$. Hence conditions (a)-(c) of the exercise are satisfied.

Conversely, suppose that conditions (a)-(c) of the exercise are satisfied. We show that Ψ solves $\mathbf{P}$ efficiently. Suppose that scientist Ψ' solves $\mathbf{P}$, and for some $\sigma \in SEQ$ for $\mathbf{P}$, either $\Psi'(\sigma)$ is undefined or $\Psi'(\sigma) \neq \Psi(\sigma)$ (if there is no such Ψ' and σ then there is nothing left to prove). Choose $\tau \in SEQ$ that extends σ and $L \in \mathbf{P}$ such that $\text{range}(\tau) \subseteq L$ and for every environment e for L that extends τ, for every $k \geq \text{length}(\sigma)$, $\Psi(e[k]) = L$. Then for every environment e for L that extends τ, $\text{length}(\sigma) \leq CP(\Psi, e) < CP(\Psi', e)$. With Definition 2.(24) this completes the proof of the exercise.   ∎

Solution to Exercise 2.(42):

Suppose that Ψ is $\mathbf{P}$-rational. The proof of Proposition 2.(31) shows that Ψ solves $\mathbf{P}$ strongly efficiently. For the opposite direction, suppose that Ψ solves $\mathbf{P}$ strongly efficiently. We show that Ψ is $\mathbf{P}$-rational. It is trivial that Ψ is $\mathbf{P}$-total, $\mathbf{P}$-bound, and $\mathbf{P}$-consistent. Let $\sigma \in SEQ$ be for $\mathbf{P}$, and suppose there exists

$n \in \Psi(\sigma)$ such that $\Psi(\sigma * n) \neq \Psi(\sigma)$. Let environment e for $\Psi(\sigma)$ begin with $\sigma * n$. Let scientist Ψ' satisfy the following, for all $\tau \in SEQ$.

(a) If $\tau \subset \sigma$ then $\Psi'(\tau)$ is undefined.

(b) If $\sigma \subseteq \tau \subseteq e$ then $\Psi'(\tau) = \Psi(\sigma)$.

(c) If $\tau \nsubseteq e$ and $\Psi(\tau)$ is defined, then $\Psi'(\tau) = \Psi(\tau)$.

Then it is easy to verify the following:

(a) Ψ' solves $\mathbf{P}$;

(b) $CP(\Psi', e) = \text{length}(\sigma)$;

(c) for every $L \in \mathbf{P}$ that contains range(σ), there exists an environment e' for L that extends σ with $CP(\Psi, e') > \text{length}(\sigma)$.

This contradicts the hypothesis that Ψ solves $\mathbf{P}$ strongly efficiently. Hence for every $n \in \Psi(\sigma)$, $\Psi(\sigma * n) = \Psi(\sigma)$. We conclude that Ψ is $\mathbf{P}$-rational. ∎

Solution to Exercise 2.(43):

For every $n \in N$, denote $\{0, n, n+1 \ldots\}$ by L_n, and set $\mathbf{P} = \{L_n \mid n \in N\} \cup \{\{0\}\}$. Define scientist Ψ as follows. Let $\sigma \in SEQ$ be given. If range(σ) $\subseteq \{0\}$ then $\Psi(\sigma) = \{0\}$. If $n \in N - \{0\}$ is smallest with $n \in$ range(σ), then $\Psi(\sigma) = L_n$. It is easy to verify that Ψ is $\mathbf{P}$-rational and solves $\mathbf{P}$. To finish the proof, it suffices to show that $\mathbf{P}$ is not solvable by enumeration. Let $\prec$ well-order $\mathbf{P}$. We show that $\Psi_\prec$ does not solve $\mathbf{P}$. Suppose that for all $n \in N - \{0\}$, $\Psi_\prec(\langle n \rangle) = L_n$. Since $n + 1 \in L_n$, this implies that $L_{n+1} \prec L_n$ for every $n \in N - \{0\}$, which contradicts the hypothesis that $\prec$ well orders $\mathbf{P}$. Using the fact that $\Psi_\prec$ is $\mathbf{P}$-rational, we infer that there is $n \in N - \{0\}$ and $m \neq n$ such that $\Psi_\prec(\langle n \rangle) = L_m$. Since $\Psi_\prec$ is $\mathbf{P}$-rational and $n \notin L_m$ for all $m > n$, it follows that $m < n$. Let e be any environment for L_n that begins with n. Since $L_n \subseteq L_m$ and m is $\prec$-least with $n \in L_m$, it is immediate that for all $k > 0$, m is $\prec$-least with range($e[k]$) $\subseteq L_m$. Hence $\Psi_\prec(e[k]) = L_m$ for all $k > 0$, and $\Psi_\prec$ fails to solve e. ∎

Section 2.4

Solution to Exercise 2.(49):

Let Ψ be a memory-limited scientist that solves problem $\mathbf{P}$. Without loss of generality, we can suppose that $\Psi(\emptyset) = \emptyset$. Let $f : \text{pow}(N) \times N \to \text{pow}(N)$ be any total function such that for all $\sigma \in SEQ$ and $n \in N$, $\Psi(\sigma * n) = f(\Psi(\sigma), n)$. Fix any function $h : N \times \text{range}(\Psi) \to \text{pow}(N)$ such that the following holds:

(8)   (a) For all $L \in (\mathrm{range}(\Psi) \cap \mathbf{P}) \cup \{\emptyset\}$ and $n \in N$, $h(n, L) = L$.

        (b) For all $L \in \mathrm{range}(\Psi) - \mathbf{P}$ and $n \in N$, $n \notin h(n, L) \notin \mathbf{P}$.

        (c) The restriction of h to $N \times [\mathrm{range}(\Psi) - (\mathbf{P} \cup \{\emptyset\})]$ is one-to-one.

By (8)a,c there exists a bijection $p : \mathrm{range}(h) \to \mathrm{range}(\Psi)$ such that for all $L \in$ $\mathrm{range}(h)$, $p(L) = L'$ iff $h(n, L') = L$ for some $n \in N$. Note that $\emptyset \in \mathrm{range}(h)$ and $h[n, f(p(L), n)]$ is well defined and belongs to $\mathrm{range}(h)$ whenever $L \in$ $\mathrm{range}(h)$. Hence we may define scientist Ψ' as follows. $\Psi'(\emptyset) = \emptyset$. For all $\sigma \in SEQ$, for all $n \in N$, $\Psi'(\sigma * n) = h[n, f(p(\Psi'(\sigma)), n)]$. It is clear that Ψ' is memory-limited. We prove the following fact.

(9)   FACT: For all $\sigma \in SEQ$, $\Psi(\sigma) = p(\Psi'(\sigma))$.

Proof of Fact (9):   Proof is by induction on $\mathrm{length}(\sigma)$. It holds for $\sigma = \emptyset$, since $\emptyset = \Psi(\emptyset) = \Psi'(\emptyset) = p(\emptyset)$. Suppose it holds for $\sigma \in SEQ$. We show that it holds for $\sigma * n$ with $n \in N$. Indeed by the induction hypothesis and the definition of f, $\Psi'(\sigma * n) = h[n, f(\Psi(\sigma), n)] = h[n, \Psi(\sigma * n)]$. Hence $p(\Psi'(\sigma * n)) = \Psi(\sigma * n)$ by the definition of p.   ∎

From Fact (9), the definition of Ψ', and the definition of f we infer that:

(10)   for all $\sigma \in SEQ$ and $n \in N$, $\Psi'(\sigma * n) = h(n, \Psi(\sigma * n))$.

From (10), (8)a, and the fact that Ψ solves $\mathbf{P}$ we infer immediately that Ψ' solves $\mathbf{P}$. So to finish the proof it suffices to show that Ψ' is $\mathbf{P}$-conservative. Let $\sigma \in SEQ$ be for $\mathbf{P}$. We distinguish two cases.

Case 1: $\Psi'(\sigma) \in \mathbf{P}$ (hence $\sigma \neq \emptyset$). By (10) and (8)a:

(11) $\Psi'(\sigma) = \Psi(\sigma)$.

Let τ be any locking-sequence for $\Psi(\sigma)$ and Ψ. Then, for all $n \in \Psi(\sigma)$, $\Psi(\sigma) = \Psi(\tau) = \Psi(\tau * n) = f(\Psi(\tau), n) = f(\Psi(\sigma), n) = \Psi(\sigma * n)$, So we have established that:

(12)   for all $n \in \Psi(\sigma)$, $\Psi(\sigma) = \Psi(\sigma * n)$.

From (10), (8)a, (12), (11), and our hypothesis on $\Psi'(\sigma)$ we infer that for all $n \in \Psi'(\sigma)$, $\Psi'(\sigma * n) = \Psi(\sigma * n)$. This with (12) and (11) again yields $\Psi'(\sigma * n) = \Psi'(\sigma)$ for all $n \in \Psi'(\sigma)$, as required.

Case 2: $\Psi'(\sigma) \notin \mathbf{P}$. If $\sigma = \emptyset$ then $\Psi'(\sigma) = \emptyset$, and for all $n \in N$, $n \notin \mathrm{range}(\Psi'(\sigma))$. Suppose $\sigma \neq \emptyset$. From (10), (8)a, and our hypothesis on $\Psi'(\sigma)$ we deduce that $\Psi(\sigma) \notin \mathbf{P}$. This with (10) and (8)b implies that $n \notin \Psi'(\sigma * n)$ for all $n \in N$. So in all cases, $n \notin \Psi'(\sigma * n)$ for all $n \in N$, which completes the proof. ■

Solution to Exercise 2.(50):

For part (a), suppose that scientist Ψ solves nonempty problem $\mathbf{P} = \{L_i \mid i \in N\}$. Let scientist Ψ' be defined as follows, for all $\sigma \in SEQ$. If there is least $i \leq \text{length}(\sigma)$ with $\Psi(\sigma) = L_i$, then $\Psi'(\sigma) = L_i$. Otherwise $\Psi'(\sigma)$ is undefined. It is immediate that Ψ' is gradualist. Let $i \in N$ and environment e for L_i be given. Since Ψ solves $\mathbf{P}$ there is $k_0 \geq i$ such that for all $k \geq k_0$, $\Psi(e[k]) = L_i$. Hence $\Psi'(e[k]) = L_i$ for all $k \geq k_0$, and Ψ' converges on e to range(e), hence solves e. So Ψ' solves $\mathbf{P}$.

For part (b), let $\mathbf{P} = \{\{0, n\} \mid n \in N\}$. It is easy to see that some memory-limited scientist solves $\mathbf{P}$. Let memory-limited, gradualist scientist Ψ be given. Since Ψ is gradualist there is $m \neq n \in N$ such that $\Psi(\langle m \rangle) = \Psi(\langle n \rangle)$. Since Ψ is memory-limited, $\Psi(\langle m, \overbrace{0 \ldots 0}^{p} \rangle) = \Psi(\langle n, \overbrace{0 \ldots 0}^{p} \rangle)$ for all $p \in N$. Hence either Ψ does not converges on environment $\langle m, 0, 0, 0 \ldots \rangle$ to $\{0, m\}$, or Ψ does not converge on environment $\langle n, 0, 0, 0 \ldots \rangle$ to $\{0, n\}$. Hence Ψ fails to solve $\mathbf{P}$. ■

Solution to Exercise 2.(51):

We first prove the right-to-left direction of the exercise. Let scientist Ψ solve problem $\mathbf{P}$ on fat environments. It suffices to show that $\mathbf{P}$ is solvable. Let function $f : SEQ \rightarrow SEQ$ be defined as follows. $f(\emptyset) = \emptyset$. For all $\sigma \in SEQ$ and $n \in N$, $f(\sigma * n) = f(\sigma) * f(\sigma) * n$. It is easy to verify the following:

(13) (a) For all $\sigma, \tau \in SEQ$, if $\sigma \subseteq \tau$ then $f(\sigma) \subseteq f(\tau)$.

     (b) For every environment e, $\bigcup_{k \in N} f(e[k])$ is a fat environment for range(e).

Let scientist Ψ' be such that for all $\sigma \in SEQ$, $\Psi'(\sigma) = \Psi(f(\sigma))$. It follows immediately from (13) and the fact that Ψ solves $\mathbf{P}$ on fat environments that Ψ' solves $\mathbf{P}$.

We now prove the other direction of the exercise. Let nonempty, solvable problem $\mathbf{P}$ be given. By Theorem 2.(16), $\mathbf{P}$ is countable, and for each $L \in \mathbf{P}$ we may choose a tip-off D'_L for L in $\mathbf{P}$. For every $L \in \mathbf{P}$ we will define a set D_L such that $D'_L \subseteq D_L \subseteq L$; hence D_L will also be a tip-off for L in $\mathbf{P}$. Fix a repetition-free enumeration $\{L_i \mid i < \kappa\}$ of $\mathbf{P}$, where $\kappa = \text{card}(\mathbf{P})$. We set $D_{L_0} = L_0$ if L_0 is finite, and $D_{L_0} = L_0 \cap \{0 \ldots, \max(D'_{L_0})\}$ otherwise. Let

$0 < i < \kappa$ be given, and suppose that D_j has been defined for all $j < i$. We set $D_{L_i} = L_i$ if L_i is finite, and $D_{L_i} = L_i \cap \{0 \dots, \sup(\max(D'_{L_i}), r)\}$ otherwise, where r is the least integer such that:

(14) $(\forall j < i)[L_i \not\subseteq L_j \rightarrow (\exists s \leq r)(s \in L_i - L_j)]$.

We rely on the following fact:

(15)   FACT: For all $i, j < \kappa$ with $i \neq j$, either $D_{L_i} - L_j \neq \emptyset$ or $D_{L_j} - L_i \neq \emptyset$.

Proof of Fact (15):   Let $j < i < \kappa$ be given. If $L_i \subset L_j$ then $D_{L_j} - L_i \neq \emptyset$ by Definition 2.(11). So suppose otherwise. If L_i is infinite then $D_{L_i} - L_j \neq \emptyset$ by (14). If L_i is finite then $D_{L_i} - L_j = L_i - L_j \neq \emptyset$.   ∎

As an immediate consequence of Fact (15), we have:

(16)   For all $i, j < \kappa$, if $i \neq j$ then $D_{L_i} \neq D_{L_j}$.

In order to define a memory-limited scientist that solves **P** on fat environments, we need some notation. Fix some one-to-one function $h : \text{pow}_{fin}(N) \rightarrow \text{pow}(N) - \textbf{P}$, where $\text{pow}_{fin}(N)$ is the class of finite subsets of N. Let function $g : \text{range}(h) \cup \textbf{P} \rightarrow \text{pow}_{fin}(N)$ be defined as follows. For all $L \in \text{range}(h)$, $g(L) = h^{-1}(L)$. For all $L \in \textbf{P}$, $g(L) = D_L$. Let function $f : \text{pow}(N) \times N \rightarrow \text{pow}(N)$ be defined as follows, for all $L \in \text{pow}(N)$ and $n \in N$:

(17)   (a)   If there is least $i < \kappa$ such that $D_{L_i} \subseteq g(L) \cup \{n\} \subseteq L_i$, then $f(L, n) = L_i$.

      (b)   Otherwise, $f(L, n) = h(g(L) \cup \{n\})$.

Finally define scientist Ψ as follows. $\Psi(\emptyset) = h(\emptyset)$. For all $\sigma \in SEQ$ and $n \in N$, $\Psi(\sigma * n) = f(\Psi(\sigma), n)$. It is trivial that Ψ is memory-limited. So it remains to show that Ψ solves **P** on fat environments.

Let $i < \kappa$ and fat environment e for L_i be given. We have to show that Ψ converges on e to L_i. In case $D_{L_i} = \emptyset$, we show that for all $k > 0$, $\Psi(e[k]) = L_i$. Proof is by induction on $k > 0$. Since $g(\Psi(\emptyset)) = g(h(\emptyset)) = \emptyset$, it follows from Fact (15) that for all $j \leq i$, $D_{L_j} \subseteq g(\Psi(\emptyset)) \cup \{e(0)\} \subseteq L_j$ iff $j = i$. Hence by (17)a, $\Psi(e[1]) = f(\Psi(\emptyset), e(0)) = L_i$. Let $k > 0$ be given, and suppose that $\Psi(e[k']) = L_i$ for all $0 < k' \leq k$. This implies that $g(\Psi(e[k])) = g(L_i) = D_{L_i} = \emptyset$, and the former argument is immediately adapted to show that $\Psi(e[k + 1]) = L_i$. Hence Ψ converges on e to L_i whenever $D_{L_i} = \emptyset$. We now examine the case: $D_{L_i} \neq \emptyset$. Proof that Ψ converges on e to L_i relies on three new facts.

(18) FACT: For all $k \in N$, $g(\Psi(e[k])) \subseteq L_i$.

Proof of Fact (18):   Proof is by induction on $k \in N$. It is trivial for $k = 0$ since $g(\Psi(\emptyset)) = \emptyset$. Let $k > 0$ be given, and suppose that $g(\Psi(e[k'])) \subseteq L_i$ for all $k' \leq k$. There are two cases.

Case 1: Suppose that there is least $j < \kappa$ such that $D_{L_j} \subseteq g(\Psi(e[k])) \cup \{e(k)\} \subseteq L_j$. Hence by (17)a, $\Psi(e[k + 1]) = L_j$. Using the induction hypothesis and the fact that $e(k) \in L_i$, we infer that $g(\Psi(e[k + 1])) = g(L_j) = D_{L_j} \subseteq g(\Psi(e[k])) \cup \{e(k)\} \subseteq L_i$.

Case 2: Suppose that there is no $j < \kappa$ such that $D_{L_j} \subseteq g(\Psi(e[k])) \cup \{e(k)\} \subseteq L_j$. Hence by (17)b, $\Psi(e[k + 1]) = h(g(\Psi(e[k])) \cup \{e(k)\})$. Using again the induction hypothesis and the fact that $e(k) \in L_i$, we infer that $g(\Psi(e[k + 1])) = g(h(g(\Psi(e[k])) \cup \{e(k)\})) = g(\Psi(e[k])) \cup \{e(k)\} \subseteq L_i$.   ∎

(19) FACT: Let $j < \kappa$ with $j \neq i$ and $k_0 \in N$ be such that for all $k \geq k_0$, $D_{L_j} \subseteq g(\Psi(e[k]))$. Then there is $k_1 \geq k_0$ such that for all $k \geq k_1$, $D_{L_j} \subset g(\Psi(e[k]))$.

Proof of Fact (19):   By Facts (15) and (18), $D_{L_i} - L_j \neq \emptyset$. Choose $n \in L_i - L_j$. Since e is fat, there exists $k_1 > k_0$ such that $e(k_1 - 1) = n$. Suppose that:

(20) $g(\Psi(e[k_1])) = D_{L_j}$.

If $\Psi(e[k_1]) \notin \mathbf{P}$, it follows from (17)b that $\Psi(e[k_1]) = h(g(\Psi(e[k_1 - 1])) \cup \{n\})$, which implies that $n \in g(\Psi(e[k_1]))$. But this contradicts Fact (18). Hence $\Psi(e[k_1]) \in \mathbf{P}$, which together with (20) and (16) implies that $\Psi(e[k_1]) = L_j$. So we showed that $D_{L_j} \subset g(\Psi(e[k_1]))$. Suppose for a contradiction that there exists a least integer $k_1' > k_1$ such that:

(21) $g(\Psi(e[k_1'])) = D_{L_j}$.

We distinguish two cases.

Case 1: $\Psi(e[k_1' - 1]) \in \mathbf{P}$. Let $s < \kappa$ be such that $\Psi(e[k_1' - 1]) = L_s$. With the definition of k_1', we infer:

(22) $g(\Psi(e[k_1' - 1])) = D_{L_s} \supset D_{L_j}$.

If $\Psi(e[k_1']) \notin \mathbf{P}$, then $g(\Psi(e[k_1' - 1])) \subseteq g(\Psi(e[k_1']))$ by (17)b, which contradicts (22) and (21). Hence $\Psi(e[k_1']) \in \mathbf{P}$, which together with (21) and (16) implies that $\Psi(e[k_1']) = L_j$. Hence $D_{L_s} = g(\Psi(e[k_1' - 1])) \subseteq L_j$ by (17)a. But $D_{L_s} \supset D_{L_j}$ and $D_{L_s} \subseteq L_j$ contradict Fact (15).

Case 2: $\Psi(e[k_1' - 1]) \notin \mathbf{P}$. By (17)b, $g(\Psi(e[k_1' - 2])) \subseteq g(\Psi(e[k_1' - 1]))$. Using (17)a and the hypothesis that $D_{L_j} \subseteq g(\Psi(e[k_1' - 2]))$, we infer that $g(\Psi(e[k_1' - 1])) \not\subseteq L_j$. Hence, using (17)a again, we infer that $\Psi(e[k_1']) \neq L_j$, which with (21) and (16) implies that $\Psi(e[k_1']) \notin \mathbf{P}$. So by (17)b, $g(\Psi(e[k_1' - 1])) \subseteq g(\Psi(e[k_1']))$. Hence by (22), $g(\Psi(e[k_1'])) \supset D_{L_j}$, which contradicts (21). ∎

Let $p \geq 1$ and $n_1 < \ldots < n_p$ be such that $D_{L_i} = \{n_1 \ldots n_p\}$.

(23)   FACT: Let $q < p$ be given. Suppose there exists $k_0 \in N$ such that:

(a) for all $k \geq k_0$, $g(\Psi(e[k])) \supseteq \{n_1 \ldots n_q\}$;

(b) there exits $k_0' \geq k_0$ with $g(\Psi(e[k_0'])) \supseteq \{n_1 \ldots n_{q+1}\}$.

Then for all $k \geq k_0' + 2$, $g(\Psi(e[k])) \supseteq \{n_1 \ldots n_{q+1}\}$.

Proof of Fact (23):   Suppose for a contradiction that $k_1 \geq k_0' + 2$ is least with $g(\Psi(e[k_1])) \not\supseteq \{n_1 \ldots n_{q+1}\}$. With (17)b, we infer that $\Psi(e[k_1]) \in \mathbf{P}$. Let $s < \kappa$ be such that $\Psi(e[k_1]) = L_s$. We will show:

(24) $D_{L_s} \neq \{n_1 \ldots n_q\}$.

Proof of (24):   Suppose for a contradiction that $D_{L_s} = \{n_1 \ldots n_q\}$. We distinguish two cases.

Case 1: $\Psi(e[k_1 - 1]) \in \mathbf{P}$. Let $r < \kappa$ be such that $\Psi(e[k_1 - 1]) = L_r$. Since $D_{L_s} = \{n_1 \ldots n_q\} \subseteq g(\Psi(e[k_1 - 1])) = D_{L_r}$, it follows from Fact (15) that $D_{L_r} - L_s \neq \emptyset$. But since $\Psi(e[k_1]) = L_s$, (17)a requires that $g(\Psi(e[k_1 - 1])) = D_{L_r} \subseteq L_s$. Contradiction.

Case 2: $\Psi(e[k_1 - 1]) \notin \mathbf{P}$. Hence by (17)b, $e(k_1 - 1) \in g(\Psi(e[k_1 - 1]))$. From $\Psi(e[k_1]) = L_s$ and (17), we deduce that $D_{L_s} \subseteq g(\Psi(e[k_1 - 1])) = g(\Psi(e[k_1 - 2])) \cup \{e(k_1 - 2)\} \subseteq L_s$. But with (17)a this implies that $\Psi(e[k_1 - 1]) \in \mathbf{P}$. Contradiction. ∎

$\Psi(e[k_1]) = L_s$ and Fact (18) imply that $g(\Psi(e[k_1])) = D_{L_s} \subseteq L_i$. Since $n_1 \ldots n_{q+1}$ are the first $q + 1$ elements of L_i, this with (24) imply the existence of $n > n_{q+1}$ such that $n \in D_{L_s}$. Suppose that $n_{q+1} \in L_s$. Then by our construction of the D_{L_j}, $n_{q+1} \in D_{L_s}$. Since $g(\Psi(e[k_1 - 1])) \supseteq \{n_1 \ldots n_q\}$, this contradicts $D_{L_s} = g(\Psi(e[k_1])) \not\supseteq \{n_1 \ldots n_{q+1}\}$. Hence $n_{q+1} \notin L_s$. Since $n_{q+1} \in g(\Psi(e[k_1 - 1]))$, this contradicts (17)a. ∎

We now end the proof of the exercise. We show that for all $q \leq p$, there exists $k_0 \in N$ such that for all $k \geq k_0$, $\{n_1 \ldots n_q\} \subseteq g(\Psi(e[k]))$. Proof is by induction on q. It is trivial for $q = 0$. Let $q < p$ be given, and suppose we have shown the property for all $q' \leq q$. By induction hypothesis, let $k_0 \in N$ be such that for all $k \geq k_0$, $g(\Psi(e[k])) \supseteq \{n_1 \ldots n_q\}$. By Fact (23) it suffices to show the existence of $k_1 \geq k_0$ such that $\{n_1 \ldots n_{q+1}\} \subseteq g(\Psi(e[k_1]))$. To this aim we choose an integer n. There are two cases:

Case 1: $\{n_1 \ldots n_q\} = D_{L_j}$ for some $j < \kappa$. Hence $D_{L_j} \subset D_{L_i}$, and relying on Fact (15) we choose $n \in L_i - L_j$.

Case 2: $\{n_1 \ldots n_q\} = D_{L_j}$ for no $j < \kappa$. Then we choose any $n \in L_i - \{n_1 \ldots n_q\}$.

Since e is fat, there exists $k_0' > k_0$ with $e(k_0' - 1) = n$. If $g(\Psi(e[k_0'])) = \{n_1 \ldots n_q\}$, then (17) and the fact that $n \notin \{n_1 \ldots n_q\}$ imply with (16) that $\Psi(e[k_0']) = L_j$ for some $j < \kappa$ with $D_{L_j} = \{n_1 \ldots n_q\}$. Hence $n \notin L_j$ by the choice of n. This together with $\Psi(e[k_0']) = L_j$ contradicts (17)a. Hence there exists $k_0' > k_0$ such that $g(\Psi(e[k_0'])) \supset \{n_1 \ldots n_q\}$. Suppose there exists $k_0'' > k_0'$ such that $g(\Psi(e[k_0''])) = \{n_1 \ldots n_q\}$. It follows from (17) that $D_{L_j} = \{n_1 \ldots n_q\}$ for some $j < \kappa$. But Fact (19) then implies that there exists $k_0^{(3)} \geq k_0''$ such that for all $k \geq k_0^{(3)}$, $g(\Psi(e[k])) \supset \{n_1 \ldots n_q\}$. So we have established that there exists $k_0^{(3)} \geq k_0$ such that for all $k \geq k_0^{(3)}$, $\{n_1 \ldots n_q\} \subset g(\Psi(e[k]))$. Since e is fat, there exists $k_1 > k_0^{(3)}$ such that $e(k_1 - 1) = n_{q+1}$. We distinguish two cases.

Case 1: $\Psi(e[k_1]) \notin \mathbf{P}$. Then $n_{q+1} \in g(\Psi(e[k_1]))$ by (17)b.

Case 2: $\Psi(e[k_1]) \in \mathbf{P}$. Suppose that $n_{q+1} \notin g(\Psi(e[k_1]))$. Since $n_1 \ldots n_{q+1}$ are the first $q + 1$ elements of L_i, $\{n_1 \ldots n_q\} \subset g(\Psi(e[k_1]))$ and Fact (18) imply that for some $n > n_{q+1}$, $n \in g(\Psi(e[k_1]))$. Then by our construction of the D_{L_j}, $n_{q+1} = e(k_1 - 1) \notin g(\Psi(e[k_1]))$. But this contradicts (17)a.

So we have proved that there exists $k_0 \in N$ such that for all $k \geq k_0$, $D_{L_i} \subseteq g(\Psi(e[k]))$. With Fact (15) and (17)a, we can thus conclude that for all $k \geq k_0$, $\Psi(e[k]) = L_i$.   ■

Solution to Exercise 2.(52):

There is nothing to prove if $m = 1$, so suppose $m \geq 2$. Let m-memory-limited scientist Ψ solve solvable problem $\mathbf{P}$. Let $f : \mathrm{pow}(N) \times N \to \mathrm{pow}(N)$ be any total function such that for all $\sigma \in SEQ$ and $n \in N$, $\Psi(\sigma * n) = f(\Psi(\sigma), n)$. From Ψ we first define another m-memory-limited scientist Ψ' that solves $\mathbf{P}$. Then we shall use Ψ' to define a 1-memory-limited scientist Ψ^* that solves $\mathbf{P}$.

Fix a repetition-free enumeration $\{H_{ij} \mid i, j \in N\}$ of nonempty subsets of $\text{pow}(N) - \mathbf{P}$ such that for all $i, j \in N$, $\min(H_{ij}) = \inf(i, j)$. Given $i, j \in N$, let one-to-one function $h_{ij} : \text{range}(\Psi) \to \bigcup_{i \in N} H_i \cup \mathbf{P}$ be such that for all $L \in \text{range}(\Psi)$:

(a) $h_{ij}(L) = L$ if $L \in \mathbf{P}$ and $\min(L) = \inf(i, j)$.

(b) $h_{ij}(L) = H_{\inf(i,j)}$ otherwise.

Let $g : \bigcup_{i,j \in N} \text{range}(h_{ij}) \to \text{range}(\Psi)$ be the (unique) function such that for all $L \in \text{domain}(g)$, $g(L) = L'$ iff there is $i, j \in N$ with $h_{ij}(L') = L$. Let $f' : \text{pow}(N) \times N^m \to \text{pow}(N)$ be any total function such that for all $L \in \text{pow}(N) - \{0\}$ and $x_1 \ldots x_m \in N$, $f'(L, x_1 \ldots x_m) = h_{\min(L)x_m}[f(g(L), x_1 \ldots x_m)]$. Define scientist Ψ' as follows.

(a) $\Psi'(\emptyset) = \emptyset$.

(b) For all $1 \le p < m$ and $x_1 \ldots x_p \in N$, $\Psi'(\langle x_1 \ldots x_p \rangle) = h_{\inf(x_1 \ldots x_p)x_p}(\Psi(\langle x_1 \ldots x_p \rangle))$.

(c) For all $\sigma \in SEQ$ and $x_1 \ldots x_m \in N$, $\Psi'(\sigma * \langle x_1 \ldots x_m \rangle) = f'(\Psi'(\sigma * \langle x_1 \ldots x_{m-1} \rangle), x_1 \ldots x_m)$.

We rely on the following two facts about Ψ'.

(25)   FACT: For all $\sigma \in SEQ$, $\min(\Psi'(\sigma)) = \min(\text{range}(\sigma))$.

Proof of Fact (25) is immediate by induction on $\text{length}(\sigma)$.

(26)   FACT: For all $\sigma \in SEQ - \{\emptyset\}$, $\Psi(\sigma) = g(\Psi'(\sigma))$.

Proof of Fact (26):   Proof is also by induction on $\text{length}(\sigma)$. It is immediate if $0 < \text{length}(\sigma) < m$. Let $p \ge m$ be given, and suppose that the property holds for all non-empty sequences of length lower than p. Let $\sigma \in SEQ$ with $\text{length}(\sigma) = p - m$ and $x_1 \ldots x_m$ be given. We show that the property holds for $\sigma * \langle x_1 \ldots x_m \rangle$.

$$g(\Psi'(\sigma * \langle x_1 \ldots x_m \rangle)) = g[f'(\Psi'(\sigma * \langle x_1 \ldots x_{m-1} \rangle), x_1 \ldots x_m)]$$

by the definition of Ψ'

$$= g[h_{\min(\Psi'(\sigma * \langle x_1 \ldots x_{m-1} \rangle))x_m}[f(g(\Psi'(\sigma * \langle x_1 \ldots x_{m-1} \rangle))), x_1 \ldots x_m)]]$$

by the definition of f'

$$= g[h_{\min(\Psi'(\sigma * \langle x_1 \ldots x_{m-1} \rangle))x_m}[f(\Psi(\sigma * \langle x_1 \ldots x_{m-1} \rangle)), x_1 \ldots x_m)]]$$

by the induction hypothesis

$$= g[h_{\min(\Psi'(\sigma * \langle x_1 \ldots x_{m-1} \rangle)) x_m}[\Psi(\sigma * \langle x_1 \ldots x_m \rangle)]]$$

by the definition of Ψ

$$= \Psi(\sigma * \langle x_1 \ldots x_m \rangle)$$

by the definition of g.   ∎

We now show that Ψ' solves **P**. Let environment e for $L \in \mathbf{P}$ be given. Since Ψ solves **P** there exists $k_0 \in N$ such that for all $k \geq k_0$:

(27) (a) $\Psi(e[k]) = L$;

       (b) $\min(L) = \min(\text{range}(e[k]))$.

Fix $k \geq \inf(k_0, m - 1)$. Then

$$\Psi'(e[k+1]) = f'(\Psi'(e[k]), e(k - m + 1) \ldots e(k))$$

by the definition of Ψ'

$$= h_{\min(\Psi'(e[k]))e(k)}[f(\Psi(e[k]), e(k - m + 1) \ldots e(k))]$$

by the definition of f' and Fact (26)

$$= h_{\min(L)e(k)}(\Psi(e[k+1]))$$

by Fact (25), (27)b, and the definition of Ψ

$$= h_{\min(L)e(k)}(L)$$

by (27)a, and the latter is equal to L by the definition of $h_{\min(L)e(k)}$. This shows that Ψ' converges on e to L, as required.

Finally we define from Ψ' a 1-memory-limited scientist Ψ^* that solves **P**. Set $\widehat{\emptyset} = \emptyset$. For all $n \in N - \{0\}$ and $\sigma = \langle x_1 \ldots x_n \rangle \in SEQ$, set:

$$\widehat{\sigma} = \langle x_1 \overbrace{y_1 \ldots y_1}^{m-1} x_2 \overbrace{y_2 \ldots y_2}^{m-1} \ldots x_n \overbrace{y_2 \ldots y_n}^{m-1} \rangle,$$

where for all $1 \leq i \leq n$, $y_i = \inf(x_1 \ldots x_i)$. Let scientist Ψ^* be such that for all $\sigma \in SEQ$, $\Psi^*(\sigma) = \Psi'(\widehat{\sigma})$. It e is an environment, then $\bigcup_{k \in N} \widehat{e[k]}$ is an environment for $\text{range}(e)$, and this with the fact Ψ' solves **P** implies that Ψ^*

solves **P**. So to finish the proof, it remains to show that Ψ^* is 1-memory-limited.

Let total function $f'' : \mathrm{pow}(N) \times N \to \mathrm{pow}(N)$ be defined as follows. For all $n \in N$, $f''(\emptyset, n) = \Psi'(\langle \overbrace{n \ldots n}^{m} \rangle)$. For all $L \in \mathrm{pow}(N) - \{\emptyset\}$ and $n \in N$,

$$f''(L, n) = f'\{ \ldots f'[f'(L, \overbrace{i \ldots i}^{m-1}, n), \overbrace{i \ldots i}^{m-2}, n, j] \ldots n, \overbrace{j \ldots j}^{m-1} \},$$

where $i = \min(L)$ and $j = \inf(i, n)$. It follows easily from Fact (25) that for all $\sigma \in SEQ$ and $n \in N$, $\Psi^*(\sigma * n) = f''(\Psi^*(\sigma), n)$. ∎

Solution to Exercise 2.(53):

We prove Part (a). Let us define an m-data selector as follows. Let $\sigma \in SEQ$ be given. Then $g(\sigma * 0) = g(\sigma)$. Let $1 \le i \le m$ and $x \in N$ be given.

(a) If $\langle i, y \rangle \in g(\sigma)$ for some $y \ge x$, then $g(\sigma * \langle i, x \rangle) = g(\sigma)$.

(b) Otherwise, $g(\sigma * \langle i, x \rangle) = (g(\sigma) \cup \{\langle i, x \rangle\}) - \{\langle i, y \rangle \in g(\sigma) \mid y \in N\}$.

Define m-buffer-limited scientist Ψ from m-data selector g together with the equality $\Psi(\emptyset) = N - \{0\}$ and the following equations, for all $L \in \mathrm{pow}(\mathcal{L}_{form})$, $D \subseteq N$ such that $\mathrm{card}(D) \le m$, and $y \in N$:

(a) If $L = N - \{0\}$ and $y \ne 0$, then $\Psi(L, D, y) = N - \{0\}$.

(b) If $L \ne N - \{0\}$ or $y = 0$, then $\Psi(L, D, y) = \{0\} \cup \bigcup_{1 \le i \le m} \{\langle i, 0 \rangle \ldots \langle i, p_i \rangle\}$ where for all $1 \le i \le m$:

(i) $p_i = 0$ if there is no $y \in N$ such that $\langle i, y \rangle \in D$;

(ii) if $\langle i, y \rangle \in D$ for some $y \in N$, then p_i is the greatest such y.

It is easy to verify that Ψ solves $\mathbf{P}_m$.

Let $(m - 1)$-buffer-limited scientist Ψ' be defined from $(m - 1)$-data selector g'. We show that Ψ' fails to solve $\mathbf{P}_m$. Suppose otherwise. Let locking-sequence σ for $N - \{0\}$ and Ψ' be given and let $p \in N$ be least such that no $\langle i, j \rangle$ with $1 \le i \le m$ and $j > p$ occurs in σ. Let $n_0 \in N$ be given. Call $\tau \in SEQ$ "n_0-particular" just in case for all $1 \le i \le m$ there is $p \le p_i \le p + n_0$ such that τ is of the form $(\langle 1, 0 \rangle \ldots \langle 1, p_1 \rangle \ldots \langle m, 0 \rangle \ldots \langle m, p_m \rangle)$. If $\tau \in SEQ$ is n_0-particular we set $L_\tau = \{0\} \cup \mathrm{range}(\tau)$. Since g' is an $(m - 1)$-data selector and at most $m(p + n_0 + 1)$ integers can occur in a n_0-particular sequence, it is easy to verify that $\{g(\sigma * \tau) \mid \tau\ n_0\text{-particular}\}$ has cardinality at most $[m(p + n_0 + 2)]^{m-1}$. Moreover there are $(n_0 + 1)^m$ n_0-particular sequences and $(n_0 + 1)^m$ corresponding distinct worlds L_τ. Choose $n_0 \in N$ such that $[m(p+1)]^{m-1} < n_0 + 1$. Then $[m(p+n_0+2)]^{m-1} < (n_0+1)^m$. Hence there are n_0-particular sequences τ_1 and τ_2 such that $\tau_1 \neq \tau_2$ and:

(28) $\quad g'(\sigma * \tau_1) = g'(\sigma * \tau_2)$.

Since σ is a locking-sequence for $N - \{0\}$,

(29) $\quad \Psi(\sigma * \tau_1) = \Psi(\sigma * \tau_2)$.

Trivially, for every n_0-particular sequence τ, $e_\tau = \sigma * \tau * 0 * 0 * \ldots$ is an environment for L_τ. From (28) and (29) we infer that $\Psi(e_{\tau_1}[k]) = \Psi(e_{\tau_2}[k])$ for all $k \geq \text{length}(\sigma * \tau)$. Hence Ψ fails to solve one of L_{τ_1} and L_{τ_2}. Hence Ψ' fails to solve $\mathbf{P}_m$.

We prove Part (b). Fix a coding of N^2 onto $N - \{0\}$, and denote by $\langle x, y \rangle$ the image of $(x, y) \in N^2$ under this coding. Let $\mathbf{P}$ be the problem consisting of $N - \{0\}$ together with every set of the form $\{0, \langle 1, 0 \rangle \ldots \langle 1, p_1 \rangle \ldots \langle n, 0 \rangle \ldots \langle n, p_n \rangle\}$, where $n \in N - \{0\}$ and $p_1 \ldots p_n \in N$. It is easy to verify that $\mathbf{P}$ is solvable. A proof similar to that of Part (a) shows that $\mathbf{P}$ is solvable by no m-buffer-limited scientist, for any $m \in N$.

We prove Part (c). Let m-buffer-limited scientist Ψ be defined from m-data selector g, and suppose that Ψ solves the problem $\mathbf{P}$ defined in the proof of Proposition (48). We show that Ψ is not $\mathbf{P}$-consistent, which allows to conclude via Exercise (40) and Lemma (29) that Ψ does not solve $\mathbf{P}$ efficiently. Let locking-sequence σ be for $N - \{0\}$ and Ψ. Let $i_0 \in N$ be least such that $\bigcup_{i \geq i_0} E_i \cap \text{range}(\sigma) = \emptyset$. Let $n_0 \in N$ be given. Call $\tau \in SEQ$ "n_0-particular" just in case τ is of the form $\gamma_0 * \ldots * \gamma_{n_0}$ where for all $j \leq n_0$, γ_j is either $\emptyset$ or $\langle \min(E_{i_0+j}) \rangle$. There are 2^{n_0+1} such sequences. Since g is an m-data selector it is easy to verify that $\{g(\sigma * \tau) \mid \tau \ n_0\text{-particular}\}$ has cardinality at most $(m + n_0 + 2)^m$. Choose $n_0 \in N$ such that $(m + n_0 + 2)^m < 2^{n_0+1}$. Hence there are particular sequences τ_1 and τ_2 such that $\tau_1 \neq \tau_2$ and:

(30) $\quad g(\sigma * \tau_1) = g(\sigma * \tau_2)$.

Since σ is a locking-sequence for $N - \{0\}$ and Ψ,

(31) $\quad \Psi(\sigma * \tau_1) = \Psi(\sigma * \tau_2)$.

Since $\tau_1 \neq \tau_2$ there is $j \leq n_0$ such that, setting $\tau_1 = \gamma_0^1 * \ldots * \gamma_{n_0}^1$ and $\tau_2 = \gamma_0^2 * \ldots * \gamma_{n_0}^2$, either $\gamma_j^1 = \emptyset$ and $\gamma_j^2 \neq \emptyset$, or $\gamma_j^1 \neq \emptyset$ and $\gamma_j^2 = \emptyset$. Suppose that $j \leq n_0$ is such that $\gamma_j^1 = \emptyset$ and $\gamma_j^2 \neq \emptyset$ (the other case is parallel). Let environment e for $N - E_{i_0+j}$ be given. It is immediate that $\sigma * \tau_1 * e$ is an environment for $N - E_{i_0+j}$. Since Ψ solves $\mathbf{P}$ there is $k_0 \geq \text{length}(\sigma) + n_0 + 1$ such that $\Psi(\sigma * \tau_1 * e[k_0]) = N - E_{i_0+j}$. From (30) and (31) we deduce that $\Psi(\sigma * \tau_2 * e[k_0]) = N - E_{i_0+j}$. Trivially, $\sigma * \tau_2 * e[k]$ is for $\mathbf{P}$, but it is not an initial segment of an environment for $N - E_{i_0+j}$ since $\gamma_j^2 \neq \emptyset$. Hence Ψ is not $\mathbf{P}$-consistent. ∎

Section 2.5

Given $i, n \in N$ and $\sigma \in SEQ$, we use $\psi_{i,n}(\sigma)$ to denote n steps in ψ_i's calculation on input σ.

Solution to Exercise 2.(73):
Let recursive function $f : N \to N$ satisfy the following, for all $i \in N$ and $\sigma \in SEQ$. If $\psi_{i,\text{length}(\sigma)}(\tau)$ is undefined for every $\tau \subseteq \sigma$ then $\psi_{f(i)}(\sigma) = 0$. Otherwise, let $\tau \subseteq \sigma$ be the initial segment of σ with maximal length such that $\psi_{i,\text{length}(\sigma)}(\tau)$ is defined; then $\psi_{f(i)}(\sigma) = \psi_i(\tau)$. It is trivial that for all $i \in N$, $\psi_{f(i)}$ underlies a total, computable scientist.

Let $i \in N$ and environment e be given. Suppose that ψ_i underlies computable scientist Ψ and that Ψ solves e. Then there exists $k_0, n \in N$ such that:

(a) for all $k \geq k_0$, $\psi_i(e[k])$ is defined and $W_{\psi_i(e[k])} = \text{range}(e)$;

(b) $\psi_{i,n}(e[k_0])$ is defined.

It is easy to verify that for all $k \geq \sup(k_0, n)$, $\psi_{f(i)}(e[k])$ is defined and $W_{\psi_i(e[k])} = \text{range}(e)$. Hence $\psi_{f(i)}$ underlies computable scientist Ψ' that solves e. This proves that $\text{scope}(\Psi) \subseteq \text{scope}(\Psi')$.

Suppose that ψ_i converges on e to index q_0 for range(e). Then there exists $k_0, n \in N$ such that:

(a) for all $k \geq k_0$, $\psi_i(e[k])$ is defined and equals q_0;

(b) $\psi_{i,n}(e[k_0])$ is defined.

It is easy to verify that for all $k \geq \sup(k_0, n)$, $\psi_{f(i)}(e[k]) = q_0$. Hence $\psi_{f(i)}$ converges on e to q_0. This proves that stable-scope(Ψ) $\subseteq$ stable-scope(Ψ'). ∎

Solution to Exercise 2.(74):
We prove Part (a). Let recursive function $f : SEQ \times N \to N$ be such that for all $\sigma \in SEQ$ and $i \in N$, $W_{f(\sigma,i)} = \text{range}(\sigma) \cup W_i$. Let partial recursive function $\psi : SEQ \to N$ underly computable scientist Ψ. Let $\psi' : SEQ \to N$

be the partial recursive function with domain domain(ψ) such that for all $\sigma \in$ SEQ, $\psi'(\sigma) = f(\sigma, \psi(\sigma))$. It is easy to verify that ψ' underlies a scope(Ψ)-consistent, computable scientist Ψ' such that scope(Ψ) $\subseteq$ scope(Ψ').

We prove Part (b). Let nonrecursive r.e. $L \subseteq N$ and $n \in N - L$ be given. Choose index i for L and index j for $N - L$. Let $f : SEQ \rightarrow N$ be such that for all $\sigma \in SEQ$, if $n \notin$ range(σ) then $\psi(\sigma) = i$; otherwise $\psi(\sigma) = j$. It is easy to verify that ψ underlies a computable scientist Ψ such that stable-scope(Ψ) = $\{L, N\}$. Let computable $\psi : SEQ \rightarrow N$ underly computable scientist Ψ' that solves $\{L, N\}$ stably. By Lemma 2.(67) let $\sigma \in SEQ$ be such that range(σ) $\subseteq L$, $W_{\psi(\sigma)} = L$ and for all $n \in L$, $\psi(\sigma * n) = \psi(\sigma)$. If Ψ' is stable-scope(Ψ)-consistent then for all $n \in N$, $\psi(\sigma * n) = \psi(\sigma)$ iff $n \in L$, which is impossible since L is nonrecursive by hypothesis. ∎

Solution to Exercise 2.(75):

Let enumeration $\{\varphi_i \mid i \in N\}$ of computable functions from N to N be such that for all $i \in N$, $W_i =$ domain(φ_i). Given $i, n, x \in N$, we use $\varphi_{i,n}(x)$ to denote n steps in φ_i's calculation on input x. Let $j \in N$ be given. Set $L_j = \{2j\} \cup \{2n + 1 \mid n \in N\}$. Given $n \in N$, denote sequence $\langle 2j, 1 \ldots 2n + 1 \rangle$ by $\sigma_{j,n}$. Set $L'_j =$ range($\sigma_{j,n}$) if $n \in N$ is least such that $\psi_j(\sigma_{j,n}) = i$ and $W_i \supset$ range($\sigma_{j,n}$), and $L'_j = \emptyset$ otherwise. Define **P** to be $\{L_j \mid j \in N\} \cup \{L'_j \mid j \in N$ and $L'_j \neq \emptyset\}$. Let $f : N \rightarrow N$ be any recursive function such that for all $j \in N$, $f(j)$ is an index of L_j. Let $g : N^2 \rightarrow N$ be any recursive function such that for all $j, n \in N$, $g(j, n)$ is an index of range($\sigma_{j,n}$). We define a partial recursive function $\psi : SEQ \rightarrow N$ as follows, for every $\sigma \in SEQ$. If range(σ) is included in no L_j, $j \in N$ then $\psi(\sigma)$ is undefined. Otherwise, let $j \in N$ be least such that range(σ) $\subseteq L_j$. Suppose there is no $n \leq$ length(σ) such that:

(32)   (a) $\psi_{j,\text{length}(\sigma)}(\sigma_{j,n}) = i$;

      (b) $\{x \leq \text{length}(\sigma) \mid \varphi_{i,\text{length}(\sigma)}(x) \text{ converges}\} \supset \text{range}(\sigma_{j,n})$.

Then $\psi(\sigma) = f(j)$. Otherwise, let n be least such that (32) holds. If range($\sigma_{j,n}$) $\supseteq$ range(σ) then $\psi(\sigma) = g(j, n)$; otherwise $\psi(\sigma) = f(j)$. It is easy to verify that ψ underlies a computable scientist Ψ such that scope(Ψ) = stable-scope (Ψ) = **P**.

Suppose that $j \in N$ is such that ψ_j underlies **P**-conservative computable scientist Ψ' that solves **P**. Hence Ψ' converges on environment $e = \langle 2j, 1, 3, 5 \ldots \rangle$ to L_j. Hence there is least $n \in N$ such that:

(33) $\psi_j(e[n + 2]) = i$ and $W_i \supset$ range($e[n + 2]$).

Since $\sigma_{j,n} = e[n+2]$, it follows from (33) that $L'_j = \text{range}(\sigma_{j,n})$. Denote by e' environment $\langle 2j, 1 \ldots 2n+1, 2n+1 \ldots \rangle$ for L'_j. With the hypothesis that Ψ' is **P**-conservative, (33) implies that $\Psi'(e[k]) = W_i$ for all $k \geq n+2$. Hence Ψ' does not converge on e' to L'_j. Contradiction.   ∎

Solution to Exercise 2.(76):

For a contradiction suppose that total computable $h : N \rightarrow N$ is such that for all $i \in N$, ψ_i underlies a scientist that fails to solve $W_{h(i)}$. Let total computable $g : N \rightarrow N$ be such that for all $i \in N$, $\psi_{g(i)}$ is the constant i-function. So, for all $i \in N$, $\psi_{g(i)}$ underlies a scientist that solves W_i. By the Recursion Theorem there is j such that $W_j = W_{h(g(j))}$. But $\psi_{g(j)}$ underlies a scientist that solves $W_j = W_{h(g(j))}$, contradicting the choice of h.   ∎

Solution to Exercise 2.(77):

We prove Part (a). Let recursive function $\psi : SEQ \rightarrow N$ be such that for every $\sigma \in SEQ$, $\psi(\sigma)$ is a Σ_0 index of $\text{range}(\sigma)$. It is immediate that ψ underlies a computable scientist that is Σ_0 and that solves $\mathbf{P}_X$ stably for all $X \subseteq N$.

We prove Part (b). Fix $n \in N$ and Σ_{n+1} set $X \subseteq N$. The predicate $P(m, x)$ defined by the formula $(x = m) \vee (x = 0 \wedge m \in X)$ is Σ_{n+1}. Hence there exists a recursive function $f : N \rightarrow N$ such that for every $m \in N$, $f(m)$ is a Σ_{n+1} index of $\{m, 0\}$ if $m \in X$, and $f(m)$ is a Σ_{n+1} index of $\{m\}$ otherwise. If $X = \emptyset$ choose Σ_{n+1} index i_0 of $\{0\}$. If $X \neq \emptyset$ choose $m_0 \in X$ and Σ_{n+1} index i_0 of $\{0, m_0\}$. Let recursive function $\psi : SEQ \rightarrow N$ be defined as follows, for every $\sigma \in SEQ$. If there is $m \in N - \{0\}$ such that $m \in \text{range}(\sigma) \subseteq \{m, 0\}$, then $\psi(\sigma) = f(m)$; otherwise, $\psi(\sigma) = i_0$. It is easy to verify that ψ underlies a computable scientist that solves $\mathbf{P}_X$ stably and efficiently.

We prove Part (c). Fix $n \in N$ and Σ_{n+1} set $X \subseteq N$ that is not a Σ_n set. Suppose that computable function $\psi : SEQ \rightarrow N$ underlies Σ_n computable scientist Ψ that solves $\mathbf{P}_X$ efficiently. Then for all $m \in N - \{0\}$, $\Psi(\langle m \rangle) = \{0, m\}$ if $m \in X$, and $\Psi(\langle m \rangle) = \{m\}$ otherwise. Hence for all $m \in N - \{0\}$:

(34)   0 belongs to the set with Σ_n index $\psi(\langle m \rangle)$ iff $m \in X$.

But "0 belongs to the set with Σ_n index $\psi(\langle m \rangle)$" is a Σ_n predicate of m, so (34) would imply that X is a Σ_n set, which contradicts the hypothesis.   ∎

Solution to Exercise 2.(78):

By Exercise 2.(73) let recursive function $g : N \rightarrow N$ be such that for all $i \in N$, if ψ_i underlies computable scientist Ψ then $\Psi_{g(i)}$ underlies total, computable scientist Ψ' such that stable-scope(Ψ) $\subseteq$ stable-scope(Ψ'). Let $\{\tau_m \mid m \in N\}$ enumerate SEQ. Let recursive function $f : N \to N$ be such that for every $\sigma \in SEQ$ and $i \in N$, the following holds. Suppose there is least $n \leq \text{length}(\sigma)$ such that:

(a) $\text{length}(\tau_n) \leq \text{length}(\sigma)$,

(b) $\text{range}(\tau_n) \subseteq \text{range}(\sigma)$, and

(c) for every $\gamma \in SEQ$ with $\text{length}(\gamma) \leq \text{length}(\sigma)$ and $\text{range}(\gamma) \subseteq \text{range}(\sigma)$, $\psi_{g(i)}(\tau_n * \gamma) = \psi_{g(i)}(\tau_n)$.

Then $\psi_{f(i)}(\sigma) = \psi_{g(i)}(\tau_n)$.

We show that f satisfies the claim of the exercise. Fix $i \in N$, and let Ψ be the computable scientist that ψ_i underlies. If scope(Ψ) is empty then there is nothing to prove, so suppose otherwise. Let environment e for $L \in$ stable-scope(Ψ) be given. By Lemma 2.(67) let $n \in N$ be least such that $\text{range}(\tau_n) \subseteq L$, $W_{\psi_{g(i)}(\tau_n)} = L$, and for all $\gamma \in SEQ$ with $\text{range}(\gamma) \subseteq L$, $\psi_{g(i)}(\tau_n * \gamma) = \psi_{g(i)}(\tau_n)$. Hence for every $m < n$, if $\text{range}(\tau_m) \subseteq L$ there is $\gamma_m \in SEQ$ such that $\text{range}(\gamma_m) \subseteq L$ and $\psi_{g(i)}(\tau_m * \gamma_m) \neq \psi_{g(i)}(\tau_m)$. Let $k_0 \geq n$ be such that for all $m \leq n$, if $\text{range}(\tau_m) \subseteq L$ then:

(a) $\text{range}(\tau_m) \subseteq \text{range}(e[k_0])$;

(b) if $m < n$ then $\text{length}(\gamma_m) \leq k_0$.

It is easy to verify that for all $k \geq k_0$, $\psi_{f(i)}(e[k]) = \psi_{g(i)}(\tau_n)$. Hence $\psi_{f(i)}$ converges on e to $\psi_{g(i)}(\tau_n)$. Since $W_{\psi_{g(i)}(\tau_n)} = L$, f satisfies the claim of the proposition.   ∎

Solution to Exercise 2.(79):

We prove Part (a). Let acceptable indexing $\{\varphi_i \mid i \in N\}$ of the computable functions from N to N be given. Given $i, n, x \in N$, we use $\varphi_{i,n}(x)$ to denote n steps in φ_i's calculation on input x. For all $i, n \in N$ set $X_{i,n} = \{x \leq n \mid \varphi_{i,n}(x) \text{ is defined }\}$. Let recursive function $f : N^2 \to N$ be such that for all $\sigma \in SEQ$, the following holds. Suppose there is no $n \in N$ such that:

(35) (a) $n \in (X_{i,\text{length}(\sigma)} - X_{j,\text{length}(\sigma)})$, or

      (b) $n \in (X_{j,\text{length}(\sigma)} - X_{i,\text{length}(\sigma)})$.

Then $\psi_{f(i,j)}(\sigma) = i$. Otherwise, let n_0 be the least $n \in N$ that satisfies (35). Suppose $n_0 \in X_{i,\text{length}(\sigma)}$. If $n_0 \in \text{range}(\sigma)$ then $\psi_{f(i,j)}(\sigma) = i$. If $n_0 \notin \text{range}(\sigma)$ then $\psi_{f(i,j)}(\sigma) = j$. Suppose $n_0 \in X_{j,\text{length}(\sigma)}$. If $n_0 \in \text{range}(\sigma)$ then $\psi_{f(i,j)}(\sigma)$

$= j$. If $n_0 \notin \text{range}(\sigma)$ then $\psi_{f(i,j)}(\sigma) = i$. We show that f satisfies the claim of (a).

Let $i, j \in N$ with $W_i \neq \emptyset$ and $W_j \neq \emptyset$ be given. Denote by Ψ the scientist that $\psi_{f(i,j)}$ underlies. It is trivial that Ψ solves $\{W_i, W_j\}$ if $W_i = W_j$. So suppose $W_i \neq W_j$. Hence there exists $n \in (W_i - W_j) \cup (W_j - W_i)$ such that for all $m < n$, $m \in W_i$ if and only if $m \in W_j$. Assume, without loss of generality, that $n \in W_i - W_j$. Let environment e for W_i or W_j be given. There exists $k_0 \in N$ such that for all $k \geq k_0$, for all $m \leq n$, $\varphi_{i,k}(m)$ is defined iff $m \in W_i$, and $\varphi_{j,k}(m)$ is defined iff $m \in W_j$. It is easy to verify that for all $k \geq k_0$, $\psi_{f(i,j)}(e[k]) = i$ if e is for W_i, and $\psi_{f(i,j)}(e[k]) = j$ if e is for W_j. Hence Ψ converges on e to range(e), and thus solves e. So Ψ solves $\{W_i, W_j\}$.

We prove Part (b). Suppose for a contradiction that recursive function $g : N \rightarrow N$ is such that for all $i \in N$, if $W_i \neq \emptyset$ then $\psi_{g(i)}$ underlies a computable scientist Ψ with $\{L, W_i\} \subseteq$ stable-scope(Ψ). Choose recursive function $f : N \rightarrow N$ that satisfies the claim of Exercise 2.(78). Let partial recursive function $\varphi : N^2 \rightarrow N$ be such that for all $i, n \in N$, $\varphi(i, n)$ is defined iff $W_i \neq \emptyset$, and $\{\varphi(i, n) \mid n \in N\} = W_i$ whenever $W_i \neq \emptyset$. Fix environment e for L, and set $X = \{i \in N \mid W_i \neq \emptyset$ and $\lim_{k\rightarrow\infty} \psi_{f(g(i))}(\langle\varphi(i, 0) \ldots \varphi(i, k)\rangle) = \lim_{k\rightarrow\infty} \psi_{f(g(i))}(e[k])$. By the limit lemma,[1] the T-degree of X is at most $\mathbf{0}'$. Moreover by the property of f, for all $i \in N$, $W_i = L$ iff $i \in X$. But $\{i \in N \mid W_i = L\}$ is of T-degree $\mathbf{0}''$.[2] Contradiction. ∎

Solution to Exercise 2.(80):

Let computable scientist Ψ be given. By Exercise 2.(73) we can suppose without loss of generality that there is total recursive function $g : SEQ \rightarrow N$ such that g underlies Ψ. Let $p : N \times SEQ \rightarrow N$ be any one-to-one recursive function such that for all $i \in N$ and $\sigma \in SEQ$, $W_{p(i,\sigma)} = W_i$. Let $\{\tau_m \mid m \in N\}$ enumerate SEQ. Let recursive function $f : SEQ \rightarrow N$ be such that for every $\sigma \in SEQ$, the following holds. Suppose there is least $n \leq \text{length}(\sigma)$ such that:

(a) length$(\tau_n) \leq$ length(σ),

(b) range$(\tau_n) \subseteq$ range(σ), and

(c) for every $\gamma \in SEQ$ with length$(\gamma) \leq$ length(σ) and range$(\gamma) \subseteq$ range(σ), $g(\tau_n * \gamma) = g(\tau_n)$.

---

1. See P. Odifreddi, *Classical Recursion Theory*, Amsterdam: North-Holland, 1989, Proposition IV.1.17, p. 373.

2. See L. Hay, "Isomorphism types of index sets of partial recursive functions," *Proceedings of the American Mathematical Society*, vol. 17, 1966, pp. 106-110.

Then $f(\sigma) = p(g(\tau_n), \tau_n)$. Denote by Ψ' the computable scientist that f underlies. It is easy to adapt the proof of Exercise 2. (78) to show that stable-scope(Ψ) $\subseteq$ stable-scope(Ψ') and:

(36) For all $r.e.$ languages $L \in \text{pow}(N) - \{\emptyset\}$, the following conditions are equivalent.

(a) Ψ' solves L stably.

(b) There exists $n \in N$ such that range(τ_n) $\subseteq L$ and for every $\gamma \in SEQ$ with range(γ) $\subseteq L$, $f(\tau_n * \gamma) = f(\tau_n)$.

Let partial recursive function $\psi : SEQ \to N$ satisfy the following, for all $\sigma \in SEQ$. If $\sigma = \emptyset$ or range(σ) $\not\subseteq W_{g(\sigma)}$, then $\psi(\sigma)$ is underfined. Suppose otherwise. We distinguish two cases.

Case 1: $N \in$ stable-scope(Ψ). Let $m \in N$ be least such that $\sigma = \tau_m$. If there is least $n \leq m$ such that $f(\tau_n) = f(\sigma)$, range(τ_n) $\subseteq$ range(σ), and $f(\tau_n * \gamma) = f(\tau_n)$ for all $\gamma \in SEQ$ with range(γ) $\subseteq W_{f(\tau_n)}$, then $\psi(\sigma) = \psi(\tau_n)$ is an index of $W_{f(\tau_n)}$. Otherwise, $\psi(\sigma)$ is an index of N.

Case 2: $N \not\in$ stable-scope(Ψ). If range(σ) $= \{0 \ldots p\}$ for some $p \in N$, then $\psi(\sigma) = \psi(\langle 0 \ldots p \rangle)$ is an index of $\{0 \ldots p\}$. Suppose otherwise. Let $m \in N$ be least such that $\sigma = \tau_m$. If there is least $n \leq m$ such that $f(\tau_n) = f(\sigma)$, range(τ_n) $\subseteq$ range(σ), and $f(\tau_n * \gamma) = f(\tau_n)$ for all $\gamma \in SEQ$ with range(γ) $\subseteq W_{f(\tau_n)}$, then $\psi(\sigma) = \psi(\tau_n)$ is an index of $W_{f(\tau_n)}$. Otherwise, there is $q \in N$ such that $\psi(\sigma)$ is an index of $\{0 \ldots q\}$.

In case $N \in$ stable-scope(Ψ'), set $\mathbf{P} =$ stable-scope(Ψ'). In case $N \not\in$ stable-scope(Ψ'), set $\mathbf{P} =$ stable-scope(Ψ') $\cup \{\{0 \ldots q\} \mid q \in N\}$. Denote by Ψ^* the scientist that ψ underlies. The implication from (36)a to (36)b together with the definition of ψ implies easily that Ψ^* solves $\mathbf{P}$. This, the implication from (36)b to (36)a, and the definition of ψ imply easily that Ψ^* is prudent. ∎

Solution to Exercise 2.(81):

Let $\langle \cdot, \cdot \rangle$ be a recursive isomorphism between N^2 and N. For every $i \in N$ such that W_i is infinite, set $L_i = \{\langle i, x \rangle \mid x \in W_i\}$. Denote by $\mathbf{P}$ the problem consisting of the sets of form L_i for all $i \in N$ such that W_i is infinite. Trivially, every member of $\mathbf{P}$ is infinite. Let computable function $\psi : SEQ \to N$ satisfy the following. $\psi(\emptyset)$ is an index for some member of $\mathbf{P}$. For all nonempty $\sigma \in SEQ$, if $\sigma = (\langle i, x_0 \rangle \ldots \langle i, x_n \rangle)$ for some $i, n, x_0 \ldots x_n \in N$, then $\psi(\sigma)$ is an index of L_i. Relying on Exercise 2.(41), it is easy to verify that ψ underlies a computable scientist that solves **P** efficiently. So it remains to prove Part (c).

Let computable $f : N \to N$ satisfy the following, for all $j \in N$. If there are $m, n, x, y \in N$ with $m \neq n$, $\langle m, x \rangle \in W_j$ and $\langle n, y \rangle \in W_j$, then $W_{f(j)} = N$. Otherwise for all $x \in N$, $x \in W_{f(j)}$ iff there is $m \in N$ such that $\langle m, x \rangle \in W_j$. It is easy to see that for all $j \in N$:

(37) (a) if W_j is infinite then $W_{f(j)}$ is infinite;

     (b) for all $i \in N$, if W_i is infinite and $W_j = L_i$, then $W_{f(j)} = W_i$.

Suppose for a contradiction that partial recursive function ψ underlies nontrivial, computable scientist Ψ that solves **P**. Set $S = \{ f(j) \mid j \in \text{range}(\psi) \}$. By (37)a, S consists of indexes of infinite, r.e. languages. Hence it is easy to verify the existence of an increasing sequence $\{ n_i \mid i \in N \}$ of integers such that $N - \{ n_i \mid i \in N \}$ is infinite and for all $i \in N$, $n_i \in W_{f(i)}$. So:

(38) S contains no index for the infinite r.e. set $N - \{ n_i \mid i \in N \}$.

Since Ψ solves **P**, $L_i \in \text{range}(\Psi)$ for all $i \in N$ such that W_i is infinite. Hence by (37)b, S contains an index of W_i for all $i \in N$ such that W_i is infinite. This contradicts (38). ∎

Solution to Exercise 2.(82):

Let acceptable indexing $\{ \varphi_i \mid i \in N \}$ of the computable functions from N to N be given. Given $i, n, x \in N$, we use $\varphi_{i,n}(x)$ to denote n steps in φ_i's calculation on input x. Let $\langle \cdot, \cdot \rangle$ be a recursive isomorphism between N^2 and N. Let $i \in N$ be such that φ_i is total and $\varphi_j \neq \varphi_i$ for all $j < i$. Let $n(i) \in N$ be least such that for all $j < i$, there exists $x \leq n(i)$ such that either $\varphi_j(x)$ is undefined or $\varphi_j(x) \neq \varphi_i(x)$. Then define L_i to be $\{ \langle 0, \langle i, n(i) \rangle \rangle \} \cup \{ \langle 1 + \langle p, x \rangle, \varphi_i(x) \rangle \mid p, x \in N \}$. Denote by **P** the problem consisting of the sets of form L_i for all $i \in N$ such that φ_i is total and $\varphi_j \neq \varphi_i$ for all $j < i$. We show that **P** satisfies the claim of the Exercise.

We prove Part (a). Let partial recursive function $\psi : SEQ \to N$ satisfy the following. Let $\sigma \in SEQ$ be given. Let $x_0 \in N$ be least such that for all $i, x \in N$, if $\langle 0, \langle i, x \rangle \rangle \in \text{range}(\sigma)$ then $x \leq x_0$. Denote by X the set of all $i \in N$ such that:

(a) for some $x \leq x_0$, $\langle 0, \langle i, x \rangle \rangle \in \text{range}(\sigma)$;

(b) for all $x \leq x_0$, $\varphi_{i,\text{length}(\sigma)}(x)$ is defined;

(c) for all $j < i$, there is $x \leq x_0$ such that either $\varphi_{j,\text{length}(\sigma)}(x)$ is undefined or $\varphi_j(x) \neq \varphi_i(x)$.

Suppose there is unique $i \in X$ such that for all $x \leq x_0$, if there is $p \in N$ and $y \neq \varphi_i(x)$ with $\langle 1 + \langle p, x \rangle, y \rangle \in \text{range}(\sigma)$, then there is $p' > p$ with $\langle 1 + \langle p', x \rangle, \varphi_i(x) \rangle \in \text{range}(\sigma)$. Then $\psi(\sigma) = i$. Denote by Ψ the computable scientist that ψ underlies. We show that Ψ solves $\mathbf{P}$ stably on noisy environments. Let $i \in N$ be such that φ_i is total and $\varphi_j \neq \varphi_i$ for all $j < i$. Let noisy environment e for L_i be given. It suffices to show that ψ converges on e to i. Since $\text{range}(e) - L_i$ is finite, let $x_0 \in N$ be least such that for all $j, x \in N$, if $\langle 0, \langle j, x \rangle \rangle \in \text{range}(e)$ then $x \leq x_0$. Again since $\text{range}(e) - L_i$ is finite, we can choose $k_0 \in N$ that satisfies the following:

(a) for all $x \leq x_0$, $\varphi_{i,k_0}(x)$ is defined;

(b) for all $x \leq x_0$, there is $p \in N$ such that $\langle 1 + \langle p, x \rangle, \varphi_i(x) \rangle \in \text{range}(e[k_0])$;

(c) for all $k \geq k_0$ and $x \leq x_0$, if there is $p \in N$ and $y \neq \varphi_i(x)$ with $\langle 1 + \langle p, x \rangle, y \rangle \in \text{range}(e[k])$, then there is $p' > p$ with $\langle 1 + \langle p', x \rangle, \varphi_i(x) \rangle \in \text{range}(e[k])$.

Using the definition of $n(i)$, it is easy to verify that for all $k \geq k_0$, $\psi(e[k]) = i$. Hence ψ converges on e to i, as required.

Now we prove Part (b). It is easy to exhibit a recursive function $f : SEQ \to SEQ$ with the following properties:

(a) For all $\sigma, \tau \in SEQ$, $\sigma \subseteq \tau$ implies $f(\sigma) \subseteq f(\tau)$.

(b) Let $k \in N$ be such that φ_k is total and $\varphi_j \neq \varphi_k$ for all $j < k$. If e is an environment for the functional language $\{\langle x, \varphi_k(x) \rangle \mid x \in N\}$, then $\bigcup_{k \in N} f(e[k])$ is an incomplete environment for L_k in which $\langle 0, \langle k, n(k) \rangle \rangle$ does not occur.

Now suppose that computable $\psi : SEQ \to N$ underlies computable scientist that solves $\mathbf{P}$ on incomplete environments. Then it is easy to verify that $\psi \circ f$ underlies a computable scientist that solves the collection of recursive functional languages. However this contradicts Theorem 2.(87).   ∎

Solution to Exercise 2.(83):

Let solvable problem $\mathbf{P}$ and computable functions $\chi_1, \chi_2 : SEQ \to N$ be such that χ_1 underlies computable scientist Ψ_1 that solves $\mathbf{P}$ efficiently and χ_2 underlies computable scientist Ψ_2 that solves $\mathbf{P}$ stably. By Exercise 2.(73) we can suppose without loss of generality that χ_2 is total. We will define a computable function $\psi : SEQ \to N$ that underlies computable scientist Ψ that solves $\mathbf{P}$ stably and efficiently. It will be clear that the definition of ψ will not depend on $\mathbf{P}$, and that an index for ψ can be effectively obtained from an index of χ_1 and from an index of χ_2. Fix a partial recursive function $\varphi : N^2 \to N$ such that for all $n \in N$:

(a) if $W_n = \emptyset$ then for all $i \in N$, $\varphi(n, i)$ is undefined;

(b) if $W_n \neq \emptyset$ then for all $i \in N$, $\varphi(n, i)$ is defined, and $\{\varphi(n, i) \mid i \in N\} = W_n$.

To define ψ, the following terminology will be used.

(39) DEFINITION: Let $\sigma, \tau \in SEQ$ and $i \in N$ be given. We say that σ i-selects τ just in case:

(a) $\psi(\sigma)$ is defined and $W_{\psi(\sigma)} \neq \emptyset$, and

(b) for all $\gamma \in SEQ$, if $\text{range}(\gamma) \subseteq \text{range}(\sigma) \cup \{\varphi(\psi(\sigma), 0) \ldots \varphi(\psi(\sigma), i)\}$ and $\text{length}(\gamma) \leq \text{length}(\sigma) + i + 1$, then $\chi_2(\tau * \gamma) = \chi_2(\tau)$.

Fix a recursive enumeration $\{\tau_n \mid n \in N\}$ of SEQ with infinitely many occurrences of $\emptyset$. We now give an algorithm for computing ψ, together with a computable function $F : SEQ \rightarrow SEQ$ useful for the construction. Set $\psi(\emptyset) = \chi_1(\emptyset)$ and $F(\emptyset) = \tau_0$. Let $\sigma \in SEQ$ be given, and suppose that $\psi(\sigma)$ and $F(\sigma)$ have been defined. Let $x \in N$ be given. If $\psi(\sigma)$ is undefined or $W_{\psi(\sigma)} = \emptyset$, then both $\psi(\sigma * x)$ and $F(\sigma * x)$ are undefined. So suppose otherwise. The values of $\psi(\sigma * x)$ and $F(\sigma * x)$ are computed as follows.

1. If $x \in \text{range}(\sigma) \subseteq \{\varphi(\psi(\sigma), 0) \ldots \varphi(\psi(\sigma), \text{length}(\sigma))\}$, then $\psi(\sigma * x) = \psi(\sigma)$ and $F(\sigma * x) = \tau_m$, where $m \leq \text{length}(\sigma)$ is greatest such that for all $i < m$, either $\text{range}(\tau_i) \not\subseteq \text{range}(\sigma)$ or σ does not 0-select τ_i. End.
   Otherwise goto 2.

2. If $\text{range}(\sigma) \not\subseteq \{\varphi(\psi(\sigma), 0) \ldots \varphi(\psi(\sigma), \text{length}(\sigma))\}$ and x is the first member of the sequence $\{\varphi(\psi(\sigma), 0) \ldots \varphi(\psi(\sigma), \text{length}(\sigma))\}$ that does not occur in σ, then $\psi(\sigma * x) = \psi(\sigma)$ and $F(\sigma * x) = \tau_m$, where $m \leq \text{length}(\sigma)$ is greatest such that for all $i < m$, either $\text{range}(\tau_i) \not\subseteq \text{range}(\sigma)$ or σ does not 0-select τ_i. End.
   Otherwise goto 3.

3. $n = 0$ and $\tau = \tau_0$.

4. Let $n' \geq n$ be least such that $\text{range}(\tau_{n'}) \subseteq \text{range}(\sigma)$. Set $n = n'$ and $\tau = \tau_n$. Goto 5.

5. If τ appears after $F(\sigma)$ in the enumeration of SEQ, then $\psi(\sigma * x) = \chi_1(\sigma * x)$ and $F(\sigma * x) = \tau$. End.
   Otherwise goto 6.

6. $i = 0$.

7. If σ does not i-select τ, then $n = n + 1$. Goto 4.
   Otherwise goto 8.

8. If $x \in \{\varphi(\psi(\sigma), 0) \ldots \varphi(\psi(\sigma), i)\}$, then $\psi(\sigma * x) = \psi(\sigma)$ and $F(\sigma * x) = \tau$. End.

Otherwise goto 9.

9. If $\chi_2[\tau * x * \varphi(\chi_1(\sigma * x), 0) * \ldots * \varphi(\chi_1(\sigma * x), i)] \neq \chi_2(\tau)$, then $\psi(\sigma * x) = \chi_1(\sigma * x)$ and $F(\sigma * x) = \tau$. End.

Otherwise $i = i + 1$. Goto 7.

Proof relies on the following lemma.

(40)   LEMMA: Let $\sigma, \tau \in SEQ$ and $x \in N$ be such that:

(a) $\sigma * x$ is for **P**.

(b) $\psi(\sigma)$ is defined and $\text{range}(\tau) \subseteq W_{\psi(\sigma)} \in \mathbf{P}$.

(c) $\text{range}(\tau) \subseteq \text{range}(\sigma)$.

(d) For all $i \in N$, σ i-selects τ.

(e) $x \notin W_{\psi(\sigma)}$.

Then there exists $i \in N$ such that $\chi_2[\tau * x * \varphi(\chi_1(\sigma * x), 0) * \ldots * \varphi(\chi_1(\sigma * x), i)] \neq \chi_2(\tau)$.

Proof of Lemma (40):   By (40)b, let e denote the environment $\tau * \varphi(\psi(\sigma), 0) * \varphi(\psi(\sigma), 1) * \ldots$ for $W_{\psi(\sigma)} \in \mathbf{P}$. It follows from (40)d that for all $k \geq \text{length}$ (τ), $\chi_2(e[k]) = \chi_2(\tau)$. This and the fact that Ψ_2 solves **P** imply that:

(41) $W_{\chi_2(\tau)} = W_{\psi(\sigma)}$.

From (40)a and the fact that Ψ_1 solves **P** efficiently, we deduce with Exercise 2.(41) that:

(42) $\text{range}(\sigma * x) \subseteq W_{\chi_1(\sigma * x)} \in \mathbf{P}$.

This with (40)e implies that:

(43) $W_{\chi_1(\sigma * x)} \neq W_{\psi(\sigma)}$.

From (40)c, (42), and the fact that Ψ_2 solves **P**, we infer that there exists $i \in N$ such that:

(44) $W_{\chi_2[\tau * x * \varphi(\chi_1(\sigma * x), 0) * \ldots * \varphi(\chi_1(\sigma * x), i)]} = W_{\chi_1(\sigma * x)}$.

We conclude with (41), (43) and (44).   ∎

We now prove a few useful facts.

(45)   FACT: For all $\sigma \in SEQ$, if σ is for **P** then:

(a) $\psi(\sigma)$ is defined and range$(\sigma) \subseteq W_{\psi(\sigma)}$.

(b) There exists $\gamma \subseteq \sigma$ such that for all $\lambda \in SEQ$, if $\gamma \subseteq \lambda \subseteq \sigma$ then $\psi(\lambda) = \chi_1(\gamma)$ (hence $W_{\psi(\sigma)} \in$ **P**).

Proof of Fact (45):   Proof is by induction on length(σ). It is trivial for $\sigma = \emptyset$. Suppose it is proved for $\sigma \in SEQ$. Let $x \in N$ be such that $\sigma * x$ is for **P**. We will prove it for $\sigma * x$. If the algorithm for computing $\psi(\sigma * x)$ does not enter a loop, then it stops in stage 1, 2, 5, 8 or 9, and it is easy to verify that the property holds for $\sigma * x$. If the algorithm for computing $\psi(\sigma * x)$ enters a loop, this means that there is $\tau \in SEQ$ such that:

(a) range$(\tau) \subseteq$ range(σ),

(b) for all $i \in N$, σ i-selects τ,

(c) $x \notin W_{\psi(\sigma)}$, and

(d) for all $i \in N$, $\chi_2[\tau * x * \varphi(\chi_1(\sigma * x), 0) * \ldots * \varphi(\chi_1(\sigma * x), i)] = \chi_2(\tau)$.

But this is in contradiction with Lemma (40). Hence the algorithm does not enter a loop, which ends the proof of the fact.   ∎

The next fact rests on the following definition.

(46)   DEFINITION:

Let *r.e.* $L \subseteq N$ and $\sigma \in SEQ$ be such that:

(a) range$(\sigma) \subseteq L$,

(b) $W_{\chi_2(\sigma)} = L$, and

(c) for all $\tau \in SEQ$ with range$(\tau) \subseteq L$, $\chi_2(\sigma * \tau) = \chi_2(\sigma)$.

We say that σ is a *stable locking-sequence* for L and χ_2.

(47)   FACT: Let environment e for $L \in$ **P** be given. Let $k_0 \in N$ be such that for all $k \geq k_0$, $W_{\psi(e[k])} = L$. Let $\tau \in SEQ$ be such that range$(\tau) \subseteq$ range$(e[k_0])$.

(a) If τ is a stable locking-sequence for L and χ_2, then for all $k \geq k_0$ and $i \in N$, $e[k]$ i-selects τ.

(b) If τ is not a stable locking-sequence for L and χ_2, then there is $k_1 \geq k_0$ such that for all $k \geq k_1$, $e[k]$ does not 0-select τ.

Proof of Fact (47):   Trivial.

(48)   FACT: Let environment e for $L \in \mathbf{P}$ be given. Let $k_0 \in N$ be such that for all $k \geq k_0$, $W_{\psi(e[k])} = W_{\psi(e[k_0])}$. Suppose $L \neq W_{\psi(e[k_0])}$. Then for all $\tau \in SEQ$ with range$(\tau) \subseteq L$, there exists $k_1 \geq k_0$ such that for all $k \geq k_1$, $e[k]$ does not 0-select τ.

Proof of Fact (48):   Suppose that $L \not\subseteq W_{\psi(e[k_0])}$, and choose $x \in L - W_{\psi(e[k_0])}$. Let $k \geq k_0$ be such that $x \in$ range$(e[k])$. With Fact (45)a we infer that $x \in W_{\psi(e[k])} = W_{\psi(e[k_0])}$, contradiction. Hence $L \subset W_{\psi(e[k_0])}$. Let $\tau \in SEQ$ be such that range$(\tau) \subseteq L$. Suppose that for all $k_1 \geq k_0$, there exists $k \geq k_1$ such that $e[k]$ 0-selects τ. This means that for all $\gamma \in SEQ$, if range$(\gamma) \subseteq W_{\psi(e[k_0])}$ then $\chi_2(\tau * \gamma) = \chi_2(\tau)$. But since range$(\tau) \subseteq L \subset W_{\psi(e[k_0])}$, it follows either that Ψ_2 does not solve L or that Ψ_2 does not solve $W_{\psi(e[k_0])}$, which is impossible since $L \in \mathbf{P}$ by hypothesis and $W_{\psi(e[k_0])} \in \mathbf{P}$ by Fact (45)b.   ∎

Now we can finish the proof of the exercise. Let e be an environment for $L \in \mathbf{P}$. First, we show that Ψ converges on e to L. By Fact (45) it suffices to show that for all $k_0 \in N$, if $W_{\psi(e[k_0])} \neq L$ then there exists $k_1 > k_0$ such that $\psi(e[k_1]) = \chi_1(e[k_1])$. Suppose for a contradiction that $k_0 \in N$ is such that $W_{\psi(e[k_0])} \neq L$ and:

(49)   for all $k > k_0$, $\psi(e[k]) \neq \chi_1(e[k])$.

By Fact (45) again, this implies that:

(50)   for all $k \geq k_0$, $\psi(e[k]) = \psi(e[k_0])$.

Hence by Fact (48), for all $k_1 \geq k_0$, there exists $k_2 \geq k_1$ such that:

(51)   for all $n \in N$, if range$(\tau_n) \subseteq L$ and τ_n does not appear after $F(e[k_1])$ in the enumeration of SEQ, then $e[k_2]$ does not 0-select τ_n.

By (49), for all $k > k_0$, the value of $\psi(e[k])$ is computed at stage 1, 2 or 8 of the algorithm. Suppose that for infinitely many $k \in N$, the value of $\psi(e[k])$ is computed at stage 1 or 2 of the algorithm. Then (50) and Fact (45)a imply easily that for all $k \geq k_0$, $\psi(e[k]) = \psi(e[k_0])$ is an index for L. So suppose otherwise, and let $k_1 \geq k_0$ be such that for all $k \geq k_1$, the value of $\psi(e[k])$ is computed at stage 8 of the algorithm. Hence for all $k \geq k_1$, the first ocurrence of $F(e[k])$ in the enumeration of SEQ does not come after $F(e[k_1])$. Considering the algorithm, this is in contradiction with (51).

Second, we show that ψ converges on e. Let $n_0 \in N$ be least such that τ_{n_0} is a stable locking-sequence for L and χ_2 (n_0 exists by Lemma 2.(67)). By Fact (47), there exists $k_0 \in N$ such that:

(a) $\text{range}(\tau_{n_0}) \subseteq \text{range}(e[k_0])$, and for all $k \geq k_0$ and $i \in N$, $e[k]$ i-selects τ_{n_0}.

(b) For all $n < n_0$, if $\text{range}(\tau_n) \subseteq L$, then $\text{range}(\tau_n) \subseteq \text{range}(e[k_0])$ and for all $k \geq k_0$, $e[k]$ does not 0-select τ_n.

It is easy to verify that for all $k \geq k_0 + n_0 + 1$, $F(e[k]) = n_0$ and the value of $\psi(e[k])$ is computed at stage 1, 2 or 8 of the algorithm. Hence for all $k \geq k_0 + n_0 + 1$, $\psi(e[k]) = \psi(e[k_0 + n_0 + 1])$.

It remains to show that Ψ solves $\mathbf{P}$ efficiently. Let $\sigma \in SEQ$ be for $\mathbf{P}$. Relying on Fact (47) and on stages 1 and 2 of the algorithm, it is easy to build $\sigma_0 \in SEQ$ that extends σ such that:

(52)   (a) $\text{length}(\sigma_0) \geq n_0$.

   (b) For all $\lambda \in SEQ$, if $\sigma \subseteq \lambda \subseteq \sigma_0$ then $\psi(\lambda)$ is computed at stage 1 or 2 of the algorithm (hence $\psi(\lambda) = \psi(\sigma)$).

   (c) $\text{range}(\tau_{n_0}) \subseteq \text{range}(\sigma_0)$ and for all $i \in N$, σ_0 i-selects τ_{n_0}.

   (d) For all $n < n_0$, if $\text{range}(\tau_n) \subseteq L$, then $\text{range}(\tau_n) \subseteq \text{range}(\sigma_0)$ and σ_0 does not 0-select τ_n.

Let e' be any environment for $W_{\psi(\sigma)}$ that extends σ_0. It follows from Fact (47)a and (52)c,d that:

(a) $\text{range}(\tau_{n_0}) \subseteq \text{range}(e'[\text{length}(\sigma_0)])$, and for all $k \geq \text{length}(\sigma_0)$ and $i \in N$, $e'[k]$ i-selects τ_{n_0}.

(b) For all $n < n_0$, if $\text{range}(\tau_n) \subseteq L$, then $\text{range}(\tau_n) \subseteq \text{range}(e'[k])$ and for all $k \geq \text{length}(\sigma_0)$, $e'[k]$ does not 0-select τ_n.

This with (52)a, b implies easily that for all $k \geq \text{length}(\sigma_0)$, $F(e'[k]) = \tau_{n_0}$ and the value of $\psi(e[k])$ is computed at stage 1, 2 or 8 of the algorithm. Hence:

(53)   for all $k \geq \text{length}(\sigma_0)$, $\psi(e'[k]) = \psi(\sigma_0)$.

We conclude with Exercise 2.(41), (52)b and (53).   ∎

Solution to Exercise 2.(84):

We prove Part (a). By Exercise 2.(73), let computable function $g : N \to N$ be such that for all $i \in N$, if ψ_i underlies computable scientist Ψ then $\psi_{g(i)}$ underlies total, computable scientist Ψ' such that stable-scope(Ψ) $\subseteq$ stable-scope(Ψ'). Let $i \in N$ be given, and denote by Ψ the scientist that $\psi_{g(i)}$ underlies. An easy variant of the proof of Lemma 2.(8) in which we substitute "Ψ solves L" by "Ψ solves L on recursive environments" and "environment" by "recursive environment" yields the following lemma:

(54) LEMMA: Suppose that Ψ stably solves every recursive environment for language L. Then there is $\sigma \in SEQ$ with the following properties.

(a) range$(\sigma) \subseteq L$,

(b) $W_{\psi_{g(i)}(\sigma)} = L$, and

(c) for all $\tau \in SEQ$ with range$(\tau) \subseteq L$, $\psi_{g(i)}(\sigma * \tau) = \psi_{g(i)}(\sigma)$.

Relying on Lemma (54), the proof of Exercise 2.(78) yields the required function f.

We prove Part (b). Let $i \in N$ be given, and denote by Ψ the scientist that ψ_i underlies. We first prove the following lemma:

(55) LEMMA: Let language $L \subseteq N$ contain at least two elements. Suppose that Ψ stably solves every nonrecursive environment for L. Then there is $\sigma \in SEQ$ with the following properties.

(a) range$(\sigma) \subseteq L$,

(b) $W_{\psi_i(\sigma)} = L$, and

(c) for all $\tau \in SEQ$ with range$(\tau) \subseteq L$, $\psi_i(\sigma * \tau) = \psi_i(\sigma)$.

Proof of Lemma (55): Suppose for a contradiction that Lemma (55) does not hold for non-singleton language L. Let $\sigma \in SEQ$ and $n, n' \in L$ with range$(\sigma) \subseteq L$ and $n \neq n'$ be given. By hypothesis there exists $\tau, \tau' \in SEQ$ such that:

(a) range$(\tau) \cup$ range$(\tau') \subseteq L$;

(b) either $W_{\psi(\sigma)} \neq L$ or $\psi(\sigma * n * \tau)$ is undefined or $\psi(\sigma * n * \tau) \neq \psi(\sigma)$;

(c) either $W_{\psi(\sigma)} \neq L$ or $\psi(\sigma * n' * \tau')$ is undefined or $\psi(\sigma * n' * \tau') \neq \psi(\sigma)$.

From this, it is easy to build uncountably many environments for L on which ψ does not converge to an index for L. This is in contradiction with the fact that there are codenumerably many nonrecursive environments for L. ∎

Relying on Lemma (55), the proof of Exercise 2.(78) yields a computable function $g : N \to N$ such that for all languages L,

(56) if L is not a singleton and Ψ stably solves every nonrecursive environment for L, then $\psi_{g(i)}$ underlies a computable scientist that stably solves L.

For all $n \in N$, denote by $h(n)$ a given index of $\{n\}$. Let recursive function $f : N \to N$ satisfy the following, for all nonempty $\sigma \in SEQ$ and $i \in N$. If range$(\sigma) = \{n\}$ for some $n \in N$, then $\psi_{f(i)}(\sigma) = h(n)$; otherwise, $\psi_{f(i)}(\sigma) = \psi_{g(i)}(\sigma)$. It is immediate that for all $i \in N$, if $\psi_{g(i)}$ underlies computable scientist Ψ' and $\psi_{f(i)}$ underlies computable scientist Ψ'', then Ψ'' stably solves all singleton languages together with all languages in stable-scope(Ψ'). We conclude with (56).   ∎

Section 2.6

Solution to Exercise 2.(95):

It is easy to explicitly define a computable function $f : SEQ \to SEQ$ with the following properties:

(a) for all $\sigma \in SEQ$, $f(\sigma)$ is the initial segment of some orderly environment;

(b) for all $\sigma, \tau \in SEQ$, $\sigma \subseteq \tau$ implies $f(\sigma) \subseteq f(\tau)$;

(c) for all environments e for functional language L, $\bigcup_{k \in N} f(e[k])$ is the orderly environment for L.

Let computable function $\psi : SEQ \to N$ underlie computable scientist Ψ, and let Ψ' denote the computable scientist that $\psi \circ f$ underlies. It is easy to verify that for every functional language L, if Ψ [stably] solves L then Ψ' [stably] solves L. The Exercise follows immediately.   ∎

Solution to Exercise 2.(96):

Let problem **P** consist of every cofinite language. By Corollary 2.(17)b, **P** is not solvable. If n_0 is an index for N, it is trivial that the constant n_0-function with domain SEQ underlies a computable scientist that FV-solves **P** stably.   ∎

Solution to Exercise 2.(97):

Let problem **P** consist of all functional languages L such that for some $i \in N$, $L = W_i$ and $\langle 0, i \rangle \in L$. It is easy to verify that some computable scientist stably solves **P**.

Suppose for a contradiction that computable $\psi : SEQ \to N$ underlies popperian, computable scientist Ψ that solves **P**. From the proof of Exercise 2.(73), it is easy to verify that ψ can be supposed to be total without loss of generality. We fist define, uniformly in $i \in N$, an environment e_i for a recursive, functional language. The construction is by induction on $k \in N$.

Construction of the e_i's

Stage 0: Set $e_i(0) = \langle 0, i \rangle$.

Stage $k + 1$: Let $y \in N$ be such that $\langle k + 1, y \rangle \in W_{\psi(e_i[k+1])}$ (such an y exists and is unique since Ψ is total and popperian). Set $e_i(k + 1) = \langle k + 1, y + 1 \rangle$.

End construction

It follows immediately from the construction that Ψ fails to solve e_i for all $i \in N$. Let recursive function $g : N \to N$ be such that for all $i \in N$, $W_{g(i)} = $ range(e_i). By the recursion theorem, there is $i_0 \in N$ such that $W_{g(i_0)} = W_{i_0}$. Since $\langle 0, i_0 \rangle \in$ range(e_{i_0}), range(e_{i_0}) $= W_{i_0} \in \mathbf{P}$. Contradiction. ∎

Solution to Exercise 2.(98):
Given $D \subseteq N$, set $L_D = \{\langle x, 0 \rangle \mid x \in D\} \cup \{\langle x, 1 \rangle \mid x \notin D\}$. Define $\mathbf{P}$ to be $\{L_D \mid D = N$ or D is a finite subset of $N\}$. Suppose for a contradiction that memory-limited scientist Ψ solves $\mathbf{P}$. By Lemma 2.(8) let $\sigma \in SEQ$ be a locking-sequence for L_N and Ψ. Let $D = \{x \in N \mid \langle x, 0 \rangle \in$ range(σ)$\}$. By Exercise 2.(18).a, let $\tau \in SEQ$ be such that $\sigma * \tau$ is a locking-sequence for L_D and Ψ. Choose $n \in N$ such that for all $y \in N$, $\langle n, y \rangle \notin$ range($\sigma * \tau$). Since σ is a locking-sequence for L_N and Ψ, $\Psi(\sigma * \langle n, 0 \rangle) = \Psi(\sigma)$. Since Ψ is memory-limited, it follows that:

(57) $\Psi(\sigma * \langle n, 0 \rangle * \tau) = \Psi(\sigma * \tau)$.

Let e be any environment for $L_{D \cup \{n\}}$ that extends $\sigma * \langle n, 0 \rangle * \tau$ and such that for all $k \geq$ length(σ) $+ 1$, $e(k) \neq \langle n, 0 \rangle$. Since $\sigma * \tau$ is locking-sequence for L_D and Ψ, we deduce from (57) and the hypothesis that Ψ is memory-limited that for all $k \geq$ length(σ) $+$ length(τ) $+ 1$, $\Psi(e[k]) = \Psi(\sigma * \tau) = L_D$. Hence Ψ does not solve e. Contradiction. ∎

Section 2.7

Solution to Exercise 2.(110):
Denote by E' the class of all environments that are not for L, and by S the class of all $\sigma \in SEQ - \{\emptyset\}$ such that:

(a) for all $i <$ length(σ) $- 1$, $\sigma(i) \in L$, and

(b) σ(length(σ) $- 1) \notin L$.

It is immediate that $E' = \bigcup_{\sigma \in S} E_\sigma$. Since $M_L(E_\sigma) = 0$ for all $\sigma \in S$, it follows that $M_L(E') \leq \Sigma_{\sigma \in S} \mu(E_S) = 0$. ∎

Solution to Exercise 2.(111):
Let memory-limited scientist Ψ be given. Let total function $f : \text{pow}(N) \times N \to \text{pow}(N)$ be such that for all $\sigma \in SEQ$ and $n \in N$, $\Psi(\sigma * n) = f(\Psi(\sigma), n)$. Suppose that Ψ **1**-solves N. We rely on the following fact.

(58) FACT: For all $n \in N$, $f(N, n) = N$.

Proof of Fact (58): Suppose that for some $n \in N$, $f(N, n) \neq N$. Then Ψ fails
to solve every environment for N in which n appears infinitely many times.
But the set of all environments in which n appears only finitely many times
has null-measure respectively to m_N. This contradicts the hypothesis that Ψ
1-solves N.   ∎

Let non-empty $\sigma \in SEQ$ be such that $\Psi(\sigma) = N$. If e is an environment for
$\text{range}(\sigma)$ that extends σ, then for every $k \geq \text{length}(\sigma)$, $\Psi(e[k]) = N$ as follows
immediately from Fact (58). Hence Ψ fails to 1-solve $\text{range}(\sigma)$.   ∎

Solution to Exercise 2.(112):

We prove part (a). Let infinite $L \subseteq N$ be given, and denote by **P** the problem
consisting of L together with every nonempty finite subset of L. Define scientist with coin Θ as follows. For every coin c, for every $\sigma \in SEQ$:

(a) if $c(0) = 0$ then $\Theta(c[\text{length}(\sigma)], \sigma) = L$;

(b) if $c(0) = 1$ then $\Theta(c[\text{length}(\sigma)], \sigma) = \text{range}(\sigma)$.

It is easy to verify that Θ solves **P** with probability $\frac{1}{2}$. Let scientist with coin
Θ be given. We have to show that Θ does not solve **P** with probability greater
than $\frac{1}{2}$. We rely on the following fact.

(59) FACT: If scientist Ψ solves L, then there exists $m \in L$ such that for all
$n \geq m$, Ψ does not solve $L \cap \{0 \ldots n\}$.

Proof of Fact (59): Suppose the conclusion of the fact does not hold. Then it
is easy to verify that there is no tip-off for L in $\mathbf{P'} = \{L\} \cup \{L \cap \{0 \ldots n\} \mid n \geq$
$\min(L)$ and Ψ solves $L \cap \{0 \ldots n\}\}$, which, via Proposition 2.(15'), contradicts
the fact that Ψ solves $\mathbf{P'}$.   ∎

To show that Θ does not solve **P** with probability greater than $\frac{1}{2}$, we first need
some notation. For every coin c, denote by Ψ_c the scientist $\lambda\sigma.\Theta(c[\text{length}(\sigma)],$
$\sigma)$. Denote by C the set of all coins c such that Ψ_c solves L. For every
$i \geq \min(L)$ denote by C_i the set of all coins c such that:

(a) If $i > \min(L)$ then Ψ_c solves $L \cap \{0 \ldots i - 1\}$;

(b) Ψ_c solves no set of form $L \cap \{0 \ldots n\}$ with $n \geq i$.

It follows from Fact (59) that:

(60) $C = \bigcup_{i \geq \min(L)} C_i$.

Let $p \in (\frac{1}{2}, 1]$ be given. Suppose for a contradiction that Θ solves $\mathbf{P}$ with probability p. Hence:

(61)   For every $L' \in \mathbf{P}$, $\Pr(\{c \mid \Psi_c \text{ solves } L'\}) \geq p$.

From (60) we infer that there exists $i_0 \geq \min(L)$ such that:

(62) $\Pr(C - \bigcup_{\min(L) \leq i \leq i_0} C_i) < p - \frac{1}{2}$.

We deduce immediately from (62) and the definition of the C_i that:

(63) $\Pr(C \cap \{c \mid \Psi_c \text{ solves } L \cap \{0 \dots i_0\}\}) < p - \frac{1}{2}$.

From (63) and (61) applied with $L' = L$ and $L' = L \cap \{0 \dots i_0\}$, we deduce that

$$\Pr(C \cup \{c \mid \Psi_c \text{ solves } L \cap \{0 \dots i_0\}\}) > 2p - (p - \frac{1}{2}) = p + \frac{1}{2} \geq 1,$$

which is impossible.

Let infinite $L \subseteq N$ be given, and denote by $\mathbf{P}$ the problem consisting of L together with every set of form $L - \{x\}$ for $x \in L$. Define scientist with coin Θ as follows. For every coin c, for every $\sigma \in SEQ$:

(a) if $c(0) = 0$ then $\Theta(c[\text{length}(\sigma)], \sigma) = L$;

(b) if $c(0) = 1$ then $\Theta(c[\text{length}(\sigma)], \sigma) = L - \{x\}$, where x is the least member of L not occurring in σ.

It is easy to verify that Θ solves $\mathbf{P}$ with probability $\frac{1}{2}$. Relying on the following fact, it is easy to adapt the proof above to show that $\mathbf{P}$ is not solvable with probability greater than $\frac{1}{2}$.

(64)   FACT: If scientist Ψ solves L, then there is $m \in N$ such that for all $n \geq m$, Ψ does not solve $L - \{n\}$.

We prove Part (b). Let $n \in N - \{0\}$ be given. Let $\mathbf{P} = \{N - D \mid D \subseteq N \text{ and } \text{card}(D) \leq n\}$. Choose $(O_0 \dots O_n)$ such that for every $i \leq n$, O_i is an open subset of the set of coins, $\Pr(O_i) = \frac{1}{n+1}$, and for every $j \leq n$, if $i \neq j$ then $O_i \cap O_j = \emptyset$. Define scientist with coin Θ as follows. For every coin c and $\sigma \in SEQ$, if $i \leq n$ is least such that $C_{c[\text{length}(\sigma)]} \subseteq O_i$, then:

(a) $\Theta(c[\text{length}(\sigma)], \sigma) = N$ if $i = 0$;

(b) $\Theta(c[\text{length}(\sigma)], \sigma) = N - \{x_1 \dots x_i\}$ otherwise, where $x_1 \dots x_i$ are the first i members of N that do not occur in σ.

It is easy to verify that for every $i \leq n$, for every $D \subseteq N$ with $\text{card}(D) = i$, and for every $c \in O_i$, $\lambda\sigma.\Theta(c[\text{length}(\sigma)], \sigma)$ solves $N - \{D\}$. Hence Θ solves $\mathbf{P}$ with probability $\frac{1}{n+1}$.

Let scientist with coin Θ be given. We have to show that Θ does not solve $\mathbf{P}$ with probability greater than $\frac{1}{2}$. We rely on the following fact.

(65) FACT: Let $q \in N - \{0\}$ and scientist Ψ that solves N be given. Then there exists $m \in N$ such that for all $D \subseteq N$ with $\text{card}(D) = q$, if $\min(D) \geq m$ then Ψ does not solve $N - D$.

Proof of Fact (65): Suppose the conclusion of the fact does not hold. Then it is easy to verify that there is no tip-off for L in $\mathbf{P}' = \{N\} \cup \{N - D \mid D \subseteq N, \text{card}(D) = q, \text{ and } \Psi \text{ solves } N - D\}$, which, via Proposition 2.(15), contradicts the fact that Ψ solves $\mathbf{P}'$. ∎

To show that Θ does not solve $\mathbf{P}$ with probability greater than $\frac{1}{2}$, we first need some notation. For every coin c, denote by Ψ_c the scientist $\lambda\sigma.\Theta(c[\text{length}(\sigma)], \sigma)$. For all $L \in \mathbf{P}$ denote by C_L the set of all coins c such that Ψ_c solves L. For all $L \in \mathbf{P}$, $q \in N - \{0\}$ and $i \in N$, denote by $C_{L,i}^q$ the set of all coins $c \in C_L$ such that:

(a) If $i \neq 0$ then Ψ_e solves some set of the form $L - D$ with $D \subseteq L, \text{card}(D) = q$, and $\min(D) < i$.

(b) Ψ_e solves no set of the form $L - D$ with $D \subseteq L$, $\text{card}(D) = q$, and $\min(D) \geq i$.

It follows from Fact (65) that for every $L \in \mathbf{P}$ and $q \in N - \{0\}$:

(66) $C_L = \bigcup_{i \in N} C_{L,i}^q$.

Let $p \in (\frac{1}{n+1}, 1]$ be given. Suppose for a contradiction that Θ solves $\mathbf{P}$ with probability p. Hence:

(67)   For every $L \in \mathbf{P}$, $\text{Pr}(\{c \mid \Psi_c \text{ solves } L\}) \geq p$.

Choose $\alpha \in (\frac{1}{p(n+1)}, 1)$, and set $\epsilon = \frac{p(1-\alpha)}{n}$. From (66), we infer that for all $L \in \mathbf{P}$ and $q \in N - \{0\}$, there exists $i_L^q \in N$ such that:

(68) $\text{Pr}(C_L - \bigcup_{0 \leq i \leq i_L^q} C_{L,i}^q) \leq \epsilon$.

Now define increasing sequence $(i_0 \dots i_n)$ of integers and sequence $(L_0 \dots L_n)$ of members of $\mathbf{P}$ as follows.

(a) $i_0 = 0$ and $L_0 = N$.

(b) For all $1 \leq q \leq n$, $i_q = \sup(i_{q-1}, i^1_{L_{q-1}} \ldots i^{n-q+1}_{L_{q-1}})$ and $L_q = L_{q-1} - \{i_q\}$.

From (68) and the former construction, we deduce that for all $0 \leq i < j \leq n$:

(69) $\Pr(C_{L_i} \cap C_{L_j}) \leq \epsilon.$

From (69) and (67) applied with $L = L_0 \ldots L_n$, we infer that for all $0 \leq i \leq n$:

$$\Pr(C_{L_i} - \bigcup_{j<i} C_{L_j}) \geq p - i\epsilon \geq p - n\epsilon = \alpha p,$$

which implies that $\Pr(\bigcup_{0 \leq i \leq n} C_{L_i}) \geq (n+1)\alpha p > 1$. Contradiction.   ∎

Section 2.8

Solution to Exercise 2.(117):
Let $P_1 = \{N\}$, $P_2 = \{N - \{n\} \mid n \in N\}$. Suppose for a contradiction that scientist Ψ solves $\{P_1, P_2\}$. Define function $f : SEQ \rightarrow \text{pow}(N)$ as follows. For all $\sigma \in SEQ$, if $\emptyset \neq \Psi(\sigma) \subseteq P_1$, then $f(\sigma) = N$; if $\emptyset \neq \Psi(\sigma) \subseteq P_2$, then $f(\sigma) = N - \{n\}$, where $n \in N$ is least such that $n \notin \text{range}(\sigma)$; otherwise, $f(\sigma)$ is undefined. It is clear that in the paradigm of Section 1, f is a scientist that solves $P_1 \cup P_2$. This contradicts Corollary 2.(17)b.   ∎

C Solutions to Exercises for Chapter 3

Solution to Exercise 3.(19):
Fix an enumeration of all the (countably many) polynomial functions with rational coefficients. Define scientist Ψ such that for all $\sigma \in SEQ$, $\Psi(\sigma) = \{S_f\}$, where f is first in the enumeration such that $\bigwedge \sigma$ is satisfiable in S_f. It is easy to see that Ψ solves the problem described in the exercise. ∎

Solution to Exercise 3.(20):
It is easy to define a computable function $f : SEQ \to SEQ$ with the following properties:

(a) for all $\sigma \in SEQ$, $f(\sigma)$ is the initial segment of some standard environment;

(b) for all $\sigma, \tau \in SEQ$, $\sigma \subseteq \tau$ implies $f(\sigma) \subseteq f(\tau)$;

(c) for all environments e for structure S, $\bigcup_{k \in N} f(e[k])$ is a standard environment for S.

Let scientist Ψ be given. We define scientist $\Psi' = \Psi \circ f$. That is, for all $\sigma \in SEQ$, $\Psi'(\sigma) = \Psi(f(\sigma))$. It is straightforward to verify that for every proposition P and for all $S \in P$, if Ψ solves P in every standard environment for S, then Ψ' solves P in every environment for S. The Exercise follows immediately. ∎

Solution to Exercise 3.(21):
It is easy (but tedious) to explicitly define a computable function $f : SEQ \to SEQ$ with the following properties:

(a) for all $\sigma \in SEQ$, $f(\sigma)$ is the initial segment of some bijective environment;

(b) for all $\sigma, \tau \in SEQ$, $\sigma \subseteq \tau$ implies $f(\sigma) \subseteq f(\tau)$;

(c) for all environments e for infinite structure S, $\bigcup_{k \in N} f(e[k])$ is a bijective environment for S.

Let scientist Ψ be given. We define scientist $\Psi' = \Psi \circ f$. That is, for all $\sigma \in SEQ$, $\Psi'(\sigma) = \Psi(f(\sigma))$. It is easy to verify that for every proposition P and for all infinite $S \in P$, if Ψ solves P in every bijective environment for S, then Ψ' solves P in every environment for S. The exercise follows immediately. ∎

Solution to Exercise 3.(22):
For the purposes of proving the Exercise, we introduce the following, temporary convention. A term or notation decorated with o indicates usage in the

sense of Chapter 1 (orders paradigm). Decoration with ℓ indicates usage in the sense of the present chapter (logical paradigm). Set $\mathbf{P} = \{\mathbf{P_K} \cap MOD(\theta), \mathbf{P_K} \cap MOD(\neg\theta)\}$.

It is easy to exhibit a total function $f : SEQ_\ell \to SEQ_o$ with the following properties.

(1)   (a) For all $\sigma, \tau \in SEQ_\ell$, if $\tau \subseteq \sigma$ then $f(\tau) \subseteq f(\sigma)$.

    (b) If $\mathcal{S} \in \mathbf{P_K}$ and environment$_\ell$ e is for $\mathcal{S}$, then $\bigcup_{k \in N} f(e[k])$ is an environment$_o$ for the unique $\prec \in \mathbf{K}$ such that for all $m, n \in N$, $\mathcal{S} \models \bar{R}\bar{m}\bar{n}$ iff $m \prec n$.

Suppose that $\mathbf{K}$ is solvable$_o$. Let scientist$_o$ Ψ_o that solves$_o$ $\mathbf{K}$ be given. We define scientist$_\ell$ Ψ_ℓ as follows, for all $\sigma \in SEQ_\ell$. If $\Psi_o(f(\sigma)) = \mathbf{y}$ then $\Psi_\ell(\sigma) = \mathbf{P_K} \cap MOD(\theta)$; otherwise $\Psi_\ell(\sigma) = \mathbf{P_K} \cap MOD(\neg\theta)$. It follows immediately from (1) that Ψ_ℓ solves$_\ell$ $\mathbf{P}$.

It is easy to exhibit a total function $g : SEQ_o \to SEQ_\ell$ with the following properties.

(2)   (a) For all $\sigma, \tau \in SEQ_o$, if $\tau \subseteq \sigma$ then $g(\tau) \subseteq g(\sigma)$.

    (b) If $\prec \in \mathbf{K}$ and environment$_o$ e is for $\prec$, then $\bigcup_{k \in N} g(e[k])$ is an environment$_\ell$ for the unique standard structure $\mathcal{S}$ such that for all $m, n \in N$, $\mathcal{S} \models \bar{R}\bar{m}\bar{n}$ iff $m \prec n$.

Suppose that $\mathbf{P}$ is solvable$_\ell$. Let scientist$_\ell$ Ψ_ℓ that solves$_\ell$ $\mathbf{P}$ be given. We define scientist$_o$ Ψ_o as follows, for all $\sigma \in SEQ_o$. If $\Psi_\ell(g(\sigma))$ is defined and $\emptyset \neq \Psi_\ell(g(\sigma)) \subseteq \mathbf{P_K} \cap MOD(\theta)$, then $\Psi_o(\sigma) = \mathbf{y}$; otherwise, $\Psi_o(\sigma) = \mathbf{n}$.

It follows immediately from (2) that Ψ_o solves$_o$ $\mathbf{K}$.   ∎

Section 3.2

Solution to Exercise 3.(41):
Let $\mathcal{B}$ be any $\subset$-chain of sets with the same order-type as $0\omega^*\omega 1$. Observe that interpreting 0 as $\emptyset$, 1 as N, $+$ as $\cup$, and $\times$ as $\cap$ makes $\mathcal{B}$ into a countable, atomless Boolean algebra. Let $\mathcal{A}$ be the collection of all finite initial segments of N. Interpreting 0, 1, $+$ and $\times$ as before, yields a countable, atomic boolean algebra. (The algebras $\mathcal{B}$ and $\mathcal{A}$ are somewhat degenerate, since both are totally ordered by the relation $\preceq$ defined as: $a \preceq b \leftrightarrow a + b = b$.)

Let P_n be the class of nonatomic Boolean algebras, and P_a the class of atomic Boolean algebras. Suppose that scientist Ψ solves P_n. It suffices to show that Ψ does not solve P_a. Since $\mathcal{B} \in P_n$, Lemma 3.(24) implies the existence of a locking pair (σ, a) for Ψ, $\mathcal{B}$, and P_n. It is easy to see that there is an environment e for $\mathcal{A}$ that extends σ. It is also clear that for every $k \geq \text{length}(\sigma)$, $\mathcal{B} \models \exists \bar{x} \bigwedge e[k][a]$, where $\bar{x}$ contains the variables in $Var(e[k]) - \text{domain}(a)$. So by Definition 3.(23), $\emptyset \neq \Psi(e[k]) \subseteq P_n$, for all $k \geq \text{length}(\sigma)$. Hence, since $\mathcal{A} \in P_y$ and $P_n \cap P_a = \emptyset$, Ψ does not solve P_a.   ∎

Solution to Exercise 3.(42):

Let $\mathcal{N}$ be the standard model of arithmetic, and let $\mathcal{M}$ be the resulting of appending another copy N' of N to $\mathcal{N}$; s, $\oplus$, and $\otimes$ are interpreted standardly on N'. Thus, the zero element of N' is not the successor of any other element, so axiom Q3 of Robinson's arithmetic is false in $\mathcal{M}$. (For Q3, see, [Boolos & Jeffrey, 1989, p. 158].) So, it is easy to see that if $\{MOD(Q), MOD(\neg Q)\}$ were solvable, then $\{\{\mathcal{N}\}, \{\mathcal{M}\}\}$ would be solvable. Suppose that scientist Ψ solves $\{\mathcal{M}\}$. It suffices to show that Ψ does not solve $\{\mathcal{N}\}$.

By Lemma 3.(24), let (σ, a) be a locking pair for Ψ, $\mathcal{M}$, and $\{\mathcal{M}\}$. It is easy to see that there is an environment e for $\mathcal{N}$ that extends σ. It is also clear that for every $k \geq \text{length}(\sigma)$, $\mathcal{M} \models \exists \bar{x} \bigwedge e[k][a]$, where $\bar{x}$ contains the variables in $Var(e[k]) - \text{domain}(a)$. So by Definition 3.(23), $\Psi(e[k]) = \{\mathcal{M}\}$, for all $k \geq \text{length}(\sigma)$. Hence, Ψ does not solve $\{\mathcal{N}\}$.   ∎

Solution to Exercise 3.(43):

By Proposition 3.(35), it suffices to show that there is no tip-off for P_1 in $\{P_1, P_2\}$. Let full assignment h to N be given. Let π-set π and nonempty, finite set of variables X be such that for all $\varphi \in \pi$, $Var(\varphi) \subseteq X$. Suppose that $N \models \pi[h]$. Since $\forall$ formulas are preserved in substructures (see [Hodges, 1993, Cor. 2.4.2]), it follows that π is satisfiable in $\{h(x) \mid x \in X\}$. Hence π is satisfiable in some member of P_2. We conclude with Definition 3.(28).   ∎

Solution to Exercise 3.(44):

By Proposition 3.(35), it suffices to show that there is no tip-off for $MOD(T \cup \{\theta\})$ in **P**. Let structure $\mathcal{S}$ be $\langle Q, \prec \rangle$, where Q is the set of rationals and $\prec$ is the usual strict order over Q. Let full assignment h to $\mathcal{S}$ be given. Let π-set π and nonempty, finite set of variables X be such that for all $\varphi \in \pi$, $Var(\varphi) \subseteq X$. Suppose that $\mathcal{S} \models \pi[h]$. Denote by $\mathcal{T}$ the substructure of $\mathcal{S}$ with $|\mathcal{T}| = \{h(x) \mid x \in X\}$. Since $\forall$ formulas are preserved in substructures (see [Hodges, 1993, Cor. 2.4.2]), it follows that π is satisfiable in $\mathcal{T}$. Since $\mathcal{T} \in MOD(T \cup \{\neg\theta\})$, π is satisfiable in some member of $MOD(T \cup \{\neg\theta\})$. We conclude with Definition 3.(28).   ∎

Solution to Exercise 3.(45):
Suppose that scientist Ψ solves P_1. By Lemma 3.(24), let (σ, a) be a locking pair for Ψ, S_0, and P_1. Let $j > 0$ be such that for every variable x that occurs free in σ, $a(x) \leq j$. It is easy to see that there is an environment e for S_j that extends σ. It is also clear that for every $k \geq \text{length}(\sigma)$, $S_j \models \exists \bar{x} \bigwedge e[k][a]$, where $\bar{x}$ contains the variables in $Var(e[k]) - \text{domain}(a)$. So by Definition 3.(23), $\Psi(e[k]) \subseteq P_2$, for all $k \geq \text{length}(\sigma)$. Hence, Ψ does not solve P_2. ∎

Solution to Exercise 3.(46):
Suppose that a sole binary predicate is the only symbol of **Sym**. Let $S = \langle N, \preceq \rangle$ be isomorphic to ω and $\mathcal{T} = \langle N, \preceq^* \rangle$ be isomorphic to ω^*. By Proposition 3.(34), $\{\{S\}, \{\mathcal{T}\}\}$ is solvable. Note that for all $\sigma \in SEQ$, if σ is for $\{S, \mathcal{T}\}$ then $\bigwedge \sigma$ is satisfiable in both S and $\mathcal{T}$. This implies immediately that for every well-ordering $\prec$ over $\{S, \mathcal{T}\}$ and every $\sigma \in SEQ$ that is for $\{\{S\}, \{\mathcal{T}\}\}$, $\Psi_\prec(\sigma)$ is the $\prec$-least member of $\{\{S\}, \{\mathcal{T}\}\}$. Hence $\{\{S\}, \{\mathcal{T}\}\}$ is not solvable by enumeration. ∎

Solution to Exercise 3.(47):
Suppose that a sole binary predicate is the only symbol of **Sym**. Let $S = \langle N, \preceq \rangle$ be isomorphic to ω and $\mathcal{T} = \langle N, \preceq^* \rangle$ be isomorphic to ω^*. By Proposition 3.(34), **P** $= \{\{S\}, \{\mathcal{T}\}\}$ is solvable. Suppose for a contradiction that **P**-conservative scientist Ψ solves **P**. Hence we can choose $\sigma \in SEQ$ such that $\bigwedge \sigma$ is satisfiable in S and $\Psi(\sigma) = \{S\}$. Note that for all $\tau \in SEQ$ that extend σ, $\bigwedge \tau$ is satisfiable in S iff $\bigwedge \tau$ is satisfiable in $\mathcal{T}$. This with the hypothesis that Ψ is **P**-conservative implies immediately that:

(a) there exists an environment for $\mathcal{T}$ that extends σ;

(b) for every environment e for $\mathcal{T}$ that extends σ and for every $k \geq \text{length}(\sigma)$, $\Psi(e[k]) = \{S\}$.

It follows immediately that Ψ does not solve $\{\mathcal{T}\}$. Contradiction. ∎

Solution to Exercise 3.(48):
We prove Part (a). Suppose that **Sym** is limited to a binary predicate. Let $S = \langle N, \preceq \rangle$ be isomorphic to ω, let $\mathcal{T} = \langle N, \preceq^* \rangle$ be isomorphic to $\omega^*\omega$. $\{S, \mathcal{T}\}$ is separated since $\mathcal{T}$ is not a substructure of S. But $\{S, \mathcal{T}\}$ is unsolvable by Proposition 3.(39).

We prove Part (b). Let problem **P** be not elementary separated. Then there are $P_1, P_2 \in \mathbf{P}$, $S \in P_1$ and $\mathcal{T} \in P_2$ with an elementary embedding s from S into $\mathcal{T}$. Let full assignment h to S be given. Hence for every π-set π, if $S \models \pi[h]$

then $\mathfrak{T} \models \pi[s \circ h]$. Definition 3.(28) then implies that there is no tip-off for P_1 in **P**. With Proposition 3.(35), we conclude that **P** is unsolvable.   ∎

Solution to Exercise 3.(49):

Suppose that **Sym** is limited to a constant $\overline{0}$ and two unary function letters s_1 and s_2. Denote by T the complete theory of all sentences true in $\langle N, S, S \rangle$, where S is the usual successor function on N. An easy adaptation of the discussion in [Enderton, 1972, Section 3.1] shows that there exist structures $\mathcal{S}$ and $\mathcal{U}$ with the following properties.

(a) $|\mathcal{S}| = |\mathcal{U}| = N$;

(b) 0 interprets $\overline{0}$ in both $\mathcal{S}$ and $\mathcal{U}$;

(c) $\mathcal{S} \models T$ and $\mathcal{U} \models T$;

(d) the interpretation of s_1 in $\mathcal{S}$ is the usual successor function on N, whereas the interpretation of s_2 in $\mathcal{S}$ contains both N and an order which is isomorphic to $\omega^*\omega$;

(e) the interpretation of s_2 in $\mathcal{U}$ is the usual successor function on N, whereas the interpretation of s_1 in $\mathcal{U}$ contains both N and an order which is isomorphic to $\omega^*\omega$.

Trivially, $\mathbf{P} = \{\{\mathcal{S}\}, \{\mathcal{U}\}\}$ is elementary. It is easy to verify that $\{\{v_i \neq s_2^n \overline{0} \mid n \in N\} \mid i \in N\}$ is a tip-off for $\{\mathcal{S}\}$ in **P**, and that $\{\{v_i \neq s_1^n \overline{0} \mid n \in N\} \mid i \in N\}$ is a tip-off for $\{\mathcal{U}\}$ in **P**. With Proposition 3.(31), we conclude that **P** is solvable.   ∎

Solution to Exercise 3.(50):

We prove Part (a). Let $\mathbf{Sym} = \emptyset$ and $P_0 = \{N\}$. Given $n > 0$, let P_n be the class of all finite subsets D of N with card$(D) = n$. Set $\mathbf{P} = \{P_n \mid n \in N\}$. It follows immediately from Exercise 3.(43) that **P** is not solvable. We show that **P** is weakly solvable. Given $n > 0$, denote by θ_n a sentence which is true in structure $\mathcal{S}$ just in case card$(|\mathcal{S}|) \leq n$. Let scientist Ψ satisfy the following, for all nonempty $\sigma \in SEQ$ and $\beta \in \mathcal{L}_{basic}$. Let $n > 0$ be least such that $\{\bigwedge \sigma, \theta_n\}$ is consistent. If $\{\bigwedge (\sigma * \beta), \theta_n\}$ is consistent, then $\Psi(\sigma * \beta) = P_n$; otherwise, $\Psi(\sigma * \beta) = P_0$. We show that Ψ weakly solves **P**. Let environment e for **P** be given. It suffices to show that:

(3)   (a) If e is for P_n with $n > 0$, then $\Psi(e[k]) = P_n$ for cofinitely many k.

     (b) If e is for P_0 then for all $n > 0$, $\Psi(e[k]) = P_n$ for finitely many k.

     (c) If e is for P_0 then $\Psi(e[k]) = P_0$ for infinitely many k.

Suppose that e is for P_n with $n > 0$. Then we can choose $k_0 \in N$ such that for all $k \geq k_0$, n is least with $\{\bigwedge e[k], \theta_n\}$ consistent. This implies (3)a. Suppose that e is for P_0. For all $n > 0$, there are only finitely many k such that $\{\bigwedge e[k], \theta_n\}$ is consistent. This implies (3)b. There is infinite $X \subseteq N$ such that for all $k \in X$, if $n > 0$ is least with $\{\bigwedge e[k], \theta_n\}$ consistent, then $\{\bigwedge e[k+1], \theta_n\}$ is inconsistent. This implies (3)c.

We prove Part (b). Let $n \in N$ and unsolvable problem $\mathbf{P} = \{P_0 \dots P_n\}$ be given. Suppose for a contradiction that scientist Ψ weakly solves $\mathbf{P}$. Let scientist Ψ' satisfy the following, for all $\sigma \in SEQ$. Given $i \leq n$, denote by $X_{\sigma,i}$ the set of all $\tau \in SEQ$ such that $\tau \subseteq \sigma$ and $\emptyset \neq \Psi(\tau) \subseteq P_i$. If $i_0 \leq n$ is such that for all $i \in \{0 \dots n\} - \{i_0\}$, $\mathrm{card}(X_{\sigma,i}) < \mathrm{card}(X_{\sigma,i_0})$, then $\Psi'(\sigma) = P_{i_0}$. To yield a contradiction, it suffices to show that Ψ' solves $\mathbf{P}$. Let $i_0 \leq n$ be given. Let environment e be for P_{i_0}. Since Ψ weakly solves $\mathbf{P}$, there exists $k_0 \in N$ such that for all $k \geq k_0$ and for all $i \in \{0 \dots n\} - \{i_0\}$, $\mathrm{card}(X_{e[k],i}) < \mathrm{card}(X_{e[k],i_0})$. Hence for all $k \geq k_0$, $\Psi'(e[k]) = P_{i_0}$. Hence Ψ' solves P_{i_0} in e. ∎

Section 3.3

Solution to Exercise 3.(68):
Denote by $+$ and $\times$ the two function symbols of **Sym**. Let

$$\theta_0(x, y) = (\bigwedge_{j>0} \overbrace{y + \dots + y}^{j} \neq x).$$

Given $i > 1$, let

$$\theta_i(x, y) = (\overbrace{y + \dots + y}^{i} = x) \wedge (\bigwedge_{1 \leq j < i} \overbrace{y + \dots + y}^{j} \neq x).$$

Then for all $i \in N$ which is either 0 or prime,

$$\exists x \exists y [\forall z ((x + z = z) \wedge (z + x = z) \wedge (y \times z = z) \wedge (z \times y = z)) \wedge \theta_i(x, y)]$$

is true in every field of characteristic i, and false in all fields of characteristic not equal not i. We conclude with Corollary 3.(52). ∎

Solution to Exercise 3.(69):
Denote by $+$ the function symbol of **Sym**. The sentence

$$\exists x[\forall z((x + z = z) \wedge (z + x = z)) \wedge \bigwedge_{i>0} \forall z(z \neq x \rightarrow \overbrace{(z + \ldots + z}^{i} \neq x))]$$

is true in every member of **F**, and false in every member of **G** − **F**. The sentence

$$\bigvee_{i>0} \exists x \exists y[\forall z((x + z = z) \wedge (z + x = z)) \wedge (x \neq y) \wedge \overbrace{(y + \ldots + y}^{i} = x)]$$

is true in every member of **G** − **F**, and false in every member of **F**. With Corollary 3.(52), we conclude that {**F**, **G** − **F**} is solvable.

We show that {**T**, **G** − **T**} is not solvable. Choose a scientist Ψ such that:

(4)   Ψ solves **G** − **T**.

It suffices to exhibit an environment e_0 for **T** such that $\Psi(e_0[k]) \subseteq \mathbf{G} - \mathbf{T}$ for infinitely many k. To define e_0 we rely on the following notation. Let sentence θ be true in structure S just in case $S \in \mathbf{G}$. Set $T = \{\theta, \forall y((v_0 + y = y) \wedge (y + v_0 = y))\}$. Let $\sigma \in SEQ$ and $n \in N$ be given. We say that σ is "complete through n" just in case:

(a) $T \cup \{\bigwedge \sigma\}$ is consistent;

(b) for all $i, j \leq n$, there is $r \in N$ such that $(v_i + v_j = v_r)$ occurs in σ;

(c) for all $i \leq n$, there is $j \in N$ such that $(v_i + v_j = v_0)$ and $(v_j + v_i = v_0)$ occur in σ;

(d) for all $i \leq n$, either $\bigwedge \sigma \models v_i = v_0$, or there is $m > 0$ such that $\bigwedge \sigma \models$ $\overbrace{v_i + \ldots + v_i}^{m} = v_0$.

It is easy to verify the following two facts.

(5)   FACT: Let $\sigma \in SEQ$ and $n \in N$ be given. If $T \cup \{\bigwedge \sigma\}$ is consistent, then there is $\tau \in SEQ$ such that $\sigma \subseteq \tau$ and τ is complete through n.

(6)   FACT: Let $\sigma \in SEQ$ be such that $T \cup \{\bigwedge \sigma\}$ is consistent. Then there is an environment e for **G** − **T** that extends σ.

We construct a sequence $\{\sigma^i \mid i \in N\}$. It will be the case that:

(7)   For all $i \in N$,

(a) $\sigma^i \subseteq \sigma^{i+1}$,

(b) $\Psi(\sigma^{2i+1}) \subseteq \mathbf{G} - \mathbf{T}$, and

(c) σ^{2i+2} is complete through i.

The compactness theorem, Conditions (7)a,c, and the definition of completion imply that $e_0 = \bigcup_{i \in N} \sigma^i$ is an environment for $\mathbf{T}$. Condition (7)b implies that it has the promised properties.

Construction of the σ^i

Stage 0: Set $\sigma^0 = \langle v_0 = v_0 \rangle$. So $T \cup \{\bigwedge \sigma_0\}$ is consistent.

Stage $2i + 1$: Suppose that σ^{2i} is defined and $T \cup \{\bigwedge \sigma^{2i}\}$ is consistent. By Fact (6), let environment e for $\mathbf{G} - \mathbf{T}$ extend σ^{2i}. By (4), let $k \geq \text{length}(\sigma^{2i})$ be such that $\Psi(e[k]) \subseteq \mathbf{G} - \mathbf{T}$. We take $\sigma^{2i+1} = e[k]$.

Stage $2i + 2$: Suppose that σ^{2i+1} is defined and $T \cup \{\bigwedge \sigma^{2i+1}\}$ is consistent. By Fact (5), let $\tau \in SEQ$ be such that τ extends σ^{2i+1} and τ is complete through i. We take $\sigma^{2i+2} = \tau$.

End construction

It is immediate that (7) is satisfied.   ∎

Solution to Exercise 3.(70):
By [Hodges, 1993, Theorem 8.3.1] if a theory is model-complete then every sentence is equivalent under the theory to a universal sentence. From this fact the exercise follows directly from Theorem 3.(55).   ∎

Solution to Exercise 3.(71):
Let distinct binary predicates R and A be given. Set $\mathbf{Sym}_o = \{R\}$, $\mathbf{Sym} = \{R, A\}$, $\theta = \forall x \exists y Rxy$, and $T = \{\forall x \exists y Rxy \leftrightarrow \exists x \forall y Axy\}$. By Theorem 3.(55), $(T, \{\theta, \neg\theta\})$ is solvable. Let structures $\mathcal{S}$ and $\mathcal{T}$ be such that:

(a) $|\mathcal{S}| = N$ and $|\mathcal{T}| = Z$, where Z is the set of integers;

(b) the interpretation of R in $\mathcal{S}$ is a total order on N isomorphic to ω;

(c) the interpretation of R in $\mathcal{T}$ is a total order on Z isomorphic to $\omega^* \omega$.

Denote by $\mathcal{S}^o$ (respect. $\mathcal{T}^o$) the restriction of $\mathcal{S}$ (respect. $\mathcal{T}$) to $\mathbf{Sym}_o$. Suppose for a contradiction that scientist Ψ solves $\{\{\mathcal{S}\}, \{\mathcal{T}\}\}$ on $\mathbf{Sym}_o$ environments. Define scientist Ψ' as follows. For all $\sigma \in SEQ$, if $\Psi(\sigma) = \{\mathcal{S}\}$ then $\Psi(\sigma) = \{\mathcal{S}^o\}$; otherwise, $\Psi(\sigma) = \{\mathcal{T}^o\}$. Definition 3.(65) then implies that Ψ' solves $\{\{\mathcal{S}^o\}, \{\mathcal{T}^o\}\}$. This contradicts Proposition 3.(39).   ∎

Section 3.4

Solution to Exercise 3.(86):

Suppose that Ψ is not dominated on $\mathbf{P}$. Let σ be for $\mathbf{P}$, and suppose for a contradiction that:

(8)   for all $P \in \mathbf{P}$ and for all environments e for P that extend σ, there is $k \geq \text{length}(\sigma)$ such that either $\Psi(e[k])$ is not defined or $\Psi(\sigma) = \emptyset$ or $\Psi(e[k]) \nsubseteq P$.

Since some environment for $\mathbf{P}$ extends σ and Ψ solves $\mathbf{P}$, there is least $k_0 \in N$ with the following property:

(9)   There is $P \in \mathbf{P}$ and environment e for P such that e extends σ and $\emptyset \neq \Psi(e[k]) \subseteq P$ for all $k \geq k_0$.

It follows from (8) and (9) that:

(10) $k_0 > \text{length}(\sigma)$.

Let scientist Ψ' satisfy the following conditions, for all $\tau \in SEQ$. If $\sigma \subseteq \tau \subseteq e[k_0]$ then $\Psi'(\tau) = P$. If $\Psi(\tau)$ is defined but either $\sigma \nsubseteq \tau$ or $\tau \nsubseteq e[k_0]$, then $\Psi'(\tau) = \Psi(\tau)$.

It is immediate that Ψ' solves $\mathbf{P}$. Let environment e' for $P' \in \mathbf{P}$ be given. If e' does not extend σ then $SP(\Psi', e', P') = SP(\Psi, e', P')$ by the definition of Ψ'. If e' extends σ and $e' \neq e$, then $SP(\Psi, e', P') \geq k_0$ by the definition of k_0, and $SP(\Psi', e', P') \leq SP(\Psi, e', P')$ by the definition of Ψ'. Finally from (9), (10), and the definition of Ψ' it is easy to see that $SP(\Psi', e, P') \leq \text{length}(\sigma) < k_0 = SP(\Psi, e, P')$. This contradicts the hypothesis. So we have shown that if Ψ is not dominated on $\mathbf{P}$, then conditions (a) and (b) are satisfied.

For the opposite suppose that conditions (a) and (b) are satisfied. Let scientist Ψ' solve $\mathbf{P}$, and let environment e for $P \in \mathbf{P}$ be such that $SP(\Psi', e, P) < SP(\Psi, e, P)$ (if there is no such Ψ', P and e, then there is nothing left to prove). Since $SP(\Psi', e, P) < SP(\Psi, e, P)$ there is $k_0 \in N$ such that:

(11)   it is not true that $\emptyset \neq \Psi(e[k_0]) \subseteq P$ and $\emptyset \neq \Psi'(e[k_0]) \subseteq P$.

By hypothesis there exists $P' \in \mathbf{P}$ and environment e' for P' such that e' extends $e[k_0]$ and $\emptyset \neq \Psi(e'[k]) \subseteq P'$ for all $k \geq k_0$. This, (11), and the fact that Ψ solves $\mathbf{P}$ imply that $SP(\Psi, e', P') \leq k_0 < SP(\Psi', e', P')$. With Definition 3.(74) we conclude that Ψ is not dominated on $\mathbf{P}$, as required.   ∎

Solution to Exercise 3.(87):

Suppose that Ψ solves $\mathbf{P}$ efficiently. Let $\sigma \in SEQ$ be for $\mathbf{P}$. We show that conditions (a)-(c) are satisfied. By Lemma 3.(75) and Exercise 3.(86) there is $P \in \mathbf{P}$ such that $\bigwedge \sigma$ is satisfiable in some member of P and $\emptyset \neq \Psi(\sigma) \subseteq P$. Suppose that for all $P' \in \mathbf{P}$, if $P' \neq P$ then $\bigwedge \sigma$ is satisfiable in no member of P'. Let scientist Ψ' have the following properties, for all $\tau \in SEQ$. If $\sigma \subseteq \tau$ then $\Psi'(\tau) = P$. If $\sigma \not\subseteq \tau$ and $\Psi(\tau)$ is defined, then $\Psi'(\tau) = \Psi(\tau)$. It is easy to see that Ψ' solves $\mathbf{P}$, and that for every $P' \in \mathbf{P}$ and for every environment e for P':

(a) $SP(\Psi', e, P') \leq SP(\Psi, e, P')$, and

(b) if e extends σ and there is $k > \text{length}(\sigma)$ such that either $\Psi(e[k])$ is undefined or $\Psi(e[k]) = \emptyset$ or $\Psi(e[k]) \not\subseteq P'$, then $SP(\Psi', e, P') < SP(\Psi, e, P')$.

It follows from Definition 3.(74) that $\emptyset \neq \Psi(e[k]) \subseteq P$ for every environment e for P that extends σ and for every $k \geq \text{length}(\sigma)$. So, conditions (a)–(c) of the exercise are satisfied. Thus, to conclude this direction of the proof, suppose that there is $P' \in \mathbf{P}$ with $P' \neq P$ such that $\bigwedge \sigma$ is satisfiable in some member of P'. Let $P' \in \mathbf{P}$ and environment e' for P' be such that $P' \neq P$ and $\sigma \subseteq e'$. Let scientist Ψ' be defined as follows, for every $\gamma \in SEQ$. If $\gamma \not\subseteq e'$ then $\Psi'(\gamma) = \Psi(\gamma)$. If $\gamma \subseteq \sigma$ then $\Psi'(\gamma) = \emptyset$. If $\sigma \subseteq \gamma \subseteq e'$ then $\Psi'(\gamma) = P'$. Trivially, Ψ' solves $\mathbf{P}$ and $\text{length}(\sigma) = SP(\Psi', e', P') < SP(\Psi, e', P')$. From this and Definition 3.(73), we deduce that there is $\mathcal{S} \in P$ and full assignment h to $\mathcal{S}$ such that $\mathcal{S} \models \bigwedge \sigma[h]$, and $\emptyset \neq \Psi(e[k]) \subseteq P$ for every environment e for $\mathcal{S}$ and h that extends σ and for every $k \geq \text{length}(\sigma)$. Hence conditions (a)-(c) of the exercise are satisfied.

Conversely, suppose that conditions (a)-(c) of the exercise are satisfied. We show that Ψ solves $\mathbf{P}$ efficiently. Suppose that scientist Ψ' solves $\mathbf{P}$. Suppose there is $P_0 \in \mathbf{P}$ and environment e_0 for P_0 such that $SP(\Psi', e_0, P_0) < SP(\Psi, e_0, P_0)$. (If there is no such Ψ', P_0 and e_0, then there is nothing left to prove.) Let $k_0 \in N$ be such that $k_0 = SP(\Psi', e_0, P_0)$. Choose $\mathcal{S} \in P$ and full assignment h to $\mathcal{S}$ such that $\mathcal{S} \models \bigwedge e_0[k_0][h]$ and for every environment e for $\mathcal{S}$ and h that extends $e_0[k_0]$, for every $k \geq k_0$, $\emptyset \neq \Psi(e[k]) \subseteq P$. Suppose that $P = P_0$. Then e_0 is for P, and $SP(\Psi, e_0, P_0) \leq k_0 = SP(\Psi', e_0, P_0)$, contradiction. Hence $P \neq P_0$. Since $\Psi'(e_0[k_0]) = P_0 \neq P$, we infer that for every environment e for $\mathcal{S}$ and h that extends $e_0[k_0]$, $SP(\Psi, e, P) \leq k_0 < SP(\Psi', e, P)$. With Definition 3.(73) this completes the proof of the exercise. ∎

Solution to Exercise 3.(88):

Suppose **Sym** is limited to two constants a and b. Let $P_1 = MOD(a = b)$,

$P_2 = MOD(a \neq b)$, and $\mathbf{P} = \{P_1, P_2\}$. Let oracle o be such that $o(\emptyset) = (a = b)$. Let scientist be defined as follows, for all $\sigma \in SEQ$. If $a = b$ occurs in σ then $\Psi(\sigma) = P_1$; otherwise, $\Psi(\sigma) = P_2$. Trivially, Ψ solves $\mathbf{P}$ using o and for all $P \in \mathbf{P}$ and environments e for P, $SP(\Psi, o(e), P) = 0$. Let environment e for P_1 and environment e' for P_2 be given. For every scientist Ψ' that solves $\mathbf{P}$, either $SP(\Psi', e, P_1) > 0$ or $SP(\Psi', e', P_2) > 0$. The Exercise follows immediately. ∎

Section 3.5

Solution to Exercise 3.(104):

Let total recursive function $G : N \times \mathcal{L}_{sen}^{<\omega} \to \mathcal{L}_{sen}^{<\omega}$ be such that for all $T = \{\delta_i \mid i \in N\}$, for all $n \in N$, and for all $\theta_0 \ldots \theta_n \in \mathcal{L}_{sen}$, the following holds. If $\theta_0 \ldots \theta_n$ are equivalent in T to $\exists\forall$ sentences, then there exists $\exists\forall$ sentences $\theta_0' \ldots \theta_n'$ and $i_0 \in N$ such that:

(a) for all $m \leq n$, $T \models \theta_m \leftrightarrow \theta_m'$, and

(b) for all $i \geq i_0$, $G(n, \delta_0 \ldots \delta_i, \theta_0 \ldots \theta_n) = (\theta_0' \ldots \theta_n')$.

Let total recursive function $F : \mathcal{L}_{sen}^{<\omega} \times SEQ \to \mathcal{L}_{sen}$ satisfy the following, for all $T = \{\delta_i \mid i \in N\}$, $n \in N$, $\theta_0 \ldots \theta_n \in \mathcal{L}_{sen}$, and $\sigma \in SEQ$. Suppose there are $\forall$ formulas $\varphi_0 \ldots \varphi_n \in \mathcal{L}_{form}$ and tuples of variables $\bar{x}_0 \ldots \bar{x}_n$ such that for all $m \leq n$, $Var(\varphi_m) = range(\bar{x}_m)$ and $G(n, \delta_0 \ldots \delta_{length(\sigma)}, \theta_0 \ldots \theta_n) = (\exists\bar{x}_0\varphi_0 \ldots \exists\bar{x}_n\varphi_n)$. Let $\{\chi_j \mid j \in N\}$ be a recursive enumeration, uniform in $n, \varphi_0 \ldots \varphi_n$, of all formulas of form $\varphi_m(\bar{y}/\bar{x}_m)$, where $m \leq n$ and $length(\bar{y}) = length(\bar{x}_m)$. Suppose there is least $j_0 \in N$ such that $\bigwedge \sigma \not\models \neg\chi_{j_0}$, and let $n_0 \leq n$ be least such that $\chi_{j_0} = \exists\bar{x}_{n_0}\varphi_{n_0}$. Then $F(\delta_0 \ldots \delta_{length(\sigma)}, \theta_0 \ldots \theta_n, \sigma) = \theta_{n_0}$.

We show that the function F satisfies the claim of the exercise. Let solvable problem of form $(T, \{\theta_0, \ldots, \theta_n\})$ be given. Let $n_0 \leq n$ and environment e for $MOD(T \cup \{\theta_{n_0}\})$ be given. It suffices to show that $\lambda\sigma . MOD(T \cup \{F(\delta_0 \ldots \delta_{length(\sigma)}, \theta_0 \ldots \theta_n, \sigma)\})$ solves $MOD(T \cup \{\theta_{n_0}\})$ in e. Since the θ_i partition the models of T, e is not an environment for $MOD(T \cup \{\theta_m\})$, for all $m < n_0$. By Theorem 3.(55) and the definition of G, there exists $i_0 \in N$, $\forall$ formulas φ_m, $m \leq n$, and tuples of variables $\bar{x}_m$, $m \leq n$, such that:

(a) for all $m \leq n$, $Var(\varphi_m) = range(\bar{x}_m)$,

(b) for all $m \leq n$, $T \models \theta_m \leftrightarrow \exists\bar{x}_m\varphi_m$, and

(c) for all $i \geq i_0$, $G(n, \delta_0 \ldots \delta_i, \theta_0 \ldots \theta_n) = (\exists\bar{x}_0\varphi_0 \ldots \exists\bar{x}_n\varphi_n)$.

Let $\{\chi_j \mid j \in N\}$, be the recursive enumeration, considered in the definition of F above, of all formulas of form $\varphi_m(\bar{y}/\bar{x}_m)$, where $m \leq n$ and $length(\bar{y}) =$

length($\bar{x}_m$). Let j_0 be least such that range(e) $\cup$ $\{\chi_{j_0}\}$ is consistent. By compactness, for all $j < j_0$, $\bigwedge e[k] \models \neg\chi_j$ for cofinitely many k. Moreover, n_0 is least with $\chi_{j_0} = \exists \bar{x}_{n_0}\varphi_{n_0}$. Hence by the definition of F, there is $i_1 \geq i_0$ such that for all $k \geq i_1$, $F(\delta_0 \ldots \delta_k, \theta_0 \ldots \theta_n, e[k]) = \theta_{n_0}$. This implies that $\lambda\sigma \, . \, MOD(T \cup \{F(\delta_0 \ldots \delta_{\text{length}(\sigma)}, \theta_0 \ldots \theta_n, \sigma)\})$ solves $MOD(T \cup \{\theta_{n_0}\})$ in e, as required. ∎

Solution to Exercise 3.(105):
We rely on the following fact.

(12) FACT: Let $T, X \subseteq \mathcal{L}_{sen}$ and scientist Ψ be such that Ψ X-solves T. Let $\mathcal{S} \in MOD(T)$ be given. Then there exists $\exists\forall$ $\chi_\mathcal{S} \in \mathcal{L}_{sen}$ such that:

(a) $\mathcal{S} \models \chi_\mathcal{S}$.

(b) $T \cup \{\chi_\mathcal{S}\} \models \theta$.

Proof of Fact (12): The proof of Lemma 3.(24) is easily adapted to show the following.

(13) LEMMA: There is $\theta \in X$, $\sigma \in SEQ$, and finite assignment $a : Var \rightarrow |\mathcal{S}|$ such that:

(a) domain(a) $\supseteq$ $Var(\sigma)$.

(b) $\mathcal{S} \models \bigwedge \sigma[a]$.

(c) For every $\tau \in SEQ$, if $\mathcal{S} \models \exists\bar{x} \bigwedge(\sigma * \tau)[a]$, where $\bar{x}$ contains the variables in $Var(\tau) - $ domain(a), then $\Psi(\sigma * \tau) = MOD(\theta)$.

Fix θ, σ and a that satisfy the claim of Lemma (13). Define π to be the collection of $\forall$ formulas φ such that $Var(\varphi) \subseteq Var(\sigma)$ and $\mathcal{S} \models \varphi[a]$. Hence:

(14) $\mathcal{S} \models \pi[a]$.

We now show that:

(15) CLAIM: $T \cup \pi \models \theta$.

Proof of Claim (15): Suppose for a contradiction that $\mathcal{U} \in MOD(T \cup \{\neg\theta\})$ and full assignment g to $\mathcal{U}$ are such that:

(16) $\mathcal{U} \models \pi[g]$.

It is clear that $\bigwedge \sigma \in \pi$, so there is an environment e for $\mathcal{U}$ and g such that $\sigma \subseteq e$. We will show that for all $k \geq$ length(σ), $\Psi(e[k]) = MOD(\theta)$. Since e is for $\mathcal{U} \in MOD(T \cup \{\neg\theta\})$, this implies that Ψ does not X-solve T, contradicting our choice of Ψ. So let $k \geq \text{length}(\sigma)$ be given. Choose $\tau \in SEQ$ such that $e[k] = \sigma * \tau$. We must show that $\Psi(\sigma * \tau) = MOD(\theta)$. By Lemma (13), it suffices to show that $\mathcal{S} \models \exists \bar{x} \bigwedge (\sigma * \tau)[a]$, where $\bar{x}$ contains the variables in $Var(\tau) - \text{domain}(a)$. For a contradiction, suppose that $\mathcal{S} \models \forall \bar{x} \neg \bigwedge (\sigma * \tau)[a]$. Then $\forall \bar{x} \neg \bigwedge (\sigma * \tau) \in \pi$, so by (16), $\mathcal{U} \models \forall \bar{x} \neg \bigwedge (\sigma * \tau)[g]$. However, this is impossible since e was chosen to be for $\mathcal{U}$ and g, and $\sigma * \tau \subseteq e$. ∎

Fact (12) follows immediately from the compactness theorem, (14), and Claim (15). ∎

We now prove the claim of the exercise. Let total recursive functions $G_1 : \mathcal{L}_{sen}^{<\omega} \to \mathcal{L}_{sen}$ and $G_2 : \mathcal{L}_{sen}^{<\omega} \to \mathcal{L}_{sen}$ be such that for all $T = \{\delta_i \mid i \in N\}$ and $X = \{\theta_i \mid i \in N\}$, there is $n \in N$ such that:

(a) for all $m \in N$, $G_1(\delta_0 \ldots \delta_m, \theta_0 \ldots \theta_m)$ and $G_2(\delta_0 \ldots \delta_m, \theta_0 \ldots \theta_m)$ are defined iff $m \geq n$;

(b) $\{G_1(\delta_0 \ldots \delta_m, \theta_0 \ldots \theta_m) \mid m \geq n\}$ enumerate all $\exists \forall$ formulas χ such that $T \cup \{\chi\} \models \theta$ for some $\theta \in X$;

(c) for all $m \geq n$, if $\chi = G_1(\delta_0 \ldots \delta_m, \theta_0 \ldots \theta_m)$, then $\theta = G_2(\delta_0 \ldots \delta_m, \theta_0 \ldots \theta_m)$ belongs to X and $T \cup \{\chi\} \models \theta$.

Let total recursive function $F : \mathcal{L}_{sen}^{<\omega} \times SEQ \to \mathcal{L}_{sen}$ satisfy the following, for all $T = \{\delta_i \mid i \in N\}$, $X = \{\theta_i \mid i \in N\}$, and $\sigma \in SEQ$. Suppose there is $n \leq \text{length}(\sigma)$ such that for all $m \leq \text{length}(\sigma)$, $G_1(\delta_0 \ldots \delta_m, \theta_0 \ldots \theta_m)$ is defined iff $m \geq n$. For all $m \leq \text{length}(\sigma) - n$, let $\chi_m = G_1(\delta_0 \ldots \delta_{m+n}, \theta_0 \ldots \theta_{m+n})$. Suppose there is least $m_0 \leq \text{length}(\sigma) - n$ such that $\bigwedge \sigma \not\models \neg \chi_{m_0}$. Then $F(\delta_0 \ldots \delta_{\text{length}(\sigma)}, \theta_0 \ldots \theta_{\text{length}(\sigma)}, \sigma) = G_2(\delta_0 \ldots \delta_{m_0}, \theta_0 \ldots \theta_{m_0})$.

We show that the function F satisfies the claim of the exercise. Let $T = \{\delta_i \mid i \in N\}$ and $X = \{\theta_i \mid i \in N\}$ be such that T is X-solvable. Let $\mathcal{S} \in MOD(T)$, full assignment h to $\mathcal{S}$, and environment e for $\mathcal{S}$ and h be given. It suffices to show that there is $\theta \in X$ with $\mathcal{S} \models \theta$ and $F(\delta_0 \ldots \delta_k, \theta_0 \ldots, \theta_k, e[k]) = \theta$ for cofinitely many k. By Fact (12), the definition of G_1, and the definition of G_2, there is least $n_0 \in N$ such that $G_1(\delta_0 \ldots \delta_{n_0}, \theta_0 \ldots \theta_{n_0})$ is defined and $\mathcal{S} \models G_1(\delta_0 \ldots \delta_{n_0}, \theta_0 \ldots \theta_{n_0})[h]$. By the definition of G_2:

(17) $\mathcal{S} \models G_2(\delta_0 \ldots \delta_{n_0}, \theta_0 \ldots \theta_{n_0}) \in X$.

Using compactness, the definition of n_0, and the definition of F, is easy to verify that for cofinitely many k, $F(\delta_0 \ldots \delta_k, \theta_0 \ldots, \theta_k, e[k]) = G_2(\delta_0 \ldots \delta_{n_0}, \theta_0 \ldots \theta_{n_0})$. We conclude with (17). ∎

Solution to Exercise 3.(106):

For the purposes of this exercise, we introduce the following convention. A term or notation decorated with η indicates usage in the sense of Chapter 2 (numerical paradigm). Decoration with ℓ indicates usage in the sense of Chapter 3 (logical paradigm). Recall the notation given in Definition 3.(98). Let $X \subseteq N$ be of T-degree greater than $\mathbf{0}'$. Hence $\mathbf{P}_\eta = \{\{0, n\} \mid n \in X\} \cup \{\{n\} \mid n \notin X\}$ is not $r.e.$-indexable. We show that $\mathbf{P}_\ell = \{P_L \mid L \in \mathbf{P}_\eta\}$ satisfies the claim of the exercise.

Trivially, some computable scientist$_\eta$ solves$_\eta$ $\mathbf{P}_\eta$. It follows from Theorem 3.(99) that some computable scientist$_\ell$ solves$_\ell$ $\mathbf{P}_\ell$. Let computable ψ_ℓ : $SEQ_\ell \to N$ underly scientist$_\ell$, and suppose for a contradiction that Ψ_ℓ respects $\mathbf{P}_\ell$ and solves$_\ell$ $\mathbf{P}_\ell$. From the proof of Theorem 3.(99), we infer that there is computable $\psi_\eta : SEQ_\eta \to N$ such that ψ_η underlies scientist$_\eta$ Ψ_η, Ψ_η solves$_\eta$ $\mathbf{P}_\eta$, and for all $\sigma \in SEQ_\eta$, $\{W^{num}_{\psi_\eta(\sigma)} \mid \sigma \in SEQ_\eta\} \subseteq \mathbf{P}_\eta$. It follows that $\{W^{num}_{\psi_\eta(\sigma)} \mid \sigma \in SEQ_\eta\} = \mathbf{P}_\eta$. This contradicts the fact that $\mathbf{P}_\eta$ is not $r.e.$-indexable.   ∎

Section 3.7

Solution to Exercise 3.(179):

Suppose that $\mathcal{S}$ and $\mathcal{T}$ are finitely isomorphic in the sense of Definition 3.(148). Lemma 3.(150)b then implies that for all $n \in N$, $I_n(\mathcal{S}, \mathcal{T}) \neq \emptyset$. With Definition 3.(148), it follows immediately that $\mathcal{S}$ and $\mathcal{T}$ are finitely isomorphic in the standard sense (taking $J_n(\mathcal{S}, \mathcal{T}) = I_n(\mathcal{S}, \mathcal{T})$, for all $n \in N$).

For the converse, suppose that $\mathcal{S}$ and $\mathcal{T}$ are finitely isomorphic in the standard sense. Let sequence $\{J_n \mid n \in N\}$ of nonempty sets of partial isomorphisms from $\mathcal{S}$ to $\mathcal{T}$ be such that for all $n \in N$, the following holds:

(a) For every $f \in J_{n+1}$ and $a \in |\mathcal{S}|$, there is $g \in J_n$ such that g extends f and $a \in \mathrm{domain}(g)$.

(b) For every $f \in J_{n+1}$ and $a \in |\mathcal{T}|$, there is $g \in J_n$ such that g extends f and $a \in \mathrm{domain}(g)$.

Let $n \in N$ be given. Since $J_m(\mathcal{S}, \mathcal{T}) \neq \emptyset$ for all $m \leq n$, we infer from Definition 3.(148) that $I_n(\mathcal{S}, \mathcal{T}) \neq \emptyset$. With Definition 3.(148) again, it follows that the empty partial isomorphism belongs to $I_n(\mathcal{S}, \mathcal{T})$. Lemma 3.(150)b then implies that $I_\omega(\mathcal{S}, \mathcal{T}) \neq \emptyset$, i.e. $\mathcal{S}$ and $\mathcal{T}$ are finitely isomorphic in the sense of Definition 3.(148).   ∎

Solution to Exercise 3.(180):

Suppose that S and T are partially isomorphic in the standard sense. Let nonempty $J \subseteq PI(S, T)$ be such that the following holds:

(a) for all $a \in |S|$ and $f \in J$, there is $g \in J$ such that g extends f and domain$(g) = $ domain$(f) \cup \{a\}$;

(b) for all $a \in |T|$ and $f \in J$, there is $g \in J$ such that g extends f and range$(g) = $ range$(f) \cup \{a\}$.

Then it is easy to verify by induction on the class of ordinals that for all ordinals α, $J \subseteq I_\alpha(S, T)$. Hence $I_\alpha(S, T) \neq \emptyset$ for all ordinals α, i.e. S and T are partially isomorphic in the sense of Definition 3.(148).

For the converse, suppose that S and T are partially isomorphic in the sense of Definition 3.(148), i.e. $I_\alpha(S, T) \neq \emptyset$ for all ordinals α. Since $PI(S, T)$ is a set and not a proper class, it follows from Lemma 3.(150)b that there is ordinal α such that $I_\alpha(S, T) = I_\beta(S, T)$ for all ordinals $\beta \geq \alpha$. Hence we may choose ordinal α such that $I_\alpha(S, T) = I_{\alpha+1}(S, T) \neq \emptyset$. Let $f \in I_\alpha(S, T)$ be given. Since $f \in I_{\alpha+1}(S, T)$, it follows that:

(a) for all $a \in |S|$, there is $g \in I_\alpha(S, T)$ such that g extends f and domain$(g) = $ domain$(f) \cup \{a\}$;

(b) for all $a \in |T|$, there is $g \in I_\alpha(S, T)$ such that g extends f and range$(g) = $ range$(f) \cup \{a\}$.

Hence S and T are partially isomorphic in the standard sense (taking $J = I_\alpha(S, T)$). ∎

D Solutions to Exercises for Chapter 4

Section 4.1

Solution to Exercise 4.(14):

Let $\phi_1 = (v_0 = v_1)$, $\phi_2 = (v_2 = v_3)$, and $\phi_3 = (v_4 = v_5)$. Set $B = \{(\phi_2 \vee \phi_3) \rightarrow \phi_1, \phi_2, \phi_3\}$. It is easy to verify that $B \perp \phi_1 = \{\{(\phi_2 \vee \phi_3) \rightarrow \phi_1\}, \{\phi_2, \phi_3\}\}$. Hence $\bigcap(B \perp \phi_1) \subseteq \{\phi_2\} \subseteq B$ and $\phi_2 \not\models \phi_1$. So there is a contraction function $\dot{-}$ such that $B \dot{-} \phi_1 = \{\phi_2\}$. Trivially, $\{\phi_2\}$ is not the intersection of some subset of $B \perp \phi_1$. The exercise follows immediately. ∎

Solution to Exercise 4.(15):

Let $\phi_1 = (v_0 = v_1)$ and $\phi_2 = (v_2 = v_3)$. Set $B_0 = \{\phi_1\}$ and $B_1 = \{\phi_1 \vee \phi_2\}$. Trivially, $B_0 \perp \phi_1 = \{\emptyset\}$ and $B_1 \perp \phi_1 = \{\phi_1 \vee \phi_2\}$. Hence $B_0 \models B_1$ but for every contraction function $\dot{-}$, $B_0 \dot{-} \phi_1 = \emptyset \not\models \{\phi_1 \vee \phi_2\} = B_1 \dot{-} \phi_1$. The exercise follows immediately. ∎

Solution to Exercise 4.(16):

We prove Part (a). By an application of Zorn's lemma there is a total ordering of $\mathrm{pow}(\mathcal{L}_{form})$ that respects inclusion (see [Stoll, 1963, p. 118]). It is easy to verify that $\dot{-}$ is definite with respect to any such ordering. To see that $\dot{-}$ is not maxichoice, note that $\{v_0 = v_1, v_0 \neq v_1\} \perp (v_0 \neq v_0) = \{\{v_0 = v_1\}, \{v_0 \neq v_1\}\}$ and $\{v_0 = v_1, v_0 \neq v_1\} \dot{-} (v_0 \neq v_0) = \emptyset$.

We prove Part (b). Before we define a maxichoice, definite contraction function that is not stringent, we need some notation and a fact. Let **two** be a sentence asserting that there are at least two individuals, **three** a sentence asserting that there are at least three individuals, **four** a sentence asserting that there are at least four individuals. Set:

(1) (a) $A_0 = \{\textbf{two}, \textbf{two} \rightarrow \textbf{three}, \textbf{two} \rightarrow \textbf{four}\}$;

    (b) $A_1 = \{\textbf{two}, \textbf{two} \rightarrow \textbf{four}\}$;

    (c) $A_2 = \{\textbf{two}\}$;

    (d) $A_3 = \{\textbf{two} \rightarrow \textbf{three}, \textbf{two} \rightarrow \textbf{four}\}$;

    (e) $A_4 = \{\textbf{two}, \textbf{two} \rightarrow \textbf{three}\}$;

    (f) $A_5 = \{\textbf{two} \rightarrow \textbf{three}\}$;

    (g) $A_6 = \{\textbf{two} \rightarrow \textbf{four}\}$;

    (h) $A_7 = \emptyset$.

Fix an enumeration $\{\varphi_i \mid i \in N\}$ of $\mathcal{L}_{form} - A_0$. Set $\varphi_\omega = \textbf{two}$, $\varphi_{\omega+1} = \textbf{two} \rightarrow$ **three**, and $\varphi_{\omega+2} = \textbf{two} \rightarrow \textbf{four}$. Let strict total order $\prec$ on $\mathrm{pow}(\mathcal{L}_{form})$ be defined as follows. For all distinct $B, B' \subseteq \mathcal{L}_{form}$, $B \prec B'$ just in case there is $n \leq \omega + 2$ such that $\varphi_n \in B$, $\varphi_n \notin B'$, and for all $m < n$, $\varphi_m \in B$ iff $\varphi_m \in B'$. Now let strict total order $\prec^{\#}$ on $\mathrm{pow}(\mathcal{L}_{form})$ be such that for all distinct $B, B' \subseteq \mathcal{L}_{form}$, $B \prec^{\#} B'$ just in case one of the following holds:

(a) $B \notin \{A_i \mid i \leq 7\}$ and $B' \in \{A_i \mid i \leq 7\}$;

(b) there is $0 \leq i < j \leq 7$ such that $B = A_i$ and $B' = A_j$;

(c) $B \notin \{A_i \mid i \leq 7\}$, $B' \notin \{A_i \mid i \leq 7\}$, and $B \prec B'$.

We show the following fact.

(2)   FACT: Let $B \subseteq \mathcal{L}_{form}$ and invalid $\phi \in \mathcal{L}_{form}$ be given. Then there is a $\prec^{\#}$-least subset Y of $\mathcal{L}_{form}$ such that $\bigcap(B \perp \phi) \subseteq Y \subseteq B$ and $Y \not\models \phi$. Moreover, $Y \in B \perp \phi$.

Proof of Fact (2):   Let $B \subseteq \mathcal{L}_{form}$ and invalid $\phi \in \mathcal{L}_{form}$ be given. We distinguish two cases.

Case 1: there is $\varphi \in \mathcal{L}_{form} - A_0$ such that $\varphi \in B$ and $\varphi \not\models \phi$. We define by induction on $\alpha \leq \omega + 2$ a subset Y^{α} of B. Suppose that $\alpha \leq \omega + 2$ and that Y^{β} is defined for all $\beta < \alpha$. Then $Y^{\alpha} = (\bigcup_{\beta < \alpha} Y^{\beta}) \cup \{\varphi_{\alpha}\}$ if $\varphi_{\alpha} \in B$ and $(\bigcup_{\beta < \alpha} Y^{\beta}) \cup \{\varphi_{\alpha}\} \not\models \phi$; otherwise $Y^{\alpha} = \bigcup_{\beta < \alpha} Y^{\beta}$. Set $Y = \bigcup_{\alpha \leq \omega + 2} Y^{\alpha}$. An easy application of compactness shows that $Y \not\models \phi$. Moreover, the construction of Y ensures that for all $\chi \in B - Y$, $Y \cup \{\chi\} \models \phi$. So $Y \in B \perp \phi$. It remains to show that Y is the $\prec^{\#}$-least subset of $\mathcal{L}_{form}$ with $\bigcap(B \perp \phi) \subseteq Y \subseteq B$ and $Y \not\models \phi$. For a contradiction suppose that $C \subseteq \mathcal{L}_{form}$ is such that:

(3)   (a) $C \prec^{\#} Y$,

        (b) $C \subseteq B$,

        (c) $C \not\models \phi$.

The hypothesis of the current case together with the construction of Y implies immediately that some member of $\mathcal{L}_{form} - A_0$ belongs to Y. From this, the definition of $\prec^{\#}$, and (3)a we infer the existence of $\alpha \leq \omega + 2$ such that $\varphi_{\alpha} \in C$, $\varphi_{\alpha} \notin Y$, and for all $\beta < \alpha$, $\varphi_{\beta} \in C$ iff $\varphi_{\beta} \in Y$. By (3)b, $\varphi_{\alpha} \in B$. Let $Z^{\alpha} = \{\varphi_{\beta} \mid \varphi_{\beta} \in C \text{ and } \beta < \alpha\} = \{\varphi_{\beta} \mid \varphi_{\beta} \in Y \text{ and } \beta < \alpha\}$. By (3)c, $Z^{\alpha} \cup \{\varphi_{\alpha}\} \not\models \phi$. These facts, along with the definition of Y imply that $\varphi_{\alpha} \in Y$, contradiction.

Case 2: for all $\varphi \in \mathcal{L}_{form} - A_0$, either $\varphi \notin B$ or $\varphi \models \phi$. Since $\{A_i \mid i \leq 7\} = \mathrm{pow}(A_0)$, we infer that every subset of B that does not imply ϕ belongs to $\{A_i \mid i \leq 7\}$. It follows immediately that there is a $\prec^{\#}$-least subset Y of $\mathcal{L}_{form}$ such that:

(4) $\bigcap(B \perp \phi) \subseteq Y \subseteq B$ and $Y \not\models \phi$.

Suppose for a contradiction that $Y \not\subseteq B \perp \phi$. By (1), for all $0 \leq i < j \leq 7$, $A_i \subset A_j$ iff $i = 2$ and $j = 4$. So it is easy to verify that:

(5)   (a) $Y = A_2$,

   (b) $A_4 \subseteq B$, and

   (c) $A_4 \not\models \phi$.

It follows from (4) and (5)a that **two** $\rightarrow$ **three** $\notin \bigcap(B \perp \phi)$. By (5)b, **two** $\rightarrow$ **three** $\in B$, so **two** $\rightarrow$ **three** $\notin \bigcap(B \perp \phi)$ implies that there is (finite) $D \subseteq B$ with:

(6) $D \not\models \phi$ and $D \cup \{\textbf{two} \rightarrow \textbf{three}\} \models \phi$.

By the hypothesis of the current case $D \subseteq A_0$. Since **two** $\rightarrow$ **four** $\models$ **two** $\rightarrow$ **three**, it follows from (6) that $D \subseteq \{\textbf{two}\}$. Hence $D \cup \{\textbf{two} \rightarrow \textbf{three}\} \subseteq \{\textbf{two}, \textbf{two} \rightarrow \textbf{three}\} = A_4$. This is in contradiction with (6) and (5)c.   ∎

By Fact (2) there is a maxichoice, definite contraction function $\dot{-}$ such that for all $B \subseteq \mathcal{L}_{form}$ and invalid $\phi \in \mathcal{L}_{form}$, $B \dot{-} \phi$ is $\prec^{\#}$-least with $\bigcap(B \perp \phi) \subseteq B \dot{-} \phi \subseteq B$ and $B \dot{-} \phi \not\models \phi$. To finish the proof of the proposition it suffices to show that $\dot{-}$ is not stringent. Suppose otherwise. Let strict total order $\prec^*$ on $\text{pow}(\mathcal{L}_{form})$ be such that for all $B \subseteq \mathcal{L}_{form}$ and invalid $\phi \in \mathcal{L}_{form}$, $B \dot{-} \phi$ is the $\prec^*$-least subset of B that does not imply ϕ. With the definition of $\prec^{\#}$ it is easy to verify that:

(7)   (a) $A_0 \perp \textbf{three} = \{A_2, A_3\}$ and $A_0 \dot{-} \textbf{three} = A_2$;

   (b) $A_0 \perp \textbf{four} = \{A_3, A_4\}$ and $A_0 \dot{-} \textbf{four} = A_3$.

From (7)a we infer immediately that $A_2 \prec^* A_3$; likewise from (7)b we infer that $A_3 \prec^* A_4$. Hence $A_2 \prec^* A_4$. Since A_4 does not imply **four** and $A_2 \subset A_4$, we deduce that A_4 is not the $\prec^*$-least subset of A_4 that does not imply **four**. This is in contradiction with the definition of $\prec^*$ and the fact that $A_4 \dot{-} \textbf{four} = A_4$.   ∎

Solution to Exercise 4.(17):
Before we define a rigid contraction function that is not definite, we need some notation and a fact. Let $\phi_1 = (v_0 = v_1)$, $\phi_2 = (v_2 = v_3)$, and $\phi_3 = (v_4 = v_5)$. Set:

(8)   (a) $A_0 = \emptyset$;

   (b) $A_1 = \{\phi_1\}$;

(c) $A_2 = \{\phi_2\}$;

(d) $A_3 = \{\phi_3\}$;

(e) $A_4 = \{\phi_1, \phi_3\}$;

(f) $A_5 = \{\phi_1, \phi_2\}$;

(g) $A_6 = \{\phi_2, \phi_3\}$;

(h) $A_7 = \{\phi_1, \phi_2, \phi_3\}$.

Fix an enumeration $\{\varphi_i \mid i \in N\}$ of $\mathcal{L}_{form} - A_7$. Set $\varphi_\omega = \phi_1$, $\varphi_{\omega+1} = \phi_2$, and $\varphi_{\omega+2} = \phi_3$. Let strict total order $\prec$ on $\mathrm{pow}(\mathcal{L}_{form})$ be defined as follows. For all distinct $B, B' \subseteq \mathcal{L}_{form}$, $B \prec B'$ just in case there is $n \leq \omega + 2$ such that $\varphi_n \in B$, $\varphi_n \notin B'$, and for all $m < n$, $\varphi_m \in B$ iff $\varphi_m \in B'$. Now let strict total order $\prec^{\#}$ on $\mathrm{pow}(\mathcal{L}_{form})$ be such that for all distinct $B, B' \subseteq \mathcal{L}_{form}$, $B \prec^{\#} B'$ just in case one of the following holds:

(a) $B \notin \{A_i \mid i \leq 7\}$ and $B' \in \{A_i \mid i \leq 7\}$;

(b) there is $0 \leq i < j \leq 7$ such that $B = A_i$ and $B' = A_j$;

(c) $B \notin \{A_i \mid i \leq 7\}$, $B' \notin \{A_i \mid i \leq 7\}$, and $B \prec B'$.

We show the following fact.

(9) FACT: Let $B \subseteq \mathcal{L}_{form}$ and invalid $\phi \in \mathcal{L}_{form}$ be given. Then there is a $\prec^{\#}$-least member of $B \perp \phi$.

Proof of Fact (9): Let $B \subseteq \mathcal{L}_{form}$ and invalid $\phi \in \mathcal{L}_{form}$ be given. We distinguish two cases.

Case 1: there is $\varphi \in \mathcal{L}_{form} - A_7$ such that $\varphi \in B$ and $\varphi \not\models \phi$. We define by induction on $\alpha \leq \omega + 2$ a subset Y^α of B. Suppose that $\alpha \leq \omega + 2$ and that Y^β is defined for all $\beta < \alpha$. Then $Y^\alpha = (\bigcup_{\beta < \alpha} Y^\beta) \cup \{\varphi_\alpha\}$ if $\varphi_\alpha \in B$ and $(\bigcup_{\beta < \alpha} Y^\beta) \cup \{\varphi_\alpha\} \not\models \phi$; otherwise $Y^\alpha = \bigcup_{\beta < \alpha} Y^\beta$. Set $Y = \bigcup_{\alpha \leq \omega + 2} Y^\alpha$. An easy application of compactness shows that $Y \not\models \phi$. Moreover, the construction of Y ensures that for all $\chi \in B - Y$, $Y \cup \{\chi\} \models \phi$. So $Y \in B \perp \phi$. It remains to show that Y is the $\prec^{\#}$-least member of $B \perp \phi$. For a contradiction suppose that $C \subseteq \mathcal{L}_{form}$ is such that:

(10) (a) $C \prec^{\#} Y$,

(b) $C \subseteq B$,

(c) $C \not\models \phi$.

The hypothesis of the current case together with the construction of Y implies immediately that some member of $\mathcal{L}_{form} - A_7$ belongs to Y. From this, the definition of $\prec^\#$, and (10)a we infer the existence of $\alpha \leq \omega + 2$ such that $\varphi_\alpha \in C$, $\varphi_\alpha \notin Y$, and for all $\beta < \alpha$, $\varphi_\beta \in C$ iff $\varphi_\beta \in Y$. By (10)b, $\varphi_\alpha \in B$. Let $Z^\alpha = \{\varphi_\beta \mid \varphi_\beta \in C \text{ and } \beta < \alpha\} = \{\varphi_\beta \mid \varphi_\beta \in Y \text{ and } \beta < \alpha\}$. By (10)c, $Z^\alpha \cup \{\varphi_\alpha\} \not\models \phi$. These facts, along with the definition of Y imply that $\varphi_\alpha \in Y$, contradiction.

Case 2: for all $\varphi \in \mathcal{L}_{form} - A_7$, either $\varphi \notin B$ or $\varphi \models \phi$. Since $\{A_i \mid i \leq 7\} = \text{pow}(A_7)$, we infer that every subset of B that does not imply ϕ belongs to $\{A_i \mid i \leq 7\}$. It follows immediately that there is a $\prec^\#$-least member of $B \perp \phi$.
∎

By Fact (2) there is a rigid contraction function $\dot{-}$ such that for all $B \subseteq \mathcal{L}_{form}$ and invalid $\phi \in \mathcal{L}_{form}$, $B \dot{-} \phi$ is the $\prec^\#$-least member of $B \perp \phi$. To finish the proof of the proposition it suffices to show that $\dot{-}$ is not definite. Suppose otherwise. Let strict total order $\prec^*$ on $\text{pow}(\mathcal{L}_{form})$ be such that for all $B \subseteq \mathcal{L}_{form}$ and invalid $\phi \in \mathcal{L}_{form}$, $B \dot{-} \phi$ is $\prec^*$-least with $\bigcap(B \perp \phi) \subseteq B \dot{-} \phi \subseteq B$ and $B \dot{-} \phi \not\models \phi$. With the definition of $\prec^\#$ it is easy to verify that:

(11)   (a) $A_5 \perp (\phi_1 \wedge \phi_2) = \{A_1, A_2\}$ and $A_5 \dot{-} (\phi_1 \wedge \phi_2) = A_1$;

     (b) $A_7 \perp (\phi_2 \wedge (\phi_1 \vee \phi_3)) = \{A_2, A_4\}$ and $A_7 \dot{-} (\phi_2 \wedge (\phi_1 \vee \phi_3)) = A_2$;

     (c) $A_7 \perp (\phi_2 \wedge \phi_3) = \{A_4, A_5\}$ and $A_7 \dot{-} (\phi_2 \wedge \phi_3) = A_4$.

From (11)a we infer immediately that $A_1 \prec^* A_2$. Likewise from (11)b we infer that $A_2 \prec^* A_4$, and from (11)c we infer that $A_4 \prec^* A_5$. Hence:

(12) $A_1 \prec^* A_4$ and $A_1 \prec^* A_5$.

By (11)c, $\bigcap(A_7 \perp (\phi_2 \wedge \phi_3)) = A_4 \cap A_5 = A_1$, which implies with (12) and the definition of $\prec^*$ that $A_7 \dot{-} (\phi_2 \wedge \phi_3) = A_1$. This is in contradiction with (11)c.
∎

Solution to Exercise 4.(18):
Let $\phi_1 = (v_0 = v_1)$, $\phi_2 = (v_2 = v_3)$, and $\phi_3 = (v_4 = v_5)$. Set $B_1 = \{\phi_1, \phi_2\}$, $B_1 = \{\phi_2, \phi_3\}$, and $B_3 = \{\phi_1, \phi_3\}$. It is easy to verify that $B_1 \perp (\phi_1 \wedge \phi_2) = \{\{\phi_1\}, \{\phi_2\}\}$, $B_2 \perp (\phi_2 \wedge \phi_3) = \{\{\phi_2\}, \{\phi_3\}\}$, and $B_3 \perp (\phi_1 \wedge \phi_3) = \{\{\phi_1\}, \{\phi_3\}\}$. We infer immediately that there exists unified, maxichoice contraction function $\dot{-}$ such that:

(13) $B_1 \dot{-} (\phi_1 \wedge \phi_2) = \{\phi_1\}$, $B_2 \dot{-} (\phi_2 \wedge \phi_3) = \{\phi_2\}$, and $B_3 \dot{-} (\phi_1 \wedge \phi_3) = \{\phi_3\}$.

Suppose for a contradiction that $\dot{-}$ is rigid. Let strict, total ordering $\prec$ of $\text{pow}(\mathcal{L}_{form})$ be such that for all $B \subseteq \mathcal{L}_{form}$ and invalid $\phi \in \mathcal{L}_{form}$, $B \dot{-} \phi$ is the $\prec$-least member of $B \perp \phi$. Suppose that $\{\phi_1\} \prec \{\phi_2\}$ and $\{\phi_1\} \prec \{\phi_3\}$ (the other cases are parallel). This implies that $B_1 \dot{-} (\phi_1 \wedge \phi_2) = \{\phi_1\}$ and $B_3 \dot{-} (\phi_1 \wedge \phi_3) = \{\phi_1\}$, which contradicts (13). ∎

Solution to Exercise 4.(19):

Let $B \subseteq \mathcal{L}_{form}$ and $\sigma \in SEQ$ be given. By hypothesis:

(14) $(B \dot{-} \neg \bigwedge \sigma) \cup \{range(\sigma)\} = (B \dot{-}' \neg \bigwedge \sigma) \cup \{range(\sigma)\}$.

It is easy to verify that for all $X \in B \perp \neg \bigwedge \sigma$, $B \cap range(\sigma) \subseteq X$. Hence $B \cap range(\sigma) \subseteq \bigcap (B \dot{-} \neg \bigwedge \sigma)$. Hence $B \cap range(\sigma) \subseteq B \dot{-} \neg \bigwedge \sigma$ and $B \cap range(\sigma) \subseteq B \dot{-}' \neg \bigwedge \sigma$. With (14), we conclude that $B \dot{-} \neg \bigwedge \sigma = B \dot{-}' \neg \bigwedge \sigma$. ∎

Solution to Exercise 4.(20):

Condition (a) implies that for every $X \in B \perp \phi$, $X \cap C = \emptyset$. This with Condition (b) implies that $B \perp \phi = \{B - C\}$. The exercise follows immediately.

∎

Solution to Exercise 4.(21):

Trivially, for every $X \in B \perp \phi$, $X \cap Y = \emptyset$. So it suffices to show that $B - Y \not\models \phi$. Suppose otherwise. By compactess there is finite $D \subseteq B - Y$ such that $D \models \phi$. By hypothesis there is $\psi \in D$ such that for all $\chi \in D$, $\psi \models \chi$. Hence $\psi \models \phi$, so $\psi \in Y$, contradiction. ∎

Section 4.2

Solution to Exercise 4.(58):

Let $\overline{0}$ be the constant, s the unary function symbol of **Sym**. For $n \in N$, let $\overline{n}$ be the result of n applications of s to $\overline{0}$. Let $\mathcal{S}$ be the structure with domain N that interprets $\overline{n}$ as n, for all $n \in N$. Following Definition 3.(11), let $\mathcal{I}(\mathcal{S})$ denote the set of all structures isomorphic to $\mathcal{S}$. Set $P = \mathcal{I}(\mathcal{S}) \cup MOD(\overline{0} = \overline{1})$.

It is immediate P is closed under isomorphism, and that the linguistic scientist Λ such that $\Lambda(\sigma) = \{\overline{0} = \overline{1}\}$ for every $\sigma \in SEQ$ solves $\{P\}$. Let $Y \subseteq \mathcal{L}_{form}$ be given. To show that $\{P\}$ is not feasible, it suffices to deduce a contradiction from:

(15) $\emptyset \neq MOD(Y \cup \{\overline{0} \neq \overline{1}\}) \subseteq P$.

Let $Z = Y \cup \{\overline{m} \neq \overline{n} \mid m \neq n\}$. It is easy to deduce from (15) that Z is consistent. If a is an auxiliary constant, then by compactness $Z \cup \{a \neq \overline{n} \mid n \in N\}$ is consistent. Hence $MOD(Z) \not\subseteq \mathcal{I}(\mathcal{S})$. So, since $Z \models \bar{0} \neq \bar{1}$, $MOD(Z) \not\subseteq P$. Since $Y \cup \{\bar{0} \neq \bar{1}\} \subseteq Z$, this implies $MOD(Y \cup \{\bar{0} \neq \bar{1}\}) \not\subseteq P$, contradicting (15). ∎

Solution to Exercise 4.(59):

Let maxichoice revision function $\dotplus$ have the following property. For all $B \subseteq \mathcal{L}_{form}$ and $\sigma \in SEQ$, if $\sigma \neq \emptyset$ and $B \cap \{\bigwedge \sigma \to \varphi \mid \varphi \in \mathcal{L}_{form}\}$ is consistent with range(σ), then $B \cap \{\bigwedge \sigma \to \varphi \mid \varphi \in \mathcal{L}_{form}\} \subseteq B \dotplus \sigma$. Let solvable and feasible problem **P** be given. There is nothing to prove if $\mathbf{P} = \emptyset$, so choose $P_0 \in \mathbf{P}$. By feasibility, there is $Y_1 \subseteq \mathcal{L}_{form}$ such that $\emptyset \neq MOD(Y_1) \subseteq P_0$. Let $Y_0 \subseteq \mathcal{L}_{form}$ be the result of doubling the index of every variable appearing in Y_1. Then also $\emptyset \neq MOD(Y_0) \subseteq P_0$. So we may choose $\mathcal{S}_0 \in P_0$ and full assignment h with $\mathcal{S}_0 \models Y_0[h]$. Let environment e_0 be for $\mathcal{S}_0$ and h. Then we have:

(16) $\mathcal{S}_0 \models Y_0 \cup \text{range}(e_0)[h]$.

Since **P** is solvable and feasible, it is easy to see that there is linguistic scientist Λ that solves **P** and satisfies:

(17)   for all $\sigma \in SEQ$, $\Lambda(\sigma)$ is consistent with range(σ).

Define $X_0 = \{\bigwedge \sigma \to \varphi \mid \emptyset \neq \text{range}(\sigma) \subseteq \text{range}(e_0) \text{ and } \varphi \in Y_0\}$. Define $X_1 = \{\bigwedge \sigma \to \varphi \mid \text{range}(\sigma) \not\subseteq \text{range}(e_0) \text{ and } \varphi \in \Lambda(\sigma)\}$. We take $X = X_0 \cup X_1$. Observe that $\mathcal{S}_0 \models X_0[h]$ because by (16) $\mathcal{S}_0 \models \varphi[h]$ for all $\varphi \in Y_0$. Also $\mathcal{S}_0 \models X_1[h]$ because by (16) again $\mathcal{S}_0 \not\models \bigwedge \sigma[h]$ for all $\sigma \in SEQ$ with range$(\sigma) \not\subseteq$ range(e_0). So X is consistent. It remains to show that $\lambda\sigma . X \dotplus \sigma$ solves **P**. Let environment e for $P_1 \in \mathbf{P}$ be given. There are two cases.

Case 1: range$(e) = $ range(e_0). Let $k > 0$ be given. By (16), $Y_0 \cup \text{range}(e[k])$ is consistent. So the definition of $\dotplus$ and X implies that $X \dotplus e[k] = \{\bigwedge e[k] \to \varphi \mid \varphi \in Y_0\} \cup \text{range}(e[k]) \models Y_0$. By Lemma 3.(7), e is for $\mathcal{S}_0 \in P_0$, so $P_0 = P_1$. Since $\emptyset \neq MOD(Y_0) \subseteq P_0$, it follows that $\lambda\sigma . X \dotplus \sigma$ solves P_1 in e.

Case 2: range$(e) \neq $ range(e_0). Let $k_0 > 0$ be least such that range$(e[k_0]) \not\subseteq$ range(e_0). Then by (17) and the definition of $\dotplus$ and X, for all $k \geq k_0$, $X \dotplus e[k] = \{\bigwedge e[k] \to \varphi \mid \varphi \in \Lambda(e[k])\} \cup \text{range}(e[k])$, hence $X \dotplus e[k] \models \Lambda(e[k])$. Since Λ solves P_1 on e, so does $\lambda\sigma . X \dotplus \sigma$. ∎

Section 4.3

Solution to Exercise 4.(88):

Let revision function $\dotplus$ and environment e for $\mathcal{S} \in MOD(T)$ be given. We distinguish two cases.

Case 1: $S \models \theta$. Then for all $k \in N$, $T \cup \{\varphi\} \not\models \neg \bigwedge e[k]$. This with Lemma 4.(13)c implies that $(T \cup \{\varphi\}) \dotplus e[k] = T \cup \{\varphi\} \cup \text{range}(e[k])$ for all $k \in N$. Hence $(T \cup \{\varphi\}) \dotplus e[k] \models T \cup \{\theta\}$ for all $k \in N$. Hence $\lambda\sigma . (T \cup \{\varphi\}) \dotplus \sigma$ solves $MOD(T \cup \{\theta\})$ in e.

Case 2: $S \models \neg\theta$. Since φ is a $\forall$ sentence there is $k_0 \in N$ such that $\varphi \models \neg \bigwedge e[k_0]$. Let $k \geq k_0$ be given. Since e is an environment for $MOD(T)$, $T \not\models \neg \bigwedge e[k]$. With Exercise 4.(20), we infer that $(T \cup \{\varphi\}) \dotplus e[k] = T \cup \text{range}(e[k])$. Since $\text{range}(e[k]) \models \neg\varphi$, it follows that $(T \cup \{\varphi\}) \dotplus e[k] \models T \cup \{\neg\theta\}$. Hence $\lambda\sigma . (T \cup \{\varphi\}) \dotplus \sigma$ solves $MOD(T \cup \{\neg\theta\})$ in e.

So we have shown that $\lambda\sigma . (T \cup \{\varphi\}) \dotplus \sigma$ solves $(T, \{\theta, \neg\theta\})$.   ∎

Solution to Exercise 4.(89):

Let T_0 be an *r.e.* axiomatization of T. By Theorem 3.(55), let $\forall$ formula $\varphi_0(\bar{x}_0)$ (with free variables $\bar{x}_0$) be such that $T \models \exists\bar{x}_0\varphi_0(\bar{x}_0) \to \theta_0$ and $T \cup \{\varphi_0(\bar{x}_0)\}$ is consistent. Let $\{\varphi_i(\bar{x}_i) \mid i \in N\}$ be a recursive enumeration of all $\forall$ formulas $\varphi(\bar{x})$ such that for some $i \leq n$, $T \models \exists\bar{x}\varphi(\bar{x}) \to \theta_i$. Note that the enumeration starts off with the formula $\varphi_0(\bar{x}_0)$ chosen to be consistent with T. Set $X = T_0 \cup \{\varphi_0(\bar{x}_0) \vee \ldots \vee \varphi_i(\bar{x}_i) \mid i \in N\}$. Since $T_0 \cup \{\varphi_0(\bar{x}_0)\}$ is consistent, so is X. We show that X satisfies the claim of the exercise.

We start with Part (a). Let partial function $f : SEQ \to N$ be defined as follows. Let $\sigma \in SEQ$ be given. If there is no $i \in N$ such that $T \cup \{\varphi_0(\bar{x}_0) \vee \ldots \vee \varphi_i(\bar{x}_i)\} \cup \text{range}(\sigma)$ is consistent, then $f(\sigma)$ is undefined; otherwise, $f(\sigma)$ is the least $i \in N$ such that $T \cup \{\varphi_0(\bar{x}_0) \vee \ldots \vee \varphi_i(\bar{x}_i)\} \cup \text{range}(\sigma)$ is consistent. Since T is decidable and since $\{\varphi_i(\bar{x}_i) \mid i \in N\}$ is a recursive enumeration, f is a partial recursive function. Let $S_0 = T_0$ and for all $i \in N$, let $S_{i+1} = \{\varphi_0(\bar{x}_0) \vee \ldots \vee \varphi_i(\bar{x}_i)\}$. Let $\mathbf{S} = \{S_i \mid i < \omega\}$, and let $\dotminus_\mathbf{S}$ be constructed from $\mathbf{S}$ as in Definition 4.(9). We take $\dotplus$ to be defined from $\dotminus_\mathbf{S}$ via Definition 4.(12). So $\dotplus$ is stringent by Proposition 4.(10). It is easy to verify that for all $\sigma \in SEQ$, the following holds. If $f(\sigma)$ is undefined then $X \dotminus \neg \bigwedge \sigma = T_0$.

If $f(\sigma)$ is defined then $X \dotminus \neg \bigwedge \sigma = T_0 \cup \{\varphi_0(\bar{x}_0) \vee \ldots \vee \varphi_i(\bar{x}_i) \mid i \geq f(\sigma)\}$. Since f is partial recursive and since T_0 is *r.e.*, it follows immediately that $\lambda\sigma . X \dotplus \sigma$ is computable.

Now we prove Part (b). Let revision function $\dotplus$, $j \leq n$, $S \in MOD(T \cup \{\theta_j\})$, full assignment h to S, and environment e for S and h be given. To finish the proof we must show that:

(18)   for cofinitely k, $X \dotplus e[k] \models T \cup \{\theta_j\}$.

By Theorem 3.(55) let i_0 be least with $S \models \varphi_{i_0}(\bar{x}_{i_0})[h]$. So, because the θ_i's are mutually exclusive in T, we have:

(19) $T \models \varphi_{i_0}(\bar{x}_{i_0}) \rightarrow \theta_j$.

By the choice of i_0, $S \models T \cup \{\varphi_0(\bar{x}_0) \vee \ldots \vee \varphi_{i_0}(\bar{x}_{i_0})\}[h]$, and for all $i < i_0$, $S \models \neg \varphi_i(\bar{x}_i)[h]$. Since the φ_i's are $\forall$ formulas, and since h is onto $|S|$, it follows that there is $k_0 \in N$ such that for all $k \geq k_0$:

(20)   (a) for all $i < i_0$, $\varphi_0(\bar{x}_0) \vee \ldots \vee \varphi_i(\bar{x}_i) \models \neg \bigwedge e[k]$, and

     (b) $T_0 \cup \{\varphi_0(\bar{x}_0) \vee \ldots \vee \varphi_i(\bar{x}_i) \mid i \geq i_0\} \not\models \neg \bigwedge e[k]$.

With Exercise 4.(20), this implies that for all $k \geq k_0$, $X \dotplus e[k] = T_0 \cup \{\varphi_0(\bar{x}_0) \vee \ldots \vee \varphi_i(\bar{x}_i) \mid i \geq i_0\} \cup \mathrm{range}(e[k])$. Hence for all $k \geq k_0$, $T_0 \cup \{\varphi_0(\bar{x}_0) \vee \ldots \vee \varphi_{i_0}(\bar{x}_{i_0})\} \cup \mathrm{range}(e[k]) \subseteq X \dotplus e[k]$. With (20)a, we infer that for all $k \geq k_0$, $X \dotplus e[k] \models T \cup \{\varphi_{i_0}(\bar{x}_{i_0})\}$. This with (19) yields (18). ∎

Solution to Exercise 4.(90):
In the proof of Theorem 4.(69), replace $X \dot{-} \neg \bigwedge e[k] = X - Y$ that occurs in the line above 4.(75) by $(X \cup \mathrm{range}(e[k])) \dot{-} \neg \bigwedge e[k] = (X \cup \mathrm{range}(e[k])) - Y$. This yields a proof of the exercise. ∎

Solution to Exercise 4.(91):
Let R be the binary predicate, $\bar{0}$ the constant, s the unary function symbol of **Sym**. For $n \in N$, let $\bar{n}$ be the result of n applications of s to $\bar{0}$. Set $\theta = \exists x \forall y R x y$. Before we define a stringent revision function $\dotplus$ such that θ, $\dotplus$ satisfy the claim of the exercise, we need some notation.

Fix an enumeration $\{\alpha_i \mid i \in N\}$ of all atomic formulas such that:

(21)   for all $i, j, m, n \in N$, if $Var(\alpha_i) \subseteq \{v_m, v_n\}$ and $\alpha_j = \alpha_i[\bar{m}/v_m, \bar{n}/v_n]$, then $j \leq i$.

Say that environment e is *standard* just in case for all $i \in N$, either $e(i) = \alpha_i$ or $e(i) = \neg \alpha_i$, and $\{v_n = \bar{n} \mid n \in N\} \subseteq \mathrm{range}(e)$. Fix an enumeration $\{\sigma_i \mid i \in N\}$ of all members of *SEQ* such that for all $i \in N$:

(22)   (a) σ_i is an initial segment of some standard environment, and

     (b) for all $j \in N$, if $i \leq j$ then $\mathrm{length}(\sigma_i) \leq \mathrm{length}(\sigma_j)$.

Now we define $\dotplus$. For all $i \in N$ let $\{S_{(\omega \times i)+n} \mid n \in N\}$ be an enumeration of all finite $D \subseteq \mathcal{L}_{form}$ such that $D \cup \mathrm{range}(\sigma_i)$ is consistent, $\mathrm{range}(\sigma_i) \cap \mathcal{L}_{sen} \subseteq D$, and $D \cup \mathrm{range}(\sigma_i) \models \theta$. Fix an enumeration $\{\varphi_i \mid i \in N\}$ of $\mathcal{L}_{form}$, and set

$S_{\omega^2+i} = \{\varphi_i\}$ for all $i \in N$. Let $\mathbf{S} = \{S_\alpha \mid \alpha < \omega^2 + \omega\}$, and let $\doteq_s$ be constructed from $\mathbf{S}$ as in Definition 4.(9). We take $\dotplus$ to be defined from $\doteq_s$ via Definition 4.(12). So $\dotplus$ is stringent by Proposition 4.(10). To prove the exercise it suffices to define $\mathcal{T} \subseteq \mathrm{pow}(\mathcal{L}_{sen})$ such that:

(23)   (a)   for all $T \in \mathcal{T}$, θ is equivalent in T to a $\forall$ sentence, and

     (b)   for all $B \subseteq \mathcal{L}_{form}$, there is $T \in \mathcal{T}$ such that $\lambda\sigma \ . \ (T \cup B) \dotplus \sigma$ does not solve $(T, \{\theta, \neg\theta\})$.

So we define $\mathcal{T} \subseteq \mathrm{pow}(\mathcal{L}_{sen})$. Given $n \in N$, define $\mathcal{T}_n$ to be the class of all $T \subseteq \mathcal{L}_{sen}$ such that for some maximally consistent $Z \subseteq \mathcal{L}_{basic} \cap \mathcal{L}_{sen}$, $T = Z \cup \{\theta \leftrightarrow \forall y R\bar{n}y\}$. Define $\mathcal{T}_\infty$ to be the class of all $T \subseteq \mathcal{L}_{sen}$ such that for some maximally consistent $Z \subseteq \mathcal{L}_{basic} \cap \mathcal{L}_{sen}$, $T = Z \cup \{\theta \leftrightarrow \forall y R\bar{n}y \mid n \in N\}$. Finally set $\mathcal{T} = \bigcup_{n \in N} \mathcal{T}_n \cup \mathcal{T}_\infty$. Trivially, (23)a is satisfied. So it remains to prove (23)b.

Let $B \subseteq \mathcal{L}_{form}$ be given. Suppose that for all $n \in N$ and $T \in \mathcal{T}_n$, $\lambda\sigma \ . \ (T \cup B) \dotplus \sigma$ solves $(T, \{\theta, \neg\theta\})$ (otherwise, there is nothing to prove). We shall exhibit a standard environment e with the following property:

(24)   (a)   there is $T \in \mathcal{T}_\infty$ such that e is for $MOD(T \cup \{\neg\theta\})$;

     (b)   there is increasing $f : N \to N$ and enumeration $\{D_n \mid n \in N\}$ of finite subsets of $T \cup B$ such that for all $n \in N$, $D_n \cup \mathrm{range}(e[f(n)])$ is consistent, $\mathrm{range}(e[f(n)]) \cap \mathcal{L}_{sen} \subseteq D_n$, and $D_n \cup \mathrm{range}(e[f(n)]) \models \theta$.

The following argument shows that the existence of e, T, f, and $\{D_n \mid n \in N\}$ satisfying (24) implies that $\lambda\sigma \ . \ (T \cup B) \dotplus \sigma$ does not solve $MOD(T \cup \{\neg\theta\})$ in e, thereby completing the proof. Let $n \in N$ be given. Let $i \in N$ be such that $\sigma_i = e[f(n)]$. From (21) and (22), we infer that for all $j \leq i$ and for all $Y \subseteq \mathcal{L}_{form}$,

(25)   if $\mathrm{range}(\sigma_j) \cap \mathcal{L}_{sen} \subseteq Y$ and $Y \cup \mathrm{range}(e[f(n)])$ is consistent, then $\sigma_j \subseteq e[f(n)]$.

By (24)b there is least $j \leq i$ such that for some finite $D \subseteq T \cup B$, $D \cup \mathrm{range}(e[f(n)])$ is consistent, $D \cup \mathrm{range}(\sigma_j)$ is consistent, $\mathrm{range}(\sigma_j) \cap \mathcal{L}_{sen} \subseteq D$, and $D \cup \mathrm{range}(\sigma_j) \models \theta$. By (25), $\sigma_j \subseteq e[f(n)]$. The choice of $\dotplus$ then implies that for all $n \in N$, $(T \cup B) \dotplus e[f(n)] \models \theta$. With (24)a we conclude that $\lambda\sigma \ . \ (T \cup B) \dotplus \sigma$ does not solve $MOD(T \cup \{\neg\theta\})$ in e, as required.

It remains to exhibit the promised standard environment e. We will build by induction on $n \in N$ a sequence $\{\tau_n \mid \in N\}$ of members of SEQ together with a sequence $\{E_n \mid n \in N\}$ of subsets of $\mathcal{L}_{basic} \cap \mathcal{L}_{sen}$ such that for all $n \in N$:

(26) (a) τ_n is an initial segment of a standard environment;

    (b) for all $m < n$, $\tau_m \subset \tau_n$;

    (c) for all $m < n$, $\text{range}(\tau_n) \models \exists y \neg R\overline{m}\, y$;

    (d) for all $m < n$, $E_m \subseteq \text{range}(\tau_n)$;

    (e) there is $q \in N$ and finite $D_n \subseteq B \cup \{\theta \leftrightarrow \forall y R\overline{q}\, y\}$ such that $D_n \cup E_n \cup \text{range}(\tau_n)$ is consistent, $\text{range}(\tau_n) \cap \mathcal{L}_{sen} \subseteq D_n \cup E_n$, and $D_n \cup E_n \cup \text{range}(\tau_n) \models \theta$.

Then we will set $e = \bigcup_{n \in N} \tau_n$. It follows from (26)a,b that e is a standard environment. It follows from (26)c that (24)a is satisfied. It follows from (26)d,e that (24)b is satisfied. Let $n \in N$ be given. Suppose that τ_p and E_p have been defined for all $p < n$ and satisfy (26) for $n = p$. We define τ_n and E_n that satisfy (26). It is easy to see that some standard environment d exists such that:

(27) (a) for all $m < n$, d extends τ_m;

    (b) for all $m < n$, $\text{range}(d) \models \exists y \neg R\overline{m}\, y$;

    (c) for all $m < n$, $E_m \subseteq \text{range}(d)$;

    (d) for some $q \in N$, $\text{range}(d) \not\models \exists y \neg R\overline{q}\, y$.

By (27)a–c we may choose $k_0 > 0$ such that for all $m < n$:

(28) (a) $k_0 > \text{length}(\tau_m)$;

    (b) $\text{range}(d[k_0]) \models \exists y \neg R\overline{m}\, y$;

    (c) $E_m \subseteq \text{range}(d[k_0])$.

Set $T = (\text{range}(d) \cap \mathcal{L}_{sen}) \cup \{\theta \leftrightarrow \forall y R\overline{q}\, y\}$. Since (i) $T \in \mathcal{T}_q$, (ii) the (isomorphic) structures for which d is an environment are models of θ by (27)d, and (iii) $\lambda\sigma . (T \cup B) \dotplus \sigma$ solves $MOD(T_q)$ by hypothesis, $(T \cup B) \dotplus d[k] \models \theta$ for cofinitely many k. Choose $k_1 \geq k_0$ such that $(T \cup B) \dotplus d[k_1] \models \theta$. Choose finite $D \subseteq (T \cup B)$ such that:

(29) $D \cup \text{range}(d[k_1])$ is consistent, $\text{range}(d[k_1]) \cap \mathcal{L}_{sen} \subseteq D$, and $D \cup \text{range}(d[k_1]) \models \theta$.

Set $\tau_n = d[k_1]$ and $E_n = D \cap \mathcal{L}_{basic} \cap T$. It follows immediately from (28) and (29) that (26) is satisfied.   ∎

Section 4.4

Solution to Exercise 4.(97):

By Theorem 3.(55), let $\forall$ formula $\varphi_0(\bar{x}_0)$ (with free variables $\bar{x}_0$) be such that $T \models \exists \bar{x}_0 \varphi_0(\bar{x}_0) \rightarrow \theta_0$ and $T \cup \{\varphi_0(\bar{x}_0)\}$ is consistent. Let $\{\varphi_i(\bar{x}_i) \mid i \in N\}$ be an enumeration of all $\forall$ formulas $\varphi(\bar{x})$ such that for some $j \leq n$, $T \models \exists \bar{x} \varphi(\bar{x}) \rightarrow \theta_j$. Note that the enumeration starts off with the formula $\varphi_0(\bar{x}_0)$ chosen to be consistent with T. Let $t \in \mathcal{L}_{sen}$ axiomatize T. For all $i \in N$, denote by ψ_i the formula $t \wedge (\varphi_0(\bar{x}_0) \vee \ldots \vee \varphi_i(\bar{x}_i))$. Set $X = \{\psi_i \mid i \in N\}$. Since $T \cup \{\varphi_0(\bar{x}_0)\}$ is consistent, so is X. We show that X satisfies the claim of the exercise.

It is easy to modify the proof of Exercise 4.(89) to show that for every revision function $\dot{+}$, $\lambda\sigma . X \dot{+} \sigma$ solves $(T, \{\theta_0, \ldots, \theta_n\})$. Let $\sigma \in SEQ$ be for $(T, \{\theta_0, \ldots, \theta_n\})$. Let $i_0 \in N$ be least such that:

(30) $\psi_{i_0} \not\models \neg \bigwedge \sigma.$

By the definition of the φ_i's, there is $j \leq n$ such that:

(31) $t \wedge \varphi_{i_0} \models T \cup \{\theta_j\}.$

By (30), let structure $\mathcal{S}$ and full assignment h to $\mathcal{S}$ be such that $\mathcal{S} \models \bigwedge \sigma[h]$ and:

(32) $\mathcal{S} \models \psi_{i_0}[h].$

Let environment e for $\mathcal{S}$ and h extend σ, and let $k \geq \text{length}(\sigma)$ be given. Exercise 4.(20), the definition of i_0, the fact that e is an environment for $\mathcal{S}$ and h, and (32) then imply that $X \dot{-} \neg \bigwedge e[k] = \{\psi_i \mid i \geq i_0\}$. Hence $\psi_{i_0} \in X \dot{-} \neg \bigwedge e[k]$. With the definition of i_0, this implies that $X \dot{+} e[k] \models t \wedge \varphi_{i_0}$. With (31), we conclude that $X \dot{+} e[k] \models T \cup \{\theta_j\}$. So we have shown that for all $k \geq \text{length}(\sigma)$, $\emptyset \neq MOD(X \dot{+} e[k]) \subseteq MOD(T \cup \{\theta_j\})$. We conclude with Exercise 3.(87).   ∎

Section 4.5

Solution to Exercise 4.(112):

We start with Part (a). By Theorem 3.(55), for all $j \leq n$, let $\forall$ formula $\varphi_j(\bar{x}_j)$ (with free variables $\bar{x}_j$) be such that $T \models \exists \bar{x}_j \varphi_j(\bar{x}_j) \rightarrow \theta_j$ and $T \cup \{\varphi_j(\bar{x}_j)\}$ is consistent. Let $\{\varphi_i(\bar{x}_i) \mid i \in N\}$ be an enumeration of all $\forall$ formulas $\varphi(\bar{x})$ such that for some $j \leq n$, $T \models \exists \bar{x} \varphi(\bar{x}) \rightarrow \theta_j$. Note that the enumeration starts off with the formulas $\varphi_0(\bar{x}_0) \ldots \varphi_n(\bar{x}_n)$ chosen above. Set $Y = \{\varphi_0(\bar{x}_0) \vee \ldots \vee \varphi_i(\bar{x}_i) \mid i \geq n\}$. Set $X = T \cup Y$. Since $T \cup \{\varphi_0(\bar{x}_0)\}$ is consistent, so is X. We show that X satisfies the claim of Part (a) of the exercise. Let foundation-like closure operator cl be given. Because the θ_j's partition $MOD(T)$, it follows from the definition of $\varphi_0(\bar{x}_0) \ldots \varphi_n(\bar{x}_n)$ that $T \models \exists \bar{x}_0 \ldots \exists \bar{x}_n (\varphi_0(\bar{x}_0) \vee \ldots \vee \varphi_n(\bar{x}_n))$. Using this fact and the definition of X, it is easy to verify the following.

(33) $\mathrm{cl}(X) \subseteq \mathrm{Cn}(T) \cup \{\psi \in \mathcal{L}_{form} \mid (\exists \psi_1 \in \mathrm{Cn}(T))(\exists \psi_2 \in Y)(\models \psi \leftrightarrow (\psi_1 \wedge \psi_2))\}$.

Let revision function $\dotplus$, $j \leq n$, $\mathcal{S} \in MOD(T \cup \{\theta_j\})$, full assignment h to $\mathcal{S}$, and environment e for $\mathcal{S}$ and h be given. We have to show that:

(34)   for cofinitely k, $\mathrm{cl}(X) \dotplus e[k] \models T \cup \{\theta_j\}$.

By Theorem 3.(55) let i_0 be least with $\mathcal{S} \models \varphi_{i_0}(\bar{x}_{i_0})[h]$. Let $i_1 = \max(i_0, n)$, and let $I = \{i \leq i_1 \mid \varphi_{i_0}(\bar{x}_{i_0}) \rightarrow \theta_j\}$. I is nonempty since $i_0 \in I$. By the definition of i_1 and I, $\mathcal{S} \models T \cup \{\varphi_0(\bar{x}_0) \vee \ldots \vee \varphi_{i_1}(\bar{x}_{i_1})\}[h]$, and for all $i \in \{0 \ldots i_1\} - I$, $\mathcal{S} \models \neg \varphi_i(\bar{x}_i)[h]$. Since the φ_i's are $\forall$ formulas, and since h is onto —$\mathcal{S}$—, it follows that there is $k_0 \in N$ such that for all $k \geq k_0$:

(35)   (a)   for all $i \in \{0 \ldots i_1\} - I$, $\varphi_i(\bar{x}_i) \models \neg \bigwedge e[k]$, and

(b) $T \cup \{\varphi_0(\bar{x}_0) \vee \ldots \vee \varphi_i(\bar{x}_i) \mid i \geq i_1\} \not\models \neg \bigwedge e[k]$.

Define Z to be the set of all $\psi \in \mathrm{cl}(X)$ such that there exists $\chi \in \mathrm{Cn}(T)$ and $i \in \{n \ldots i_0 - 1\}$ with $\models \psi \leftrightarrow (\chi \wedge (\varphi_0 \vee \ldots \vee \varphi_i))$. From the definition of i_0, (33), (35), and Exercise 4.(20), we infer that for all $k \geq k_0$, $\mathrm{cl}(X) \dotplus e[k] = (\mathrm{cl}(X) - Z) \cup \mathrm{range}(e[k])$. Hence for all $k \geq k_0$, $T \cup \{\varphi_0(\bar{x}_0) \vee \ldots \vee \varphi_{i_1}(\bar{x}_{i_1})\} \cup \mathrm{range}(e[k]) \subseteq \mathrm{cl}(X) \dotplus e[k]$. With (35)a, we infer that for all $k \geq k_0$, $\mathrm{cl}(X) \dotplus e[k] \models T \cup \{\varphi_i(\bar{x}_i) \mid i \in I\}$. This with the definition of (nonempty) I yields (34).

Now we consider Part (b). Our proof relies on the following fact about contraction, which is proved via trivial modifications to the proof of Exercise 4.(21).

(36)   FACT: Let $B' \subseteq B \subseteq \mathcal{L}_{form}$, $\phi \in \mathcal{L}_{form}$ and contraction function $\dotminus$ be given. Suppose that:

(a) $B' \not\models \phi$,

(b)   for all $\psi \in B - B'$, $\psi \models B$, and

(c) for all $\psi, \chi \in B - B'$, either $\psi \models \chi$, or $\chi \models \psi$.

Then $B \dot{-} \phi = B - Y$, where $Y = \{\psi \in B - B' \mid \psi \models \phi\}$.

First we define X, using some notation. By Theorem 3.(55), we can fix strictly increasing $f : N \to N$ with $f(0) = 0$ and enumeration $\{\varphi_n \mid n \in N\}$ of $\forall$ formulas such that the following holds:

(37) (a) for all $m \in N$, the existential closure of $\varphi_{f(m)} \vee \ldots \vee \varphi_{f(m+1)-1}$ is a logical consequence of T;

     (b) for every finite disjunction φ of $\forall$ formulas, if the existential closure of φ is a logical consequence of T, then there is $m \in N$ such that $\varphi = \varphi_{f(m)} \vee \ldots \vee \varphi_{f(m+1)-1}$.

For all $m \in N$, denote by ψ_m the formula $\varphi_{f(m)} \vee \ldots \vee \varphi_{f(m+1)-1}$. Directly from (37)a:

(38) for all $m \in N$, the existential closure of ψ_m is a logical consequence of T.

Recall from the proof of Proposition 4.(76) the notion of "formula associated with some nonempty, increasing sequence of natural numbers," which was defined from a fixed enumeration of all $\forall$ formulas. Here we use the same notion, but defined from the enumeration $\{\psi_m \mid n \in N\}$. For example, the formula associated with $(2, 5, 7)$ is:

$$(\psi_0 \vee \psi_1 \vee \psi_2) \wedge (\psi_0 \vee \psi_1 \vee \psi_3 \vee \psi_4 \vee \psi_5) \wedge (\psi_0 \vee \psi_1 \vee \psi_3 \vee \psi_4 \vee \psi_6 \vee \psi_7).$$

Let $t \in \mathcal{L}_{sen}$ axiomatize T. Then we define X to be the set of all consistent formulas of form $t \wedge \chi$, where χ is any formula associated with some nonempty, increasing sequence of natural numbers. We will show that X satisfies the claim of Part (b) of the exercise. First, we note some useful properties of X. The proof of Fact 4.(77) can be immediately adapted to show the following:

(39) FACT: For every $\delta, \delta' \in X$, $\delta \models \delta'$ or $\delta' \models \delta$.

Since every member of X is consistent, it follows from Fact (39) that X is consistent. Let foundation-like closure operator cl be given. Using Fact (39), (38), and the definition of X, it is easy to verify the following.

(40) $\mathrm{cl}(X) \subseteq \mathrm{Cn}(T) \cup \{\phi \in \mathcal{L}_{form} \mid (\exists \phi' \in X)(\models \phi \leftrightarrow \phi')\}$.

Now let solvable problem of form $(T, \{\theta_0, \ldots, \theta_n\})$, revision function $\dotplus$, and $i \leq n$ be given. We have to show that $\lambda \sigma . \text{cl}(X) \dotplus \sigma$ solves $MOD(T \cup \{\theta_i\})$. Let $\mathcal{S} \in MOD(T \cup \{\theta_i\})$, full assignment h to $\mathcal{S}$, and environment e for $\mathcal{S}$ and h be given. By Theorem 3.(55) and (37)b, let $m_0 \in N$ be such that:

(41) $f(m_0 + 1) - f(m_0) = n + 1$ and for all $j \leq n$, $T \models \varphi_{f(m_0)+j} \to \theta_j$.

From Fact (36), Fact (39), and (40), we deduce that:

(42)   for all $k \in N$, $\text{cl}(X) \dotplus e[k]$ includes the set of all formulas in $\text{cl}(X)$ that are consistent with $\bigwedge e[k]$.

Denote by γ the increasing sequence of integers that ends with m_0 and such that for all $m \leq m_0$, m occurs in γ if and only if $\mathcal{S} \models \psi_m[h]$. Denote by χ the formula associated with γ. It is immediate that $t \wedge \chi \in X$, and that for all $k \in N$, $t \wedge \chi$ is consistent with $\bigwedge e[k]$. We can thus deduce from (42) that:

(43)   for all $k \in N$, $t \wedge \chi$ belongs to $X \dotplus e[k]$.

Since h is onto —$\mathcal{S}$—, there is $k_0 \in N$ such that for all $k \geq k_0$ and for all $m \leq m_0$, if $m \notin \text{range}(\gamma)$ then $\bigwedge e[k] \models \neg \psi_m$. From the definition of χ this can be seen to imply that for all $k \geq k_0$ and for all $m \in \text{range}(\gamma)$, $\text{range}(e[k]) \cup \{\chi\} \models \psi_m$. In particular:

(44)   for all $k \geq k_0$, $\text{range}(e[k]) \cup \{\chi\} \models \psi_{m_0}$.

Since h is onto —$\mathcal{S}$—, there is $k_1 \geq k_0$ such that for all $k \geq k_1$ and $j \leq n$:

(45)   if $\mathcal{S} \models \neg \theta_j$ then $\bigwedge e[k] \models \neg \varphi_{f(m_0)+j}$.

From (41), (43), (44), and (45), we deduce that for all $k \geq k_1$, $\text{cl}(X) \dotplus e[k] \models T \cup \{\theta_i\}$. Hence $\lambda \sigma . \text{cl}(X) \dotplus \sigma$ solves $MOD(T \cup \{\theta_i\})$ in e.   ■

Solution to Exercise 4.(113):

Let R be the binary predicate of **Sym**. Let $T = \{\exists x \forall y Rxy \leftrightarrow \neg \exists y \forall x Rxy\}$, and $\theta = \exists x \forall y Rxy$. Fix an enumeration $\{\theta_i \mid i \in N\}$ of all sentences that imply θ, and set $S_i = \{\theta_i\}$ for all $i \in N$. Fix an enumeration $\{\chi_i \mid i \in N\}$ of $\mathcal{L}_{form}$, and set $S_{\omega+i} = \{\chi_i\}$ for all $i \in N$. Set $\mathbf{S} = \{S_\alpha \mid \alpha < \omega \times 2\}$, let $\dotdiv_S$ be constructed from $\mathbf{S}$ as in Definition 4.(9), and take $\pm$ to be defined from $\dotdiv_S$ via the definition above with $\dotdiv = \dotdiv_S$. So Proposition 4.(10) implies that $\pm$ is stringent. We show that $(T, \{\theta, \neg \theta\})$ and $\pm$ satisfy the claim of the exercise.

$(T, \{\theta, \neg \theta\})$ is solvable by Proposition 3.(54). Let $B \subseteq \mathcal{L}_{form}$ be given. If for all $\sigma \in SEQ$ and $B' \subseteq B$, either $B' \cup \text{range}(\sigma)$ is inconsistent or $B' \cup$

range$(\sigma) \not\models T \cup \{\theta\}$, then it is easy to verify that $\lambda\sigma \,.\, B \pm \sigma$ does not solve $MOD(T \cup \{\theta\})$, hence does not solve $(T, \{\theta, \neg\theta\})$. So suppose otherwise. By compactness choose finite $D \subseteq B$ and $\sigma_0 \in SEQ$ such that $D \cup$ range(σ_0) is consistent and $D \cup$ range$(\sigma_0) \models T \cup \{\theta\}$. Let ψ be the existential closure of $\bigwedge D \wedge \bigwedge \sigma_0$. With our hypothesis on cl, we deduce that:

(46) (a) $\psi \models T \cup \{\theta\}$, and

(b) $\psi \in$ cl$(B \cup$ range$(\sigma_0))$.

Let structure S be such that $S \models \psi$ and $\bigwedge \sigma_0$ is satisfiable in S. The following is an immediate consequence of (46)b and the fact that $S \models \psi$.

(47) For all $\tau \in SEQ$, if τ extends σ_0 and τ is satisfiable in S, then $\psi \in$ cl$(B \cup$ range$(\tau))$.

Let structure U be such that $|U| = |S|$ and for all $x, y \in |U|$, $(x, y) \in R^U$ if and only if $(y, x) \in R^S$. Since $S \models \psi$, it follows from (46)a that:

(48) (a) $U \models T \cup \{\neg\theta\}$.

(b) For all $\sigma \in SEQ$, U satisfies $\bigwedge \sigma$ if and only if S satisfies $\bigwedge \sigma$.

Let e be an environment for U that extends σ_0. By (48)b, for all $k \in N$, $\bigwedge e[k]$ is satisfiable in S. This, (46)a, (47), and the definition of $\dot{-}_S$ then imply that for all $k \geq$ length(σ_0), there is $\varphi \in B \pm e[k]$ such that $\varphi \models \theta$. With (48)a, we deduce that $\lambda\sigma \,.\, B \pm \sigma$ does not solve $MOD(T \cup \{\neg\theta\})$, hence does not solve $(T, \{\theta, \neg\theta\})$. ∎

Section 4.6

Solution to Exercise 4.(125):
Up to obvious changes in notation, the proof of Proposition 4.(116) (respectively, Theorem 4.(117)) is a proof of Part (a) (respectively, Part (b)) of the exercise. ∎

References

Alchourrón & Makinson, 1985 C. E. Alchourrón & D. Makinson. On the logic of theory change: Safe contraction. *Studia Logica*, 44:405–422, 1985.

Alchourrón *et al.*, 1985 C. E. Alchourrón, P. Gärdenfors, & D. Makinson. On the logic of theory change: Partial meet contraction and revision functions. *Journal of Symbolic Logic*, 50:510–530, 1985.

Angluin & Smith, 1983 D. Angluin & C. Smith. A survey of inductive inference: Theory and methods. *Computing Surveys*, 15:237–289, 1983.

Angluin, 1980 D. Angluin. Inductive inference of formal languages from positive data. *Information and Control*, 45:117–135, 1980.

Angluin, 1988 D. Angluin. Identifying languages from stochastic examples. Technical Report 614, Yale University, 1988.

Anthony & Biggs, 1992 M. Anthony & N. Biggs. *Computational Learning Theory*. Cambridge University Press, New York, 1992.

Baird, 1992 D. Baird. *Inductive Logic: Probability and Statistics*. Prentice Hall, Englewood Cliffs, NJ, 1992.

Baliga *et al.*, 1996 G. R. Baliga, J. Case, & S. Jain. Synthesizing Enumeration Techniques for Language Learning, 1996.

Barwise & Feferman, 1985 J. Barwise & S. Feferman, editors. *Model theoretic logics*. Springer-Verlag, New York, NY, 1985.

Barwise, 1975 J. Barwise. *Admissible Sets and Structures*. Springer-Verlag, Berlin, 1975.

Barzdin, 1974 J. M. Barzdin. Two theorems on the limiting synthesis of functions. *Latv. Gos. Univ. Uce. Zap.*, 210:82–88, 1974.

Billingsley, 1986 P. Billingsley. *Probability and Measure (Second Edition)*. John Wiley and Sons, New York, 1986.

Blum & Blum, 1975 L. Blum & M. Blum. Toward a mathematical theory of inductive inference. *Information and Control*, 28:125–155, 1975.

Boolos & Jeffrey, 1989 G. Boolos & R. Jeffrey. *Computability and Logic (Third Edition)*. Cambridge University Press, 1989.

Boutilier, 1996 C. Boutilier. Iterated revision and minimal change of conditional beliefs. *Journal of Philosophical Logic*, 25:263–305, 1996.

Carnap, 1950 R. Carnap. *Logical Foundations of Probability (2nd Edition)*. University of Chicago Press, Chicago, IL, 1950.

Case & Ngo-Manguelle, 1979 J. Case & S. Ngo-Manguelle. Refinements of inductive inference by popperian machines. Technical report, SUNY Buffalo, Dept. of Computer Science, 1979.

Case & Smith, 1983 J. Case & C. Smith. Comparison of identification criteria for machine inductive inference. *Theoret. Comput. Sci.*, 25:193–220, 1983.

Case *et al.*, 1993 J. Case, S. Jain, & A. Sharma. Complexity issues for vacillatory function identification. *Inform. Comput.*, 1993. To appear.

Case *et al.*, 1994 J. Case, S. Jain, & S. Ngo Manguelle. Refinements of inductive inference by popperian and reliable machines. *Kybernetika*, 30–1:23–52, 1994.

Chang & Keisler, 1977 C. C. Chang & H. J. Keisler. *Model Theory (2nd Edition)*. North Holland, Amsterdam, 1977.

Cohen, 1992 L. J. Cohen. *Belief & Acceptance*. Oxford University Press, Oxford, UK, 1992.

Craig, 1953 W. Craig. On axiomatizability within a system. *Journal of Symbolic Logic*, 18:30–32, 1953.

Daley & Smith, 1986 R. Daley & C. Smith. On the complexity of inductive inference. *Information and Computation*, 69:12–40, 1986.

Daley, 1986 R. Daley. Inductive inference hierarchies: Probabilistic vs. pluralistic. In *Lecture Notes in Computer Science 215*, pages 73–82. Springer-Verlag, 1986.

Darwiche & Pearl, 1994 A. Darwiche & J. Pearl. On the Logic of Iterated Belief Revision, 1994.

Davis & Weyuker, 1983 M. D. Davis & E. J. Weyuker. *Computability, Complexity, and Languages*. Academic Press, New York, 1983.

Dorling, 1979 J. Dorling. Bayesian personalism, the methodology of scientific research programmes, and Duhem's problem. *Studies in History and Philosophy of Science*, 10:177–87, 1979.

Earman & Glymour, 1980 J. Earman & C. Glymour. Relativity and eclipses: The British eclipse expeditions of 1919 and their predecessors. *Historical Studies in the Physical Sciences*, 11:49–85, 1980.

Earman, 1992 J. Earman. *Bayes or Bust?* MIT Press, Cambridge MA, 1992.

Ebbinghaus et al., 1994 H. D. Ebbinghaus, J. Flum, & W. Thomas. *Mathematical Logic, Second Edition*. Springer-Verlag, Berlin, 1994.

Enderton, 1972 H. Enderton. *A Mathematical Introduction to Logic*. Academic Press, New York, 1972.

Enderton, 1977 H. Enderton. *Elements of Set Theory*. Academic Press, New York, NY, 1977.

Feyerabend, 1975 P. Feyerabend. *Against Method*. Verso Press, London, 1975.

Fraissé, 1972 R. Fraissé. *Cours de logique mathématique, Tome 2*. Gauthier-Villars, Paris, 1972.

Freivalds et al., 1995 R. Freivalds, E. Kinber, & C. H. Smith. On the Intrinsic Complexity of Learning. *Information and Computation*, 123(1):64–71, 1995.

Fuhrmann, 1991 A. Fuhrmann. Theory contraction through base contraction. *Journal of Philosophical Logic*, 20:175–203, 1991.

Fulk & Jain, 1994 M. Fulk & S. Jain. Approximate inference and scientific method. *Information and Computation*, 114–2:179–191, 1994.

Fulk et al., 1994 M. Fulk, S. Jain, & D. Osherson. Open Problems in *systems that learn*. *Journal of Computer and System Sciences*, 49(3):589 – 604, 1994.

Fulk, 1988 M. Fulk. Saving the phenomenon: Requirements that inductive machines not contradict known data. *Inform. Comput.*, 79:193–209, 1988.

Fulk, 1990 M. Fulk. Prudence and other conditions on formal language learning. *Information and Computation*, 85(1):1–11, 1990.

Gaifman & Snir, 1982 H. Gaifman & M. Snir. Probabilities over rich languages. *Journal of Symbolic Logic*, 47:495–548, 1982.

Gaifman et al., 1990 H. Gaifman, D. Osherson, & S. Weinstein. A reason for theoretical terms. *Erkenntnis*, 32:149–159, 1990.

Gärdenfors, 1988 P. Gärdenfors. *Knowledge in Flux: Modeling the Dynamics of Epistemic States*. MIT Press, Cambridge MA, 1988.

Gasarch & Pleszkoch, 1989 W. Gasarch & M. Pleszkoch. Learning via queries to an oracle. In R. Rivest, D. Haussler, & M. K. Warmuth, editors, *Proceedings of the Second Annual Workshop on Computational Learning Theory*, pages 214–229. Morgan Kaufmann Publishers, Inc., 1989.

Gasarch et al., 1992 W. Gasarch, M. Pleszkoch, & R. Solovay. Learning via Qeries in [+, <]. *Journal of Symbolic Logic*, 57(1):53–81, 1992.

Gemes, 1994 K. Gemes. A New Theory of Content I: Basic Content. *Journal of Philosophical Logic*, 23(6):595–620, 1994.

Gibson & Wexler, 1994 T. Gibson & K. Wexler. Triggers. *Linguistic Inquiry*, 25(4), 1994.

Gleitman & Liberman, 1995 L. Gleitman & M. Liberman. *Invitation to Cognitive Science: Language (2nd Edition)*. MIT Press, Cambridge, MA, 1995.

Glymour, 1980 C. Glymour. *Theory and Evidence*. Princeton University Press, Princeton, NJ, 1980.

Glymour, 1985 C. Glymour. Inductive inference in the limit. *Erkenntnis*, 22:23–31, 1985.

Gold, 1967 E. M. Gold. Language identification in the limit. *Information and Control*, 10:447–474, 1967.

Goldman, 1986 A. Goldman. *Epistemology and Cognition*. Harvard University Press, Cambridge MA, 1986.

Gustason, 1994 W. Gustason. *Reasoning from Evidence: Inductive Logic*. Macmillan, New York City, 1994.

Hansson, 1992 S. O. Hansson. A dyadic representation of belief. In P. Gärdenfors, editor, *Belief Revision*, pages 89–121. Cambridge University Press, New York, 1992.

Hansson, 1993a S. O. Hansson. Changes of disjunctively closed bases. *Journal of Logic, Language, and Information*, 2:225–284, 1993.

Hansson, 1993b S. O. Hansson. Reversing the Levi Identity. *Journal of Philosophical Logic*, 22(6):637–669, 1993.

Hansson, 1994 S. O. Hansson. Kernel contraction. *Journal of Symbolic Logic*, 59(3):845–859, 1994.

Hinman, 1978 P. G. Hinman. *Recursion-Theoretic Hierarchies*. Springer-Verlag, Berlin, 1978.

Hodges, 1993 W. Hodges. *Model Theory*. Cambridge University Press, Cambridge, England, 1993.

Horwich, 1982 P. Horwich. *Probability and Evidence*. Cambridge University Press, Cambridge, 1982.

Howson & Urbach, 1993 C. Howson & P. Urbach. *Scientific Reasoning: The Bayesian Approach (2nd Edition)*. Open Court, La Salle, Illinois, 1993.

Jain & Sharma, 1990a S. Jain & A. Sharma. Finite learning by a *team*. In *Proc. 3rd Annu. Workshop on Comput. Learning Theory*, pages 163–177, San Mateo, CA, 1990. Morgan Kaufmann.

Jain & Sharma, 1990b S. Jain & A. Sharma. Language learning by a team. In *Proc. 17th International Colloquium on Automata, Languages, and Programming*, pages 153–166, Berlin, 1990. Springer-Verlag.

Jain & Sharma, 1993a S. Jain & A. Sharma. Learning with the knowledge of an upper bound on program size. *Inform. Comput.*, 102–1:118–166, 1993. Presented at the Workshop on Computational Learning Theory and Natural Learning Systems.

Jain & Sharma, 1993b S. Jain & A. Sharma. Prudence in vacillatory language identification. *Math. Syst. Theory*, 1993.

Jain & Sharma, 1994 S. Jain & A. Sharma. On monotonic strategies for learning r.e. languages. In *Algorithmic Learning Theory, Proceedings of the 4th International Workshop on Analogical and Inductive Inference, AII'94 and 5th International Workshop on Algorithmic Learning Theory, ALT'94, Reinhardsbrunn Castle, Germany*, pages 349–364. Springer Verlag, October 1994. Lecture Notes in AI 872.

Jain & Sharma, 1995 S. Jain & A. Sharma. The Structure of Instrinsic Complexity of Learning. In Paul Vitányi, editor, *Computational Learning Theory: Second European Conference, EuroCOLT '95 (Barcelona, Spain)*. Springer: Lecture Notes in Artificial Intelligence 904, Berlin, 1995.

Jain et al., forthcoming S. Jain, D. Osherson, J. Royer, & A. Sharma. *Systems that Learn, 2nd Edition*, forthcoming.

Juhl, 1993 C. Juhl. Bayesianism and reliable scientific inquiry. *Philosophy of Science*, 60:302–319, 1993.

Kanazawa, 1994 M. Kanazawa. *Learnable classes of categorial grammars*. PhD thesis, Department of Linguistics, Stanford University, 1994.

Kearns & Vazirani, 1994 M. J. Kearns & U. V. Vazirani. *An Introduction to Computational Learning Theory*. MIT Press, Cambridge, MA, 1994.

Keisler, 1977 H. J. Keisler. Fundamentals of model theory. In Jon Barwise, editor, *Handbook of Mathematical Logic*, pages 47–104. North Holland, Amsterdam, 1977.

Kelly & Glymour, 1992 K. Kelly & C. Glymour. Inductive inference and theory-laden data. *Journal of Philosophical Logic*, 21(4), 1992.

Kelly *et al.*, 1995 K. Kelly, O. Schulte, & V. Hendricks. Reliable belief revision. *Proceedings of the X International Joint Congress for Logic, Methodology and the Philosophy of Science*, 1995.

Kelly, 1996 K. T. Kelly. *The Logic of Reliable Inquiry*. Oxford University Press, New York, NY, 1996.

Kinber & Stephan, 1995 E. Kinber & F. Stephan. Language learning from texts: mind changes, limited memory and monotonicity. *Information and Computation*, 123:224–241, 1995.

Kornblith, 1985 H. Kornblith, editor. *Naturalizing Epistemology*. MIT Press, Cambridge MA, 1985.

Kugel, 1977 P. Kugel. Induction, pure and simple. *Information and Control*, 33:276–336, 1977.

Kuhn, 1957 T. S. Kuhn. *The Copernican Revolution: planetary astronomy in the development of Western thought*. Harvard University Press, Cambridge MA, 1957.

Kurtz & Royer, 1988 S. A. Kurtz & J. S. Royer. Prudence in language learning. In D. Haussler & L. Pitt, editors, *Proceedings of the Workshop on Computational Learning Theory*, pages 143–156. Morgan Kaufmann Publishers, Inc., 1988.

Lange & Zeugmann, 1993 S. Lange & T. Zeugmann. Language learning with a bounded number of mind changes. In *Proc. 10th Annual Symposium on Theoretical Aspects of Computer Science*, pages 682–691. Springer–Verlag, 1993. Lecture Notes in Computer Science 665.

Langley *et al.*, 1987 P. Langley, H. A. Simon, G. L. Bradshaw, & Z. M. Zytkow. *Scientific Discovery*. MIT Press, Cambridge MA, 1987.

Levi, 1980 I. Levi. *The Enterprise of Knowledge*. MIT Press, Cambridge MA, 1980.

Levy, 1979 A. Levy. *Basic Set Theory*. Springer-Verlag, Berlin, 1979.

Lindström & Rabinowicz, 1989 S. Lindström & W. Rabinowicz. On Probabilistic Representation of Non-Probabilistic Belief Revision. *Journal of Philosophical Logic*, 18(1):69–102, 1989.

Maas & Turán, 1996 W. Maas & Gy. Turán. On learnability and predicate logic. *NeuroCOLT Technical Report*, NC-TR-96-023, 1996.

Machtey & Young, 1978 M. Machtey & P. Young. *An Introduction to the General Theory of Algorithms*. North-Holland, New York, 1978.

Martin & Osherson, 1995a E. Martin & D. Osherson. A note on the use of probabilities by mechanical scientists. In Paul Vitányi, editor, *Computational Learning Theory: Second European Conference, EuroCOLT '95 (Barcelona, Spain)*. Springer: Lecture Notes in Artificial Intelligence 904, Berlin, 1995.

Martin & Osherson, 1995b E. Martin & D. Osherson. Scientific discovery via rational hypothesis revision. In Maria Luisa Dalla Chiara, editor, *Proceedings of the 10th International Congress of Logic, Methodology, and Philosophy of Science (Florence, Italy)*, 1995.

Martin & Osherson, in press E. Martin & D. Osherson. Scientific discovery based on belief revision. *Journal of Symbolic Logic*, in press.

Matthews & Demopoulos, 1989 R. J. Matthews & W. Demopoulos, editors. *Learnability and Linguistic Theory*. Kluwer, 1989.

Matthews, 1990 R. Matthews, editor. *Language Acquisition and Learnability*. Reidel, 1990.

Montagna, 1996a F. Montagna. Investigations on measure one identification of classes of languages with respect to fixed probability distributions, 1996.

Montagna, 1996b F. Montagna. Two theorems on measure one identification of r.e. classes of languages. Technical Report 301, University of Siena, 1996.

Myrvold, 1996 W. C. Myrvold. Bayesianism and Diverse Evidence: A Reply to Andrew Wayne. *Philosophy of Science*, 63:661 – 665, 1996.

Nayak, 1994a A. C. Nayak. Foundational Belief Change. *Journal of Philosophical Logic*, 23(5): 495–534, 1994.

Nayak, 1994b A. C. Nayak. Iterated Belief Change Based on Epistemic Entrenchment. *Erkenntnis*, 41:353–390, 1994.

Odifreddi, 1989 P. Odifreddi. *Classical recursion theory*. North-Holland, Amsterdam, 1989.

Odifreddi, 1997 P. Odifreddi. *Classical recursion theory, Vol II*. North-Holland, Amsterdam, 1997.

Osherson & Weinstein, 1982a D. Osherson & S. Weinstein. Criteria of language learning. *Information and Control*, 52:123–138, 1982.

Osherson & Weinstein, 1982b D. Osherson & S. Weinstein. A note on formal learning theory. *Cognition*, 11:77–88, 1982.

Osherson & Weinstein, 1989 D. Osherson & S. Weinstein. Identifiable collections of countable structures. *Philosophy of Science*, 56(1):94–105, 1989.

Osherson & Weinstein, 1993 D. Osherson & S. Weinstein. Relevant consequence and scientific discovery. *Journal of Philosophical Logic*, 22:437–448, 1993.

Osherson & Weinstein, 1995 D. Osherson & S. Weinstein. On the study of first language acquisition. *Journal of Mathematical Psychology*, 39(2):129 – 145, 1995.

Osherson *et al.*, 1982 D. Osherson, M. Stob, & S. Weinstein. Learning strategies. *Information and Control*, 53(1):32–51, 1982.

Osherson *et al.*, 1984 D. Osherson, M. Stob, & S. Weinstein. Learning theory and natural language. *Cognition*, 17(1):1–28, 1984.

Osherson *et al.*, 1986a D. Osherson, M. Stob, & S. Weinstein. Aggregating inductive expertise. *Information and Control*, 70(1):69–95, 1986.

Osherson *et al.*, 1986b D. Osherson, M. Stob, & S. Weinstein. Analysis of a learning paradigm. In W. Demopoulos & A. Marras, editors, *Language Learning and Concept Acquisition*. Ablex Publ. Co., New Jersey, 1986.

Osherson *et al.*, 1986c D. Osherson, M. Stob, & S. Weinstein. *Systems that Learn*. M.I.T. Press, Cambridge MA, 1986.

Osherson *et al.*, 1988a D. Osherson, M. Stob, & S. Weinstein. Mechanical learners pay a price for Bayesianism. *Journal of Symbolic Logic*, 53(4):1245–1251, 1988.

Osherson *et al.*, 1988b D. Osherson, M. Stob, & S. Weinstein. Synthesizing inductive expertise. *Information and Computation*, 77(2):138–161, 1988.

Osherson *et al.*, 1989 D. Osherson, M. Stob, & S. Weinstein. On approximate truth. In R. Rivest, D. Haussler, & M. Warmuth, editors, *Proceedings of the Workshop on Computational Learning Theory*. Morgan Kaufmann Publishers, Inc., 1989.

Osherson *et al.*, 1991a D. Osherson, M. Stob, & S. Weinstein. New directions in automated scientific discovery. *Information Sciences*, 1991.

Osherson *et al.*, 1991b D. Osherson, M. Stob, & S. Weinstein. A universal inductive inference machine. *Journal of Symbolic Logic*, 56(2):661–672, 1991.

Osherson *et al.*, 1996 D. Osherson, S. Weinstein, D. de Jongh, & E. Martin. A first-order framework for learning. In J. van Bentham & Alice ter Meulen, editors, *Handbook of Logic and Language*. Elsevier Science Publishers, New York, 1996.

Osherson, 1985 D. Osherson. Computer Output. *Cognition*, 20:261–4, 1985.

Papadimitriou, 1994 C. H. Papadimitriou. *Computational Complexity*. Addison-Wesley Publishing Company, Reading, MA, 1994.

Pappas, 1979 G. Pappas, editor. *Justification and Knowledge*. Reidel, Dordrecht, 1979.

Partee *et al.*, 1990 B. H. Partee, A. ter Meulen, & R. E. Wall. *Mathematical Methods in Linguistics*. Kluwer Academic Publishers, Dordrecht, 1990.

Pitt & Smith, 1988 L. Pitt & C. Smith. Probability and plurality for aggregations of learning machines. *Information and Computation*, 77:77–92, 1988.

Pitt, 1989 L. Pitt. Probabilistic inductive inference. *J. ACM*, 36(2):383–433, 1989.

Popper, 1959 K. Popper. *The Logic of Scientific Discovery*. Hutchinson, London, 1959.

Putnam, 1975 H. Putnam. Probability and confirmation. In *Mathematics, Matter, and Method*. Cambridge University Press, 1975.

Rogers, 1987 H. Rogers. *Theory of Recursive Functions and Effective Computability*. MIT Press, New York, 1987.

Rosenkrantz, 1977 R. Rosenkrantz. *Inference, Method, and Decision*. Reidel, Boston MA, 1977.

Royer, 1986 J. Royer. Inductive inference of approximations. *Information and Control*, 70:156–178, 1986.

Schurz & Weingartner, 1987 G. Schurz & P. Weingartner. Verisimilitude defined by relevant consequence-elements. A new reconstruction of Popper's idea. In T. A. Kuipers, editor, *What is Closer-to-the-Truth?* Rodopi, Amsterdam, 1987.

Schurz, 1996 G. Schurz. The role of relevance in deductive reasoning, 1996.

Smith & Velauthapillai, 1986 C. Smith & M. Velauthapillai. On the inference of programs approximately computing the desired function. *Lecture Notes in Computer Science*, 265:164–176, 1986.

Solomonoff, 1964 R. J. Solomonoff. A formal theory of inductive inference. *Information and Control*, 7:1–22, 224–254, 1964.

Stephan, 1996a F. Stephan. Learning via Queries and Oracles, 1996.

Stephan, 1996b F. Stephan. Noisy Inference and Oracles, 1996.

Stoll, 1963 R. R. Stoll. *Set Theory and Logic*. W. H. Freeman and Company, San Francisco, 1963.

Tzouvaras, 1996 A. Tzouvaras. Aspects of analytic deduction. *Journal of Philosophical Logic*, 25:581–596, 1996.

Wexler & Culicover, 1980 K. Wexler & P. Culicover. *Formal Principles of Language Acquisition*. M.I.T. Press, Cambridge MA, 1980.

Wiehagen *et al.*, 1984 R. Wiehagen, R. Freivald, & E. Kinber. On the power of probabilistic strategies in inductive inference. *Theoretical Computer Science*, 28:111–133, 1984.

Wiehagen, 1977 R. Wiehagen. Identification of formal languages. In *Lecture Notes in Computer Science 53*, pages 571–579. Springer-Verlag, 1977.

Zeugmann *et al.*, 1995 T. Zeugmann, S. Lange, & S. Kapur. Characterizations of monotonic and dual monotonic language learning. *Inform. Comput.*, 120, No. 2:155–173, 1995.

Symbol Index

Index